Springer Proceedings in Physics 68

Springer Proceedings in Physics

Managing Editor: H. K. V. Lotsch

---

44 *Optical Fiber Sensors*
Editors: H. J. Arditty, J. P. Dakin, and R. Th. Kersten

45 *Computer Simulation Studies in Condensed Matter Physics II: New Directions*
Editors: D. P. Landau, K. K. Mon, and H.-B. Schüttler

46 *Cellular Automata and Modeling of Complex Physical Systems*
Editors: P. Manneville, N. Boccara, G. Y. Vichniac, and R. Bidaux

47 *Number Theory and Physics*
Editors: J.-M. Luck, P. Moussa, and M. Waldschmidt

48 *Many-Atom Interactions in Solids*
Editors: R .M. Nieminen, M. J. Puska, and M. J. Manninen

49 *Ultrafast Phenomena in Spectroscopy*
Editors: E. Klose and B. Wilhelmi

50 *Magnetic Properties of Low-Dimensional Systems II: New Developments*
Editors: L. M. Falicov, F. Mejía-Lira, and J. L. Morán-López

51 *The Physics and Chemistry of Organic Superconductors*
Editors: G. Saito and S. Kagoshima

52 *Dynamics and Patterns in Complex Fluids: New Aspects of the Physics–Chemistry Interface*
Editors: A. Onuki and K. Kawasaki

53 *Computer Simulation Studies in Condensed Matter Physics III*
Editors: D. P. Landau, K. K. Mon, and H.-B. Schüttler

54 *Polycrystalline Semiconductors II*
Editors: J. H. Werner and H. P. Strunk

55 *Nonlinear Dynamics and Quantum Phenomena in Optical Systems*
Editors: R. Vilaseca and R. Corbalán

56 *Amorphous and Crystalline Silicon Carbide III, and Other Group IV-IV Materials*
Editors: G. L. Harris, M. G. Spencer, and C. Y.-W. Yang

57 *Evolutionary Trends in the Physical Sciences*
Editors: M. Suzuki and R. Kubo

58 *New Trends in Nuclear Collective Dynamics*
Editors: Y. Abe, H. Horiuchi, and K. Matsuyanagi

59 *Exotic Atoms in Condensed Matter*
Editors: G. Benedek and H. Schneuwly

60 *The Physics and Chemistry of Oxide Superconductors*
Editors: Y. Iye and H. Yasuoka

61 *Surface X-Ray and Neutron Scattering*
Editors: H. Zabel and I. K. Robinson

62 *Surface Science: Lectures on Basic Concepts and Applications*
Editors: F. A. Ponce and M. Cardona

63 *Coherent Raman Spectroscopy: Recent Advances*
Editors: G. Marowsky and V. V. Smirnov

64 *Superconducting Devices and Their Applications*
Editors: H. Koch and H. Lübbig

65 *Present and Future of High-Energy Physics*
Editors. K.-I. Aoki and M. Kobayashi

66 *The Structure and Conformation of Amphiphilic Membranes*
Editors: R. Lipowsky, D. Richter, and K. Kremer

67 *Nonlinearity with Disorder*
Editors: F. Abdullaev, A. R. Bishop, and S. Pnevmatikos

68 *Time-Resolved Vibrational Spectroscopy V*
Editor: H. Takahashi

69 *Evolution of Dynamical Structures in Complex Systems*
Editors: R. Friedrich and A. Wunderlin

Volumes 1–43 are listed at the end of the book

H. Takahashi (Ed.)

Time-Resolved Vibrational Spectroscopy V

Proceedings
of the 5th International Conference
on Time-Resolved Vibrational Spectroscopy,
Tokyo, Japan, June 3–7, 1991

With 187 Figures

Springer-Verlag
Berlin Heidelberg New York
London Paris Tokyo
Hong Kong Barcelona
Budapest

Professor Hiroaki Takahashi

Department of Chemistry, School of Science & Engineering,
Waseda University, Tokyo 169, Japan

ISBN-13:978-3-642-84773-8 e-ISBN-13:978-3-642-84771-4
DOI: 10.1007/978-3-642-84771-4

Softcover reprint of the hardcover 1st edition 1992

Typesetting: Camera ready by authors/editors
54/3140 - 5 4 3 2 1 0 - Printed on acid-free paper

Preface

The work contained in this volume is representative of the presentations made by the participants at the Fifth International Conference on Time-Resolved Vibrational Spectroscopy, which was held at Waseda University, Tokyo, Japan, from June 3 to 7, 1991. The conference was the fifth in a biennial series initiated in 1982 by Prof. George H. Atkinson (University of Arizona) at Lake Placid, USA, and subsequently convened by Prof. Alfred Laubereau (University of Bayreuth, Germany) and Dr. Manfred Stockburger (Max-Planck Institut, Göttingen, Germany) at Bayreuth-Bischofsgrün, Germany, in 1985, by Prof. Joop D.W. Van Voorst (University of Amsterdam) at Amersfoort, The Netherlands, in 1987, and by Prof. Thomas G. Spiro (Princeton University) at Princeton, USA, in 1989.

The purpose of the conference is to bring together researchers from various disciplines and provide a forum for discussion of the latest advances in time-resolved spectroscopies concerned with transient vibrational phenomena and their application to fundamental scientific and engineering studies. The 167 registered participants, including 46 students, from 14 different countries, represented a wide range of scientific disciplines, and clearly indicated that the field continues to expand into new areas of physics, chemistry, biology, and materials science. Their enthusiasm and the originality and quality of the contributions presented produced a very successful and enjoyable conference.

Although the space available for each contribution does not permit a very detailed discussion of any particular topic, I believe that this volume makes it feasible for scientists interested in this field of research to learn about the current state-of-the-art results. This volume also serves as an official record of the conference.

The conference was supported by grants from Waseda University and the Commemorative Association for the Japan World Exposition (1970). Financial support received from the following companies is gratefully acknowledged: Tokyo Instruments, Seki Technotron, Hamamatsu Photonics, Jasco, Jeol, Hitachi, Atago Bussan, Leonix, Rikei, Spectra-Physics, Marubun, Seishin Trading, Nicolet Japan, Shimadzu, Showa Denko, Toray Research Center, Nippon Bio Rad Laboratories, and Toray Techno. The contributions of Miss Takayo Naoi and Miss Masako Yonemitsu to the organization of the conference and the preparation of this volume are also greatly appreciated.

But my thanks are mainly due to the authors, for providing their latest results for publication, and to the participants for vigorous and good-natured discussions which enriched the contributions to these proceedings. This book is their work.

Tokyo
August 1991

Hiroaki Takahashi

Contents

Part I Biological Systems

Part II Metal Complex and Organometallic Systems

Part III Photochemical Reactions

Part IV Transient Species

Part V Excited State Dynamics

Part VIII Theoretical Aspects

Part I

Biological Systems

Protein Dynamics From UV Resonance Raman Spectroscopy

T.G. Spiro, K. Rodgers, and C. Su

Princeton University, Department of Chemistry,
Princeton, NJ 08544-1009, USA

Introduction

The advent of reliable pulsed lasers whose frequency can be shifted into the ultraviolet region has made possible the application of ultraviolet resonance Raman (UVRR) spectroscopy to proteins. Appropriate tuning of the laser yields selective enhancement of vibrational modes of the aromatic side chains phenylalanine, tyrosine and tryptophan, or of the amide bonds [1-3]. The amide UVRR bands are responsive to the protein secondary structure. Enhancements of the classic amide II and III bands decrease with increasing α-helical content, [4] as does an additional UVRR band, amide S [5-8]. The aromatic side chain UVRR bands show frequency and/or intensity changes in response to changes in their local environment, particularly to H-bonding changes [9-12]. They can therefore monitor protein conformation changes that alter the tertiary or quaternary structure.

The UVRR experiment can be applied in a time-resolved mode to monitor structural changes associated with transient processes in proteins. The R-T quaternary transition in hemoglobin is currently being studied in this manner, using a pair of nanosecond pulsed lasers, one to photolyze the CO adduct, and the other to generate UVRR spectra [13-15]. Changes in the tyrosine and tryptophan UVRR signals are observed at time delays on the microsecond scale. These changes provide indicators of protein motions associated with the tertiary and quaternary rearrangements on the R-T pathway.

Probes of Aromatic Residue Environment

UVRR enhancements for the aromatic residues phenylalanine (Phe), tyrosine (Tyr) and tryptophan (Trp) are sensitive to the polarity of the environment due to alterations in the electronic transition frequencies and intensities [9-12]. In the case of Trp, H-bonding from the indole NH group has specific effects on the excitation profiles [13,15], and, for some of the bands, there are perceptible frequency shifts as well [11]. The 880 cm^{-1} mode is particularly sensitive to H-bonding since it has a significant contribution from the N-H bending coordinate. The Trp W3 mode, at ca. 1555 cm^{-1}, is sensitive to variations of the dihedral angle, $\chi 1,2$, about the bond connecting the indole ring to the Cβ atom of the residue [11]. In the case of Tyr, the OH group can be either an H-bond donor or acceptor. These interactions produce opposite shifts in the excitation profiles, and also in some of the vibrational frequencies themselves [9,15]. The ν_{8a} and ν_{8b} ring modes, at ca. 1615 and 1600 cm^{-1}, are particularly useful as monitors of H-bonding. They are strongly enhanced at 230 nm, the optimum excitation wavelength for discriminating Tyr from Phe (whose ν_{8a} and ν_{8b} modes overlap those of Tyr) [9]. The ν_{8b} frequency shifts down significantly when p-cresol, a model for Tyr, interacts with H-bond acceptors [9], and increases when it interacts with H-bond donors [15]. Experiments with ovomucoid proteins containing a Tyr residue, H-bonded to a carboxylate side chain, showed a clear upshift in ν_{8b} when the H-bond was broken at low pH [9].

Springer Proceedings in Physics, Vol. 68 Time Resolved Vibrational Spectroscopy V
Ed.: H. Takahashi

Protein Dynamics of Hemoglobin

When the 230 nm-excited UVRR spectrum of the CO adduct of Hb, which is in the R state, is subtracted from that of deoxy-Hb, which is in the T state, well-defined difference signals are obtained for several bands of Tyr and Trp [13,15]. The same difference signals, albeit with lower amplitude, are obtained when the spectrum of HbCO is subtracted from that of the HbCO photoproduct, provided that the delay between the photolysis and probe pulses is 10-20 μs. If there is no delay, there is no difference spectrum, showing that the Tyr and Trp signals are unaffected by the ligation of the heme per se. 20 μs is also the time constant determined by Gibson and coworkers [16] for the transition from fast to slow recombining Hb, and suggested by them to be the R-T quaternary transition. The Hb UVRR data provide direct structural support for this suggestion. Kitagawa and coworkers have likewise determined a 20 μs time constant for a perturbation of the 880 cm^{-1} Trp band in 218 nm-excited UVRR spectra of the HbCO photolysis product. [14]

The UVRR difference signals are due to the changes in the environments of the six inequivalent tyrosine residues and the three inequivalent tryptophan residues in the Hb tetramer. Among the tyrosines, the most significant change is the formation of an H-bond between Tyr α42 and Asp β99, across the $\alpha_1\beta_2$ interface, in the T state. Removal of this H-bond, via mutation of either residue, lowers the cooperativity of ligand binding by Hb. The interaction has always been assumed to involve H-bond donation from the Tyr OH to the Asp carboxylate group. However, the UVRR difference signal does not show a downshift for ν_{8b}, as would be expected for H-bond donation, and as is seen in the ovomucoid proteins with carboxylate H-bond acceptors [9]. Instead, ν_{8b} as well as ν_{8a}, shift up slightly, suggesting that the Tyr OH is acting as an H-bond acceptor. Examination of the x-ray coordinates of the T state show the Asp β99 side chain to be surrounded by hydrophobic residues except for Tyr α42. Even though it is at the subunit interface, it is inaccessible to water. The UVRR spectra therefore indicate that the carboxylate group is protonated, and donates an H-bond to Tyr α42.

The evolution of the UVRR difference spectrum reflects both tertiary and quaternary protein motions on the R-T reaction coordinate. These can best be distinguished via the Trp W3 signal, which can be interpreted with the aid of the correlation between the W3 frequency and the indole χ1,2 dihedral angle discovered by Harada and coworkers [11]. When this correlation is applied to the χ1,2 angles calculated from the x-ray coordinates for the three inequivalent Trp residues of Hb, it is found that Trp β37, which is located at the interface between the α_1 and β_2 subunits, is predicted to have a frequency that coincides with a shoulder on the low-frequency side of the broad W3 band, while Trp α14 and Trp β15, which are located on the A helices in the subunit interiors, are predicted to have frequencies in the main part of the W3 band [15]. This assignment of the Trp β37 frequency was confirmed with a mutant protein, Hb Rothschild, in which Trp β37 is replaced by an arginine residue. The low-frequency shoulder was missing for this protein.

In comparing the R and T state crystal structures, one finds that the Trp β37 indole ring is H-bonded to the carboxylate group of Asp α94 in the T state but not in the R state. At 230 nm, H-bonding is expected to increase the W3 intensity, since the excitation profile, normally peaked at 220 nm, shifts to the red. Consequently, the Trp β37 H-bond in the T state should produce a positive difference signal on the low-frequency side of the W3 band, and this is what is observed. The interior residues, Trp α14 and β15, are H-bonded to the OH groups of Thr and Ser residues, respectively, on the E helices . Changes in these H-bonds, due to a slight realignment of the A and E helices, can account for the complex shape of the W3 difference signal in the main part of the band. It may be significant that the E helix forms the distal side of the heme binding pocket, and it is therefore plausibly on the pathway from ligand dissociation to the R-T switch.

References

1. I. Harada, H. Takeuchi in "Spectroscopy of Biological Systems," R. T. Clark and R. E. Hester, Ed., J. Wiley & Sons, New York, 1986, pp. 112-175.
2. B. Hudson and L. C. Mayne in "Biological Applications of Raman Spectroscopy," T. G. Spiro, Ed., J. Wiley & Sons, New York 1987, Vol. 2, pp. 181-210.
3. S. A. Asher, Ann. Rev. Phys. Chem., 1988, 39, 537-588.
4. R. A. Copeland and T. G. Spiro, Biochemistry, 1985, 26, 2134.
5. Y. Wang, R. Purrello, and T. G. Spiro, J. Am. Chem. Soc. 1989, 111, 8274-8276.
6. Wang, Y., R. Purrello, T. Jordan, and T. G. Spiro, J. Am. Chem. Soc. 1990, in press.

7a. S. Song, S. A. Asher, S. Krimm, and J. Vandekar, J. Am. Chem. Soc. 1988, 110, 8548-8550.

7b. S. Krimm, S. Song, and S. A. Asher, J. Am. Chem. Soc., 1988, 111, 4290-4294.

8. S. Song, S. A. Asher, and K. D. Shaw, J. Am. Chem. Soc., 1991, 113, 1155-1163.
9. P. Hildebrandt, R. A. Copeland, T. G. Spiro, J. Otlewski, M. Laskowski, and F. G. Prendergast, Biochemistry, 1988, 27, 5426.
10. H. Takeuchi, N. Watanabe, Y. Satoh, and I. Harada, J. Raman Spectrosc., 1989, 20, 233-237.
11. T. Miura, H. Takeuchi, and I. Harada, J. Raman Spectrosc., 1989, 20, 667.
12. G.-Y. Liu, C. A. Grygon, and T. G. Spiro, Biochemistry, 1989, 28, 5046.
13. C. Su, Y. D. Park, G.-Y. Liu, and T. G. Spiro, J. Am. Chem. Soc., 1989, 111, 3457.
14. S. Kaminaka, T. Ogura, and T. Kitagawa, J. Am. Chem. Soc., 1990, 112, 23.
15. C. Su, K. Rodgers, S. Subramaniam, and T. G. Spiro (in preparation).
16. C. Sawicki and Q. H. Gibson, J. Biol. Chem., 1976, 251, 1533.

Ultrafast Infrared Spectroscopic Studies of Protein Dynamics

R. Diller, M. Iannone, B. Cowen, S. Maiti, J. Owrutsky, M. Li, M. Sarisky, Y.R. Kim, R.B. Locke, T. Lian, and R.M. Hochstrasser

Department of Chemistry, University of Pennsylvania, Philadelphia, PA 19104, USA

Introduction: In the last few years there was progress in creating experimental configurations to carry out subnanosecond transient infrared spectroscopy (1-3). Although it is known how to generate infrared from optical sources by nonlinear mixing, the adaptation of these technologies to infrared pump/probe experiments with ultrashort time resolution is a recent occurrence.

Fundamental understanding of chemical dynamics in condensed media requires the results of experiments which expose pathways of energy flow as well as the identity of the products and the overall kinetics. Infrared spectroscopy is ideally suited for these purposes. What is needed are tunable optical and IR pulses to pump the system and a tunable IR source to probe: The optical pulses can initiate chemical reactions while those in the IR can be used to initiate known non-thermal distributions whose character can be followed in time by the probe.

A gated IR detector system that utilizes upconversion of the IR to the visible with a stabilized dye laser pulse followed by detection with a PMT, presented a particularly high detectivity system (1-3) suitable for both optically and infrared pumped experiments. The IR pump method was used to determine T_1 relaxation times in a variety of molecules and also to measure the orientational relaxation times of chemical bonds. These examples include N_3^- in various solvents (4) and CO vibrational excitation dynamics in metal carbonyls (5). These measurements are important for understanding solvent collision induced relaxation processes which are intimately involved in the full picture of the photoreactions discussed in the second part of this paper. Photoprocesses generally yield products that are vibrationally hot and the cooling of these states can be an important part of the chemical dynamics. Obviously the T_1 relaxation times for individual modes sets the limit on thermalization times for complex molecules.

These methods were used in this laboratory for the study of metal carbonyls, hemoglobin dynamics and bacteriorhodopsin (6).

Dynamics of BR_{570} with Picosecond IR-Spectroscopy: It was shown that the IR difference bands around 1610, 1650 and 1658 cm^{-1}, in the region of the amide I absorption are present from ca. 25 ps and remain unaltered out to 14 ns after excitation (see Figure 1). Thus if protein structural changes cause these transitions (7), they would have to be due to amino acids close to the chromophore, directly influenced by the primary isomerisms.

Infrared Pump/Infrared Probe Measurements of T_1 Relaxation: We recently generated IR pulses of ca. 350 fs duration and a capability to adjust the pulsewidth to be larger than the T_2 time of the vibrational excitation (otherwise unwanted coherent effects are observed (8)).

To study N_3^- in H_2, D_2O, CH_3OH and C_2H_5OH we use ca. 1 ps pulses while the dephasing times are in the range 200 fs. The T_1 relaxation times were ca. 3 ps in all the solvents (see Fig. 2) suggesting either that the dominant pathway for relaxation is internal or that the solvent mode coupling the electropositive central atom to solvent O atoms is allowing energy to couple directly to solvent modes (5). The latter explanation is preferred since no evidence of intrinsic or solvent induced Fermi resonances are seen in the IR spectrum. In the N_3^- asymmetric stretch, the central atom displaces twice as

Springer Proceedings in Physics, Vol. 68 Time Resolved Vibrational Spectroscopy V
Ed.: H. Takahashi

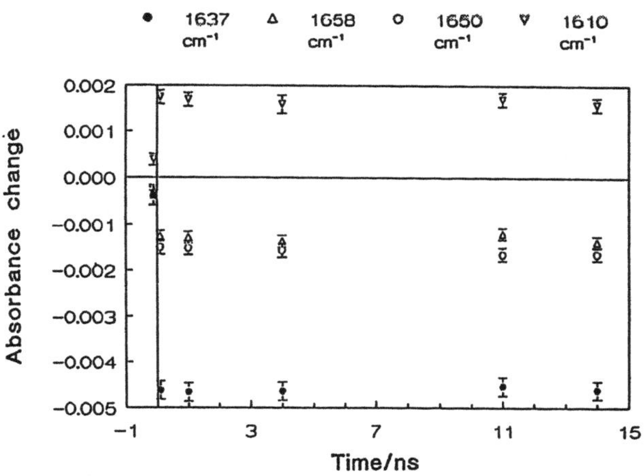

Figure 1
IR - absorbance kinetics of purple membrane in the amide I region after excitation at 550 nm.

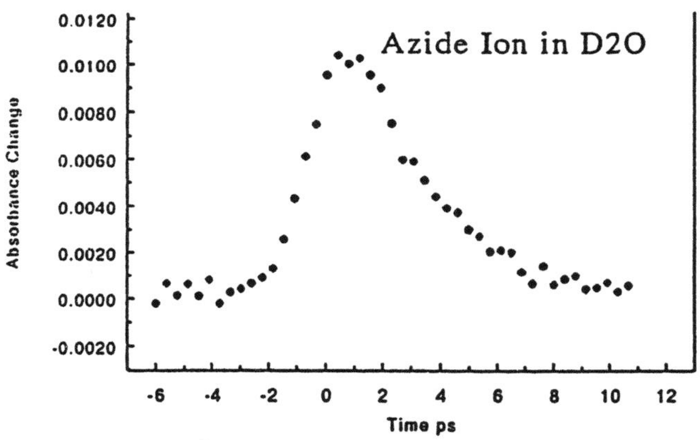

Figure 2
Decay of the vibrationally excited asymmetric stretch of azide ion in solution demonstrating the fastest T1 relaxation yet observed.

much as the end atoms. Even CN^- that has no opportunity for internal relaxation displays a T_1 in the few ps regime (9) again emphasizing the strong forces that couple vibrational states of ions to H_2O and other hydroxyllic solvents.

High frequency vibrations in neutral molecules such as metal carbonyls (10) relax much more slowly than ions, but a wide range of T_1 times occur. For example WCO_6 excited in the CO region shows T_1 = 700 ps (3,10) whereas the CO's in $Fe_2\,CO_4\,Cp_2$ relax on the 15 - 30 ps timescale (5). In the latter case there are both terminal Fe-CO and bridged Fe-CO-Fe bonds formed. By means of IR pump / IR probe experiments we have seen that the terminal CO relaxes slower (ca. 32 ps) than the bridges (ca. 18 ps). These processes may involve intramolecular energy flow between symmetric and asymmetric stretch vibrations of CO. The bridge relaxation is not much slower than found for CO adsorbed on metals (11).

Of course the cooling of molecules, heated as a result of chemical reactions, may involve much lower frequency modes than those discussed here and not much is yet known about their T_1 relaxation.

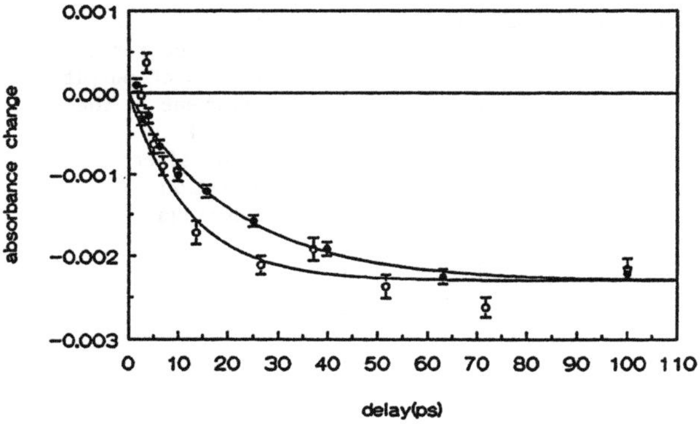

Figure 3
Absorbance changes monitored at 1809 cm^{-1} for solutions of deoxy Hb (closed circles) and deoxy Mb (open circles) following excitation at 580 nm.

Infrared Probe Measurements of Cooling and Heat Transport in Proteins: Internal conversion processes following optical or infrared excitation, or photoreactions invariably will leave the molecules or reaction products with excess vibrational energy. For small molecules in the gas phase the energy distributions of reactions can be measured directly, but for larger molecules and/or condensed phases, no such information yet exists. The accepted view is that IVR is sufficiently effective in large molecules, that it is reasonable to consider the excess energy distributed according to a Boltzman distribution, so that the product distribution is defined by a temperature. Coupling of the molecules to the surrounding solvent will cool this distribution as a result of T_1 relaxation processes.

The effective temperature for the protoporphyrins of hemoglobin excited with photons at 580 nm is ca. 500 K. The T_1 relaxation heats the surrounding protein and, ultimately, the water solvent. This process was studied using transient IR spectroscopy of D_2O in the range 1840 cm^{-1} to 2000 cm^{-1}. The temperature dependence of the D_2O spectrum in this range is known (12) and our IR probe technique is capable of recording temperature changes as small as 10^{-2}K. The results for Hb and Mb in D_2O are given in Figure 3. They represent an increase in transmission at 1840 cm^{-1} corresponding to a limiting change of 0.5 K in the D_2O. The solid lines are calculated from a classical thermal diffusion treatment in which a time dependent source of temperature is located at the center of the protein. From these results we can conclude that both the response of D_2O to the heat impulse and the evolution of the source temperature (cooling of the heme) must be faster than ca. 10 ps. This is in the range predicted by calculations (13) and estimated from other experiments on heme cooling (14).

Infrared-Optical Two Photon Absorption (TPA) of Hemes: The sharp spike in IR absorption of Hb and HbCO reported earlier (15) was thought to indicate an ultrafast (<1ps) component of molecular cooling. We have now carried out a detailed study of this response and have shown it is due to TPA: The signal depends on the optical laser power, whereas the D_2O heating signal (Figure 3) depends on the energy in the laser pulse. The TPA signal (Figure 4) anisotropy shows that it derives from the electronic polarizability and not the nuclear response. In all likelihood both IR-optical and optical-IR time orderings contribute about equally since the Q-band heme absorption involved in the optical step is so broad. This effect should be ubiquitous in IR/Optical two pulse experiments, even when the IR pulse if off-resonance but more so when it is also resonant, causing possible confusion with IR spectra of short-lived excited states.

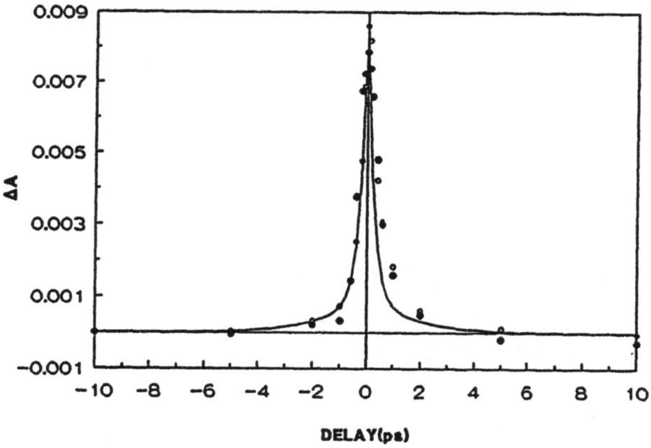

Figure 4
TPA for solutions of deoxy Mb (closed circles) and doxy Hb (open circles) at 1855 cm^{-1}. The solid line is the autocorrelation of the visible excitation pulse.

Acknowledgements: Supported by NSF and NIH.

1. P. Anfinrud, C. Han, P. A. Hansen, J. N. Moore and R. M. Hochstrasser, "Picosecond and femtosecond infrared spectroscopy with CW diode lasers," in Ultrafast Phenomena VI; eds. T. Yajima, K. Yoshihara, C. B. Harris and S. Shionoya (Springer-Verlag, Berlin-Heidelberg, 1988), p. 442.
2. R. M. Hochstrasser, P. A. Anfinrud, R. Diller, C. Han, M. Iannone, T. Lian and B. Locke, "Femtosecond infrared spectroscopy of complex molecules using CW IR lasers in Ultrafast Phenomena VII; eds. C. B. Harris, E. P. Ippen, G. A. Mourou and A. H. Zewail (Springer-Verlag, Berlin-Heidelberg, 1990), p. 429.
3. M. Iannone, B. R. Cowen, R. Diller, S. Maiti and R. M. Hochstrasser, "A high repetion-rate infrared pump, infrared probe spectrometer," Optics Letters, in press.
4. J. C. Owrutsky, Y. R. Kim, M. Li, M. J. Sarisky and R. M. Hochstrasser, "Determination of the vibrational energy relaxation time of the N_3^- ion in protic solvents", in process of publication.
5. J. C. Owrutsky, M. Iannone and R. M. Hochstrasser, "Vibrational energy relaxation of bridged and terminal CO bonded to iron", in process of publication.
6. R. Diller, M. Iannone, R. Bogomolni and R. M. Hochstrasser, Biophysical Journal, in press.
7. M. S. Braiman, P. L. Ahl and K. J. Rothschild, Proc. Natl. Acad. Sci. USA 84, 5221 (1984).
8. P. A. Anfinrud, C. Han, T. Lian and R. M. Hochstrasser, J. Phys. Chem. 94, 1180 (1990).
9. E. J. Heilweil, F. Doany, R. Moore and R. M. Hochstrasser, J. Chem. Phys. 80, 201 (1984).
10. E. J. Heilweil, Ann. Rev. Phys. Chem. 40, 143 (1990).
11. A. L. Harris, L. Rothberg, L. H. Dubois, N. J. Levinos and L. Dhar, Phys. Rev. Lett. 64, 2086 (1990); J. D. Beckerle, M. P. Cassassa, R. Cavanaugh, E. J. Heilweil and J. C. Stephenson, Phys. Rev. Lett. 64, 2090 (1990).
12. D. A. Draegert, N. W. B. Stone, B. Curnutte and D. Williams, J. Opt. Soc. Amer. 56, 64 (1966).
13. E. R. Henry, W. A. Eaton and R. M. Hochstrasser, Proc. Natl. Acad. Sci. USA 83, 8982 (1986).
14. J. W. Petrich, J. L. Martin, D. Houde, C. Poyart and A. Orzag, Biochem. 26, 7914 (1987); R. Lingle, Jr., X. Xu, H. Zhu, S.-C. Yu and J. B. Hopkins, J. Amer. Chem. Soc. (in press).
15. P. A. Anfinrud, C. Han and R. M. Hochstrasser, Proc. Natl. Acad. Sci. 86, 8387 (1989).

Time-Resolved Resonance Raman Investigation of Photodissociation Dynamics of Liganded Hemoglobin and Myoglobin

T. Kitagawa[1], *Y. Sakan*[2], *S. Kaminaka*[1], *T. Ogura*[1], *M. Nagai*[3], *and M. Ikeda-Saito*[4]

[1]Institute for Molecular Science,
Okazaki National Research Institutes, Okazaki 444, Japan
[2]Graduate University for Advanced Studies, Myodaiji, Okazaki 444, Japan
[3]Kanazawa University School of Allied Medical Professions,
Kanazawa 920, Japan
[4]Case Western Reserve University School of Medicine,
Cleveland, Ohio, OH 44106-4970, USA

1. Introduction

Dynamical properties of proteins are essential for a given protein to accomplish physiological functions but have not been clarified sufficiently. It would be informative to investigate a single protein throughly. Accordingly, dynamics of proteins upon photodissociation of carbonmonoxide from hemoglobin (Hb) and myoglobin (Mb) have been investigated with various techniques. Resonance Raman (RR) spectroscopy is powerful for this purpose since the interaction time of a photon with a molecule is short and the response time of a detector is sufficiently fast. In the previous TRVS conference, we reported time-resolved ultraviolet (UV) RR spectra about the quaternary structure change of COHb upon its photolysis [1]. It is interesting to examine how the protein dynamics are affected by protonation to a key residue as well as by replacement of an amino acid residue. In the present study, we extended the time resolved UV RR measurements to study the Bohr effect of Hb, and also pursued the recombination process of CO to the photodissociated heme of some mutant Hb and Mb by using visible TR^3 spectroscopy.

2. Experimental Procedures

The visible TR^3 experiment was carried out with the second harmonic (532 nm) of a Nd:YAG laser as a pump beam (4.4 mJ/pulse) and with the first H_2 Stokes shifted Raman line (416 nm, 0.13 mJ/pulse) of the third harmonic as a probe beam [2]. The sample was contained in a spinning cell (50 rpm) filled with CO. The system for observing the UV RR spectra was described previously [1]. In order to prevent the photodissociated Hb from recombining with CO, a new sample circulation system [3] with an N_2-filled sample-illuminating chamber was constructed and applied successfully to determine the end point of the quaternary structure change.

Springer Proceedings in Physics, Vol. 68 Time Resolved Vibrational Spectroscopy V
Ed.: H. Takahashi

3. Results and Discussion

Visible TR^3 Study CO adduct of Hb gives three CO-associated Raman bands; CO stretching (ν_{CO}) at 1952 cm^{-1}, Fe-CO stretching (ν_{Fe-CO}) at 507 cm^{-1} and Fe-C-O bendig (δ_{FeCO}) at 576 cm^{-1}. However, the fully reduced CO-bound form of a natural mutant Hb, in which the distal histidine (His-E7) is replaced by tyrosine (Tyr) in either the α (Hb M Boston) or β (Hb M Saskatoon) subunit, gave another set of the CO-associated Raman bands in addition to the normal set mentioned above [4]; ν_{CO} = 1972 cm^{-1}, ν_{Fe-CO} = 488-490 cm^{-1}, and δ_{FeCO} = too weak to be identified. These new bands arise from the abnormal subunit. On the basis of the observed isotopic frequency shifts for $^{13}C^{16}O$, $^{12}C^{18}O$, and $^{13}C^{18}O$, and the normal coordinate calculations for an isolated Fe-C-O molecule with an arbitrary bend angle (θ), the new species was deduced to adopt an upright geometry ($\theta = 176^o$)[4], while the normal COHb is reported to adopt a bent structure ($\theta = 160^o$) [5].

In the CW-laser excited RR experiments for COHb M Boston, the two sets of the CO-associated RR bands were observed with the spinning cell (1800 rpm), but only the new set was observable when the spinning of the cell was abolished [4]. In order to clarify the difference between the two species, we carried out visible TR^3 experiments [2]. For the delay time (Δt_d) of -200 ns, the two ν_{Fe-CO} bands were observed at 505 and 490 cm^{-1}. However, for $\Delta t_d = 10$ ns, the 505 cm^{-1} band was very weak, while the 490 cm^{-1} band remained unchanged. The intensity of the 505 cm^{-1} band was restored in $\Delta t_d = 500$ ns. Even when the laser power of the pump beam was raised to the maximum (11 mJ/pulse), this phenomenon was unaltered. If the 490 cm^{-1} species were photodissociated but very fast recombination took place, a photosteady state should be established quickly and the equilibrium would be shifted toward the dissociated form upon the increase of the pump laser power. Insensitivity of the 490 cm^{-1} intensity to the pump power suggests that the 490 cm^{-1} species is not photodissociable.

It is known that COMb at acidic pH exhibits two sets of the CO-associated Raman bands [6] spectroscopically similar to the fully reduced CO-bound form of Hb M Boston. When the CO adduct of horse Mb at pH 4.5 was examined, the time resolved measurements for $\Delta t_d = -200$ ns yielded two ν_{Fe-CO} bands at 506 and 492 cm^{-1}, but for $\Delta t_d = 50$ ns, the 506 cm^{-1} band decreased in intensity significantly, while the 492 cm^{-1} band remained unchanged. The 506 cm^{-1} band was restored in 1 μs. In these measurements, the Fe-C-O bending mode was too weak to be identified.

For human COMb, which has an amino acid sequence different from that of horse Mb, and its Gln mutant of distal His, the ν_{Fe-CO} band was observed at 509 cm^{-1} but at 492 cm^{-1} for the Val and Leu mutants of distal His [7]. For the two kinds of Mbs, the δ_{FeCO} band was noticeably intense for $\Delta t_d = -20$ ns. The recombination process of human Mb was quite different from that for horse Mb. For native human Mb, the ν_{Fe-CO} RR band became considerably broad but the δ_{FeCO} band was still intense for $\Delta t_d = 0$. This feature was retained even for $\Delta t_d = 100$ μs. On the other hand, for the distal Val mutant of human Mb, the ν_{Fe-CO} band became broad and the δ_{FeCO} band became somewhat stronger at

Δt_d=0 similar to native human Mb, but the ν_{Fe-CO} band was restored faster than that for native one. Since the Raman intensity of δ_{FeCO} is known to increase as the bend angle becomes smaller, it is likely that the bend angle is quite small at the initial stage of recombination and has appreciable distribution, which results in the broad feature of the ν_{Fe-CO} band. This initial stage may correspond to a photoexcited state of the CO bound form, if non-photodissociable, or to the geminate recombined state.

UV TR^3 Spectra With the new device of the sample flowing system, the UV RR spectrum probed at 217.8 nm for Δt_d=200 μs was similar to that of the equilibrium deoxyHb; the 878 cm^{-1} band of Trp in the spectrum for -100 μs was partly shifted to 883 cm^{-1} in the spectrum for Δt_d=200 μs and the 1011 cm^{-1} band of Trp decreased in intensity. There are three Trp residues per $\alpha\beta$ dimer. Two of them are contained in a helix part and would not change upon the quaternary structure change. The remaining is Trp-β37, which is located at the α_1 and β_2 subunit interface, and its interaction with Tyr-α140 is reported to be altered upon quaternary structure change [8]. Accordingly, this Trp is deduced to be responsible for the UV RR spectral change after photolysis.

When the 1011 cm^{-1} intensity of Trp relative to the 1003 cm^{-1} intensity of Phe was plotted against Δt_d, it exhibited a smooth two state change; the value for Δt_d=5 μs was the same as that for Δt_d=-100 μs and it changed between Δt_d=5 and 20 μs but remained unchanged after 20 μs. Consequently, it became evident that the subunit interface begins to change around 5 μs after photolysis of CO at the heme and ends it around 20 μs after it. Similar experiments were carried out for COHb at pH 5.8 where Bohr residues are protonated. In this case the spectrum had already been altered at Δt_d=10 μs and after that it did not change. The values of the relative intensity before and after the structural change at pH 5.8 were close to the corresponding values for pH 7.4. Therefore, it is inferred that the structural change which occurred after photolysis, is scarcely affected by protonation of Bohr residues but only its rate is accerelated by protonation.

There are four Bohr residues [9]; His-β146, His-β143, His-α89, and His-α20. The similarity of the structural change between pH 5.8 and 7.4 suggests that His-α89 (pK_a=7.2 for deoxyHb and 5.6 for COHb)[9], which is hydrogen bonded to Lys-α139, seems to contribute significantly to the structural change observed here. The ligand dissociation causes a change in its trans ligand (His-α87) and thus the F-helix in 10 ns - 1 μs. Next, Lys-α139 is changed through the hydrogen bond with His-α89, and this leads to small movement of Tyr-α140. Finally, Trp-β37 is altered at 5 - 20 μs after photolysis of CO due to the interaction with Tyr-α140. Thus, time-resolved UV RR spectroscopy demonstrated that the ligand dissociation in the heme moiety is communicated to another heme through a mechanical change of higher order structures of proteins.

This study was supported by Grant-in-Aids for Priority Area (Bioinorganic chemistry) of the Ministry of Education, Science, and Culture to T.K. (03241105)

References

1. Kaminaka, S. Ogura, T., and Kitagawa, T., J. Am. Chem. Soc. 112, 23-27 (1990).
2. Sakan, Y., Ogura, T., Nagai, M., and Kitagawa, T., to be published
3. Kaminaka, S. and Kitagawa, T., to be published.
4. Nagai, M., Yoneyama, Y., and Kitagawa, T., Biochemistry, in press.
5. Moore, J. N., Hansen, P. A., and Hochstrasser, R. M., Proc. Natl. Acad. Sci. U.S.A. 85, 5062-5066 (1988).
6. Han, S., Rousseau, D. L., Giacometti, G., and Brunori, M., Proc. Natl. Acad. Sci. U.S.A. 87, 205-209 (1990).
7. Sakan, Y, Ikeda-Saito, M. Ogura, T., and Kitagawa, T., to be published.
8. Baldwin, J. and Chothia, C., J. Mol. Biol. 129, 175-220 (1979).
9. Ohe, M. and Kajita, A., Biochemistry 19, 4443-4450 (1980).

Time-Resolved Resonance Raman Study of Photodissociated Oxy-Myoglobin and Carbonmonoxy-Myoglobin

Y. Sakan, T. Ogura, and T. Kitagawa

The Graduate University for Advanced Studies and
Institute for Molecular Science, Okazaki National Research Institutes,
Myodaiji, Okazaki 444, Japan

1. Introduction

Both oxy-myoglobin(MbO_2) and carbonmonoxy-myoglobin(MbCO) are photodissociable upon laser illumination in the blue region. It is known from the transient resonance Raman spectroscopy [1, 2] that relaxation of the photodissociated MbCO to the equilibrium deoxy-myoglobin(deoxyMb) takes place in a ps time scale, but some differences between MbO_2 and MbCO are noted about the relaxation behavior after photolysis [3, 4]. If liganded myoglobin can be photolyzed by UV light, it is expected that the rate of the photolysis becomes slower than that for the direct photolysis by visible light, because the energy of UV light must be transferred to the heme through the protein. Then the difference between MbCO and MbO_2 might be revealed more explicitly. Accordingly, we first determined the action spectrum (250 - 700 nm) for photodissociation of MbCO. For hemoglobin, the iron-histidine stretching ($\nu_{Fe\text{-}His}$) mode is reported to be a good marker of protein structure. Therefore, in this syudy, the UV light photolyzed time-resolved resonance Raman spectra in the $\nu_{Fe\text{-}His}$ region of MbCO and MbO_2 were investigated.

2. Experimental Section

The action spectra for photodissociation of MbCO were determined with the facility of Okazaki Large Spectrograph [5]. The pump/probe time-resolved resonance Raman spectra were obtained by using a 442.0 nm pulse as a probe beam and a 223.1 nm pulse as a pump beam. The 442.0 nm pulse was generated with a Nd;YAG (10 ns, 10 Hz)/dye (coumarine-440) laser system, and the 223.1 nm pulse was obtained as the fourth anti-Stokes Raman shift of H_2 from the 3rd harmonic of a Nd;YAG laser(10 ns, 10 Hz). In the laser power used, the probe beam partially photodissociates the ligand. The delay-time of the probe pulse from pump pulse was changed from -20 ns to 100 μs. A slow spinning cell (50 rpm) was used for the present measurements. The scattered light at right angle was dispersed with a triple-monochromator (Spex 1877) and detected by an intensified photodiode array (PAR 1420B).

Springer Proceedings in Physics, Vol. 68 Time Resolved Vibrational Spectroscopy V
Ed.: H. Takahashi

3. Results and Discussion

The action spectrum for photodissociation of MbCO was in close agreement with the absorption profile of MbCO. This indicates facile photodissociation of CO upon UV illumination. The UV light should be absorbed by some aromatic amino-acids first. The excited aromatic residues fluorescence in a ps time scale, but only tryptophane is expected to fluorescence around 400 nm region where the absorption maximum of the heme is located. Therefore, the absorbed UV light energy would be transferred to the heme through tryptophane residues. Unexpectedly, the $\nu_{Fe\text{-}His}$ frequencies of the photodissociated species observed did not exhibit any differences between MbCO, MbO_2 and equilibrium deoxyMb in all time regions that we examined (0 sec - 100 μs). However, the temporal behavior of intensities of $\nu_{Fe\text{-}His}$ bands relative to the intensity of 370 cm^{-1} exhibited some differences; MbCO gave a maximum at 50 ns after photolysis, whereas MbO_2 gave a constant intensity. This suggests two alternative interpretations; one is that the energy transfer rate is different between MbCO and MbO_2; it takes place in 50 ns for MbCO but within 10 ns for MbO_2 and another is that MbO_2 is not photolyzed by UV light and only the probe beam photolyzed species is observed. Although more experiments are required, we can conclude at the present stage that the relaxation of the protein structure after photolysis is so fast irrespective of UV or visible photolysis that the Fe-His bond is accommodated to the equilibrium structure within 10ns.

4. References and Acknowledgment

[1] S. Dasgupta, T. G. Spiro, C. K. Johnson, G. A. Dalickas, and R. M. Hochstrasser, Biochemistry, 24, 5295 (1985).

[2] E. W. Findsen, T. W. Scott, M. R. Chance, J. M. Friedman and M. R. Ondrias, J. Am. Chem. Soc., 107, 3355 (1985).

[3] F. Sato, Y. Shiro, Y. Sakaguchi, T. Iizuka, and H. Hayashi, J. Biol. Chem., 265, 2004 (1990).

[4] M. R. Chance, S. H. Courtney, M. D. Chavez, M. R. Ondrias, and J. M. Friedman, Biochemistry, 29, 5537 (1990).

[5] This study was carried out under the NIBB Cooperative Research Program for the Okazaki Large Spectrograph (90-534).

Ultrafast Measurements of Energy Flow, Geminate Recombination, and Fast Structural Rearrangements in Photoexcited Heme Proteins

R. Lingle, Jr., H. Zhu, X. Xu, and J.B. Hopkins

Louisiana State University, Department of Chemistry,
Baton Rouge, LA 70803, USA

SUMMARY. Transient picosecond Raman spectroscopy is capable of differentiating vibrational relaxation from conformational changes by comparing the Stokes and anti-Stokes dynamics. We have accomplished pump-probe picosecond Raman experiments on deoxy- and oxyhemoglobin (deoxyHb and oxyHb, respectively) using 8 ps 532 nm pump pulses and 8 ps 355 nm probe pulses. Heme-to-protein vibrational cooling has been directly observed in deoxyHb for the first time, and the deconvolved cooling time constant is measured to be 2 - 5 ps.[1-2] By applying our mode-specific Stokes and anti-Stokes technique to oxyHb, we find that any geminate recombination of photodeligated O_2 must occur in either less than two picoseconds or longer than a nanosecond. The relative resonant enhancements for the initial oxyHb bands and the transient deoxyHb bands must be carefully considered in order to correctly interpret the picosecond dynamics data.

1. VIBRATIONAL RELAXATION IN DEOXYHEMOGLOBIN

Spectroscopists need to be able to differentiate between electronic, vibrational, conformational, and ligation dynamics in order to correctly interpret ultrafast spectra. We have directly studied vibrational energy relaxation (VER) in a heme protein for the first time by measuring the picosecond Stokes and anti-Stokes dynamics of photoexcited deoxyHb. Previous Raman experiments[3-4] did not probe the transient anti-Stokes spectrum and were interpreted using indirect analyses. Our experiment is unique in that it *directly* and *unambiguously* detects hot vibrations in both low- and high-frequency porphyrin modes.

We studied deoxyHb by pumping at 532 nm and probing at 355 nm. A pure transient Raman spectrum is produced by subtracting the background probe-only components from the spectrum obtained with the pump-probe sequence. We observe the VER dynamics of photoexcited deoxyHb by monitoring the *Stokes* and *anti-Stokes* transient signals for specific deoxyHb vibrational modes as the time delay between pump and probe pulses is varied.[2] Heme Stokes Raman bands deplete promptly and return to the baseline with a 2-5 ps deconvolved time constant. The dynamics at heme

Springer Proceedings in Physics, Vol. 68 Time Resolved Vibrational Spectroscopy V
Ed.: H. Takahashi

anti-Stokes band positions rise promptly and then decay to the baseline with a 2-5 ps deconvolved time constant. Negative transient Stokes bands indicate removal of population (by the 532 nm pump photons) from those levels thermally populated at room temperature. Positive transient anti-Stokes bands indicate that there is more population in the low-lying excited vibrational modes of the photoexcited heme than in unexcited heme at room temperature. We estimate that the photoexcited deoxy heme is approximately $\approx$ 36 K above room temperature, averaged over the laser pulse width. A 532 nm photon can produce a heme internal temperature jump in excess of 400 K. Therefore the 2-5 ps VER time constant represents the loss of most of the heme internal energy.

2. HEME-LIGAND DYNAMICS IN OXYHEMOGLOBIN

Following photoexcitation, heme-oxygen deligation and geminate recombination may occur in addition to vibrational relaxation. We can separate the heme-ligand dynamics from the vibrational relaxation dynamics by comparing *Stokes* and *anti-Stokes* Raman dynamics. VER with a time constant $\approx$ 5 ps is observed for both hot oxyHb and hot deoxyHb. Our results indicate that O_2 recombination occurs either faster than 2 ps or slower than 1 ns.

The resonant Raman spectrum of oxyHb differs from that of the deoxy photoproduct due to band frequency shifts and/or band intensity changes. These differences in the resonant enhancement of oxyHb and deoxyHb will dramatically affect the dynamics observed for the photodeligation of oxyHb. For example, the oxyHb ν_4 band appears at 1376 cm^{-1}, while the deoxyHb band is shifted to 1359 cm^{-1}. The oxyHb and deoxyHb ν_{16} bands are at roughly the same Raman shift. For Stokes Raman scattering, the ν_4 band is almost a factor of three more intense for oxyHb than for deoxyHb, while ν_{16} is about a factor of three stronger for deoxyHb than for oxyHb. In the anti-Stokes spectrum, both ν_4 and ν_{16} are more intense in deoxyHb than in oxyHb.

Figure 1 shows both Stokes and anti-Stokes dynamics of the ν_4 vibrational mode at 1376 cm^{-1}. Following a pulse-width limited bleaching of the Stokes signal, we observe two dynamics as the transient returns toward the baseline (established by the signal at negative time). The anti-Stokes dynamics rises promptly and then decays rapidly to a level *below* the baseline. Taken together, these dynamics indicate both vibrational cooling and slow recombination of the O_2 ligand. Photodeligation will remove a certain number of oxyHb molecules from the sample volume, leaving deoxy-like Hb in their place. Because the deoxyHb ν_4 band is shifted 20 cm^{-1} off of the position we are observing, *the oxyHb signal dominates the observed dynamics*. Thus photodeligation will result in negative transient signals for both the

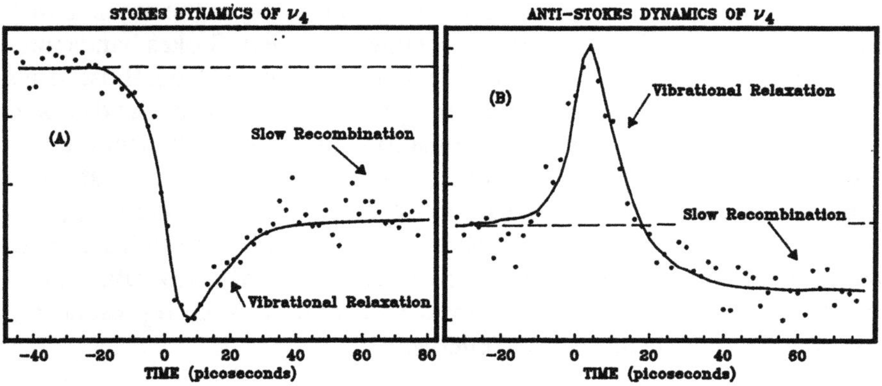

Figure 1

Dynamics of oxyHb for the ν_4 vibrational mode (1376 cm^{-1}). (A) Stokes dynamics; (B) anti-Stokes dynamics. The dots are data points; the solid line is the fit.

Stokes and anti-Stokes bands. This is clearly observed in the long time dynamics, which represents heme-ligand recombination on the nanosecond time scale.

The early time dynamics of Figure 1 is assigned to VER. The complementary negative/positive behavior in the Stokes/anti-Stokes early time dynamics is identical to that observed in deoxyHb. Because the oxyHb resonant Raman signal dominates the spectrum of Figure 1A, the early time dynamics is believed to result from VER in hot *oxyHb* which either never lost its ligand or else underwent geminate recombination faster than 2 ps. Both the Stokes and anti-Stokes dynamics yield a deconvolved time constant for VER of $\approx$ 5 ps. On the other hand, fast recombination would appear as a transient moving in the positive direction in both the Stokes and anti-Stokes dynamics. This is not observed.

Both the Stokes and anti-Stokes dynamics of the oxyHb ν_{16} band in Figures 2A and 2B show *positive* transient signal at long times, because the resonant enhancements for the ν_{16} Stokes and anti-Stokes bands are larger for deoxyHb than for oxyHb. That is, *the observed dynamics of ν_{16} are dominated by the deoxy photoproduct*. These deligated hemes will be vibrationally hot, as shown in the early time dynamics. The ν_{16} anti-Stokes transient signal rises as the deoxyHb low-lying excited levels become populated and then decays to the long time offset (which represents deoxyHb at room temperature). The ν_{16} Stokes transient provides a window to observe the cold vibrational levels of deoxyHb; this signal rises as the deoxyHb vibrationless level is populated. We do not observe fast geminate recombination in the Stokes dynamics, since no negative-going feature appears in Figure 2A.

We see no evidence of ligand recombination on a 2.5 ps time scale, as inferred from transient absorption.[3] A 2.5 ps geminate recombination is

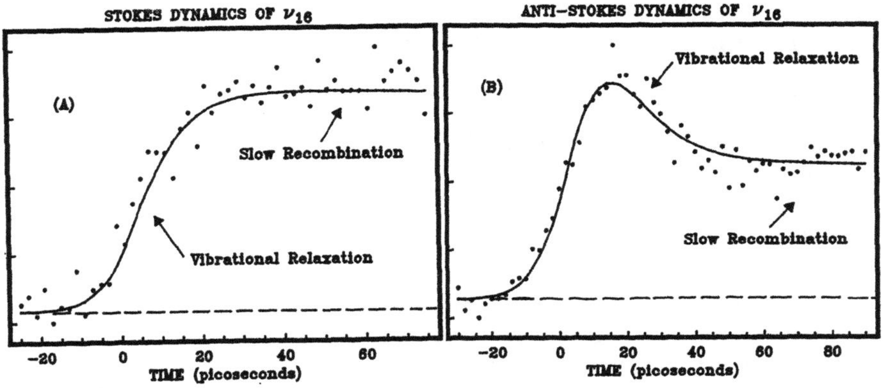

Figure 2

Dynamics of oxyHb for the ν_{16} vibrational mode (755 cm^{-1}). (A) Stokes dynamics; (B) anti-Stokes dynamics. The dots are data points; the solid line is the fit.

on the same time scale as the VER we observe in deoxyHb, so it should appear in the dynamics of Figures 1 and 2 if it exists.

3. REFERENCES

[1.] Robert Lingle, Jr., Xiaobing Xu, Huiping Zhu, Soo-Chang Yu, J. B. Hopkins, J. Am. Chem. Soc., **113**, 3992 (1991).

[2.] Robert Lingle, Jr., Xiaobing Xu, Huiping Zhu, Soo-Chang Yu, J. B. Hopkins, J. Phys. Chem., in press

[3.] J. W. Petrich and J. L. Martin, Chemical Physics, **131**, 31 (1989).

[4.] R. G. Alden, et al., J. Am. Chem. Soc., **112**, 3241 (1990).

Intermediates Formed in the Reaction of Cytochrome *c* Oxidase with Oxygen

Sanghwa Han, Sunho Song, Yuan-chin Ching, and D.L. Rousseau*

AT & T Bell Laboratories, Murray Hill, NJ 07974, USA
*Current address: Department of Biochemistry, School of Natural Sciences, Kangwon National University, Chuncheon, Korea

Cytochrome *c* oxidase is the terminal enzyme in the electron transport chain. In this role, it catalyzes the transfer of four electrons from cytochrome *c* to di-oxygen, the physiological substrate. Oxygen binds to cytochrome *c* oxidase on the microsecond time scale and is fully reduced to water in milliseconds. An understanding of the molecular basis for the catalytic mechanism of the oxygen reduction has been sought after for many years but has remained elusive due to the difficulty in establishing the identity of each intermediate on this time scale by optical spectroscopic techniques. Recently, we [1-5] and others [6-11] have applied resonance Raman spectroscopy to the study of the oxygen reduction process and achieved considerable success. In this paper, we discuss the findings from the resonance Raman scattering experiments and based on these results present a model for the catalytic pathway.

To probe the microsecond-millisecond time scale of the reaction of oxygen with cytochrome *c* oxidase by resonance Raman spectroscopy, we have built a continuous flow apparatus. In this technique a sample of the carbon monoxide-bound enzyme is mixed with an oxygen saturated buffer and then passed into a small capillary or a free standing liquid jet. In the flowing CO-bound enzyme-oxygen mixture one laser beam photodissociates the carbon monoxide from the enzyme allowing the oxygen to react with it. A second laser beam is used to probe the spectrum. The separation between the beams determines the time by which the reaction has progressed. At flow rates of 3.2 ml/min, velocities of 0.85 um/us are reached in the spectroscopic cell and 3.0 um/us in the free jet leading to a time resolution as small as 5 microseconds with our geometry. The scattered light from the second beam is directed onto the entrance slit of a monochromator where it is dispersed and detected by a linear photodiode array.

Cytochrome *c* oxidase contains four redox centers: two heme groups and two copper atoms. One heme group, cytochrome a_3, and one copper atom, Cu_B, form a binuclear center which binds oxygen. The other heme, cytochrome *a*, and the other copper atom, Cu_A, are involved in electron transfer from cytochrome *c* to the binuclear oxygen binding site. The objective of our resonance Raman experiments is to follow the spectra of both of the heme groups in order to identify and characterize the intermediates, measure their kinetic properties, and determine when the electron transfers take place.

To identify the intermediates in the reaction, measurements were made comparing the reaction of the enzyme with $^{16}O_2$ and $^{18}O_2$. Oxygen-isotope sensitivity was detected in three lines (Fig. 1). At the earliest time with $^{16}O_2$, a line is located at 568 cm^{-1} [1,2,5,8,9]. We assign this line as the Fe-O_2 stretching mode of the primary intermediate. As this line decays a new oxygen isotope-sensitive line is detected at 786 cm^{-1} [5,10,11]. This line has hydrogen isotope sensitivity [5] and, in analogy to other proteins, we assign it as the iron-oxygen stretching mode of a ferryl ($Fe^{+4}=O^{-2}$)

Springer Proceedings in Physics, Vol. 68 Time Resolved Vibrational Spectroscopy V
Ed.: H. Takahashi

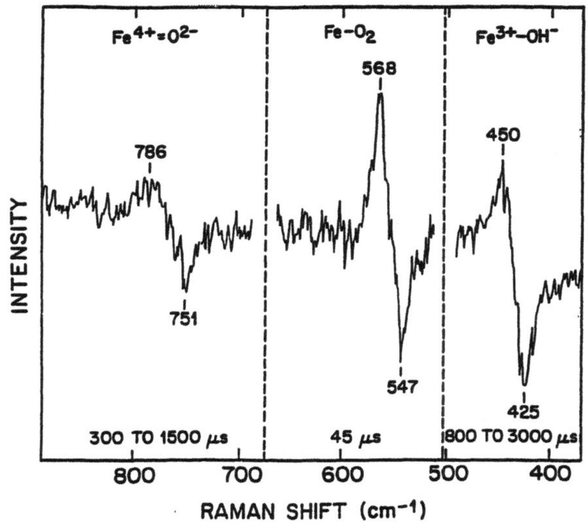

Figure 1. Resonance Raman spectra of the oxygen isotope-sensitive lines detected in the reaction of oxygen with cytochrome *c* oxidase. As the reaction proceeds the Fe-O_2, the $Fe^{4+}=O^{2-}$, and then the Fe^{3+}-OH^- stretching modes are detected sequentially.

intermediate. The hydrogen isotope shift demonstrates that the oxo atom of the ferryl moiety is hydrogen bonded. At the longest times a mode is detected at 450 cm^{-1} which, based on its oxygen and hydrogen isotope shifts, we assign as the Fe-OH stretching mode of a hydroxide intermediate [5]. By monitoring the high frequency region of the resonance Raman spectrum, the rate of oxidation of cytochrome *a* may be followed, thereby allowing for the determination of the rate at which electrons are transferred to the binuclear oxygen binding site. This oxidation occurs very rapidly (over 50% oxidation of cytochrome *a* in less than 100 microseconds) [3,10]. Owing to the equilibrium between cytochrome *a* and Cu_A this result demonstrates that one electron is rapidly transferred from the low potential sites, not from Cu_B, to the binuclear site [3].

Based on the Raman scattering results, we are able to postulate the intermediates in the catalytic pathway. Our results confirm the major intermediates proposed in the past by Wikstrom [12,13] and by Witt and Chan [14], but the electron transfer events, especially the early electron transfer from cytochrome *a* are quite different and instead confirm the branched type of mechanism proposed by Hill and Greenwood [15]. The total scheme is illustrated in Fig. 2. On the right side the oxidation level of cytochrome *a* and Cu_A are shown, and to their left are the oxygen intermediates at the binuclear site.

The starting form of the enzyme under our conditions has CO bound to cytochrome a_3 and all of the redox centers reduced (E_{CO}). Photodissociation yields the ligand free enzyme fully reduced enzyme (E_{red}). Reaction with oxygen gives the primary oxygen-bound intermediate (I). We write this intermediate in the superoxide form based on analogy with oxy complexes of hemoglobin and myoglobin [16] and on our studies of the

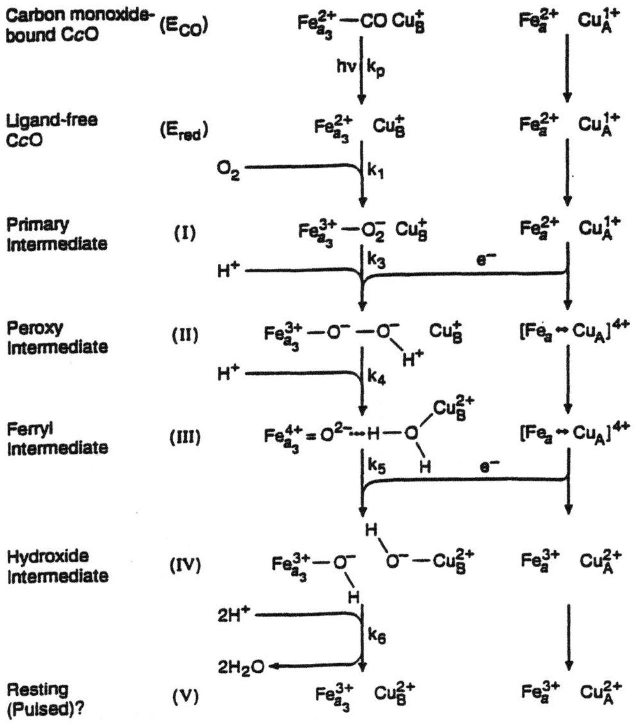

Figure 2. A proposed model for the catalytic mechanism of cytochrome c oxidase. See text for details.

reaction of oxygen with the mixed valence form of cytochrome *c* oxidase which are most easily interpreted as forming a superoxide [4].

The decay of the primary intermediate is generally regarded as a consequence of electron transfer from Cu_B which forms the oxygen binding pocket [12-14]. However, we find that the oxidation of cytochrome *a* proceeds so rapidly that electron transfer from this center can account for the decay of the primary intermediate without the need for any electron transfer from Cu_B [3]. This leads to a peroxy intermediate (II) in which cytochrome *a* and Cu_A are in a redox equilibrium consistent with the reported rapid electron transfer between these centers [17]. In our studies, we have not been able to identify the peroxy intermediate spectroscopically, consistent with observations made by others [18].

As the peroxy intermediate decays the di-oxygen is split with the formation of a ferryl intermediate (III) as confirmed by the Raman data identifying the Fe=O stretching mode. Isotope substitution studies demonstrate that the oxo atom is hydrogen bonded [5]. We speculate that this hydrogen bond is between the oxo atom and a proton on a water molecule bound to Cu_B. Electron transfer to the ferryl intermediate leads to the hydroxy intermediate (IV) which we have identified. Additional protons convert this intermediate to the resting (or the pulsed) form of the enzyme (V).

The model which we have presented here should serve as a starting point for obtaining a quantitative understanding of the full catalytic function of cytochrome *c* oxidase. At present we do not have data to determine which steps in the mechanism involve protonation or play a role in proton translocation. Further studies are necessary to confirm these steps and establish each of the kinetic constants as well as the linkage between proton translocation and oxygen reduction.

References

1. S. Han, Y.-c. Ching, D. L. Rousseau: Biochemistry **29**, 1380 (1990)
2. S. Han, Y.-c. Ching, D. L. Rousseau: Proc. Nat. Acad. Sci. USA **87**, 2491 (1990)
3. S. Han, Y.-c. Ching, D. L. Rousseau: Proc. Nat. Acad. Sci. USA **87**, 8408 (1990)
4. S. Han, Y.-c. Ching, D. L. Rousseau: J. Am. Chem. Soc. **112**, 9445 (1990)
5. S. Han, Y.-c. Ching, D. L. Rousseau: Nature **348**, 89 (1990)
6. G. T. Babcock, J. M. Jean, L. N. Johnson, G. Palmer, W. H. Woodruff: J. Am. Chem. Soc. **106**, 8305 (1984)
7. G. T. Babcock, J. M. Jean, L. N. Johnson, W. H. Woodruff, G. Palmer: J. Inorg. Biochem. **23**, 243 (1985)
8. C. Varotis, W. H. Woodruff, G. T. Babcock: J. Am. Chem. Soc. **111**, 6439 (1989); 112, 1297 (1990)
9. T. Ogura, S. Takahashi, K. Shinzawa-Itoh, S. Yoshikawa, T. Kitagawa: J. Am . Chem. Soc. **112**, 5630 (1990)
10. C. Varotis, G. T. Babcock: Biochemistry **29**, 7359 (1990)
11. T. Ogura, S. Takahashi, K. Shinzawa-Itoh, S. Yoshikawa, T. Kitagawa: J. Biol. Chem. **265**, 14721 (1990)
12. M. Wikstrom: Proc. Nat. Acad. Sci. USA **78**, 4051 (1981)
13. M. Wikstrom: Nature **338**, 776 (1989)
14. S. N. Witt, S. I. Chan: J. Biol. Chem. **262**, 1446 (1987)
15. B. C. Hill, C. Greenwood: Biochem. J. **218**, 913 (1984)
16. J. B. Wittenberg, B. A. Wittenberg, J. Peisach, W. E. Blumberg: Proc. Nat. Acad. Sci. USA **67**, 1846 (1970)
17. J. E. Morgan, P. M. Li, D.-J. Jang, M. A. El-Sayed, S. I. Chan: Biochemistry **28**, 6975 (1989)
18. Y. Orii: Ann. N. Y. Acad. Sci. **550**, 105 (1988)

Picosecond Infrared Study of Carbonmonoxy Cytochrome *c* Oxidase: Ligand Transfer Dynamics and Binding Orientations

K.A. Peterson, P.O. Stoutland, R.B. Dyer, and W.H. Woodruff

Los Alamos National Laboratory, Los Alamos, NM 87545, USA

1. Introduction

Cytochrome *c* oxidase (CcO), an enzyme which catalyzes the reduction of dioxygen to water in the terminal step of the respiratory chain, combines several fundamental chemical processes in performing its function; electron, proton and ligand transfers.[1] The coordination chemistry and ligation dynamics of the cytochrome a_3-Cu_B site, where O_2 and other small molecules such as CO, NO and isocyanates can bind, are essential to the function of the enzyme.[2] The sensitivity of the vibrational frequencies and bandwidths of small molecules to changes in coordination and environment makes infrared spectroscopy uniquely useful as a probe for these processes, particularly at Cu_B^+, which generally is not observable by other spectroscopies.[1,2] Recent time-resolved infrared (TRIR) and visible absorption measurements have shown that coordination to Cu_B^+ is an obligatory mechanistic step for CO entering the cytochrome a_3 heme site and departing the protein after photodissociation.[2] The timescale (> 10^{-7} s) of the TRIR measurements, however, precluded observation of the ligation dynamics immediately following photodissociation. Here we report a picosecond timescale TRIR study of these events. The results reveal that the photoinitiated ligand transfer of CO from Fe_{a3}^{2+} to Cu_B^+, which are believed to lie 4-5 Å apart [1], occurs within 1 ps.

2. Experimental

The experiments reported here were performed on the fully reduced enzyme (Fe_a^{2+}, Fe_{a3}^{2+}, Cu_A^+, Cu_B^+). Beef heart CcO was exchanged into D_2O, reduced with dithionate (pH 7.4) and exposed to ca. 1 atm. CO in an O_2 free environment. The samples were ca. 1 mM in protein, in a 200 µm IR cell. TRIR measurements were obtained in visible pump-IR probe experiments using optical delay. The pump-probe cross-correlation is generally well described by a $sech^2$ function of FWHM = 3.5 ps. The visible pump pulse (ca. 595 nm) is in resonance with the alpha band of the hemes and induces rapid dissociation of CO from Fe_{a3}^{2+}. The IR pulse is tuned to the CO vibrational absorption when bound to Fe_{a3}^{2+} (1963 cm^{-1}) or Cu_B^+ (2062 cm^{-1}). [3] Induced linear dichroism measurements were obtained by rotating the pump polarization relative to the probe polarization. Laser power dependence measurements ensured there was no distortion of the data due to high pump pulse powers. Experimental details are given elsewhere.[4]

Springer Proceedings in Physics, Vol. 68 Time Resolved Vibrational Spectroscopy V
Ed.: H. Takahashi

3. Results and Discussion

The bleach in the absorption at 1963 cm^{-1} is instantaneous, as expected since photodissociation occurs within 150 fs.[6] The formation of the CO complex of Cu_B^+ was directly observed by monitoring the appearance of the absorption at 2062 cm^{-1} (Figure 1). Surprisingly, this risetime is also instrument response limited. The experimental instrument response and a convolution of this with a 1 ps exponential rise are also shown in Fig. 1. The failure of this convolution to fit the data clearly indicates that the rise of the IR transient is less than 1 ps. In scans out to 250 ps, no further changes in absorption were seen at either probe wavelength. Thus there is no significant recombination of CO with Fe_{a3}^{2+} or further development of the Cu_B^+-CO complex within this time period. This agrees with previous TRIR measurements which determined a 1.5 µs lifetime for the Cu_B^+-CO complex and rebinding to cytochrome a_3 on the ms timescale.[2]

The low temperature, static extinction coefficient of the CO-Cu_B^+ absorption is 7 times less than for CO coordinated to Fe_{a3}^{2+}.[3] The ratio of the transient ΔA's at 1963 and 2062 cm^{-1} is equal to the static ratio when the two measurements are performed under the same optical conditions (beam overlap, percent photodissociation, etc.) and suggests that complete formation of the Cu_B^+-CO complex occurs within one picosecond. This interpretation was tested by obtaining transients at various probe frequencies within the Cu_B^+-CO absorption band. The spectrum generated from these transients is identical to both the low temperature spectrum and that obtained at 1 µs [6], except that it is broadened by the 10 cm^{-1} width of the probe pulse (Figure 2). No wavelength dependence to the absorption rise was observed. We conclude that CO is quantitatively transferred and the Cu_B^+-CO complex is formed in 1 ps or less.

Induced linear dichroism experiments can provide a measure of the average angle of the CO dipole relative to the heme plane normal.[7,8] In the limit of 0% photodissociation, the polarization ratio at 1963 cm^{-1} is 1.75±0.06, yielding an average angle of 20±3° for CO bound to Fe_{a3}^{2+}. This result was obtained

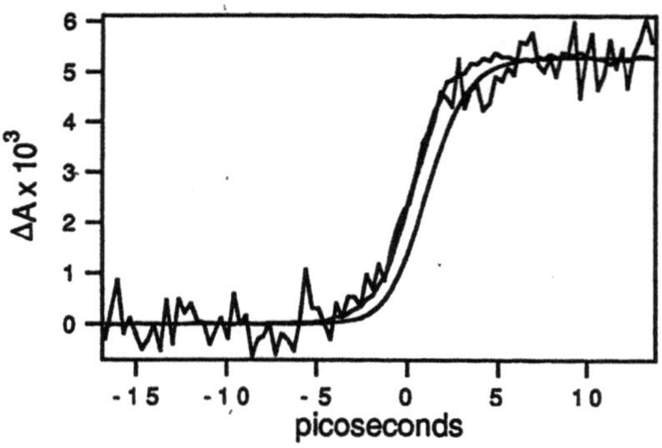

Figure 1. Time-resolved IR absorption for photodissociated CO-CcO monitored at the peak of the Cu_B^+-CO absorption at 2062 cm^{-1}. The smooth traces are; leftmost, the experimental instrument response (3.2 ps $sech^2$) and rightmost, the convolution of this with a 1 ps exponential rise.

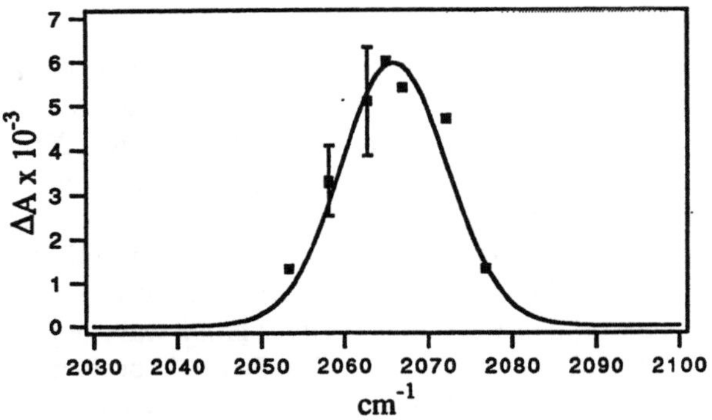

Figure 2. Picosecond transient IR spectrum of the 2062 cm^{-1} stretch of CO when bound to Cu_B^+. The gaussian (FWHM = 12 cm^{-1}) is not fit to the data.

previously on the μs timescale [8] and is similar to values obtained for Hb-CO (18°) and one of the Mb-CO conformers (20°).[7] A polarization ratio of 1.00±.05 was obtained at 2062 cm^{-1}. This corresponds mathematically to either an average angle of 55±3° or an isotropic orientation for CO when bound to Cu_B^+. An isotropic distribution could result either from molecular reorientation on the experimental timescale or an orientation of the CO species formed at 2062 cm^{-1} which is uncorrelated to the orientation of the photoselected hemes. The rotation time for CcO is on the microsecond timescale and cannot contribute significantly to depolarization in a few picoseconds. Furthermore, the narrow bandwidth of for the CO vibration at 2062 cm^{-1} indicates that the CO is in a single, unique environment. For example, "free" CO in the hemoglobin pocket has a 30 cm^{-1} wide band [9] as compared to 7 cm^{-1} when coordinated to Cu_B^+. Consequently, we conclude that the measured polarization ratio of 1.0 means that the CO dipole in the Cu_B^+-CO complex is oriented at an angle near 55° to the cytochrome a_3 heme normal.

The remarkable rate of CO transfer from Fe_{a3}^{2+} to Cu_B^+ provides new insight into the structure of CcO and suggests an unhindered pathway or channel is required to facilitate ligand transfer from one metal center to the other. The rapidity of this transfer does not allow for any barriers to CO translation or rotation nor for any ligand reorganization at Cu_B^+. It is possible that the ligand tranfers via a concerted-like mechanism. The close proximity of the two metals (4-5 Å) means that the CO is potentially within van der Waals contact with Cu_B while bound to Fe_{a3}. When CO is coordinated to Fe_{a3}^{2+}, the CO vibrational frequency, bandwidth, and insensitivity to changes at Cu_B^+ suggest that the Cu_B^+-O interaction is weak. As the CO dissociates, however, it can begin to interact with Cu_B^+ and form a new bond. In any event, the heme pocket must be constructed in such a manner that it restricts the motion of bound CO but expedites rapid transfer of the photodissociated CO between metal centers. This feature of the protein is significant to the role of Cu_B as a ligand shuttle to Fe_{a3} in the functional dynamics of the protein.

Acknowledgement

This work was supported by LANL ISR Grant X15B and NIH Grant DK36263 and performed at Los Alamos National Laboratory under the auspices of the U. S. Department of Energy.

References

1. B. G. Malmstrom, Chem. Rev. **90**, 1247 (1990) and references therein.
2. W.H. Woodruff, O. Einarsdottier, R. B. Dyer, K. A. Bagley, G. Palmer, S. J. Atherton, R. A. Goldbeck, T. D. Dawes, and D. S. Kliger, Proc. Natl. Acad. Sci. USA. **88**, 2588 (1991) and references therein.
3. F. G. Fiamingo, R. A. Altschuld, P. P. Moh, and J. O. Alben, J. Biol. Chem. **257**, 1639 (1982).
4. R. B. Dyer, K. A. Peterson, P. O Stoutland, and W. H. Woodruff, (a) J. Am. Chem. Soc. (1991) in press, (b)Proc. SPIE Soc. Opt. Eng. **1432**, 502 (1991).
6. P. O. Stoutland, J.-C. Lambry, J.-L. Martin and W. H. Woodruff, J. Phys. Chem. (1991) submitted.
7. J. N. Moore, P. A. Hansen, and R. M. Hochstrasser, Proc. Natl. Acad. Sci. USA. **85**, 5062 (1988).
8. R. B. Dyer, J. J. Lopez-Garriga, O. Einarsdottier, W. H. Woodruff, J. Am. Chem. Soc. **111**, 8962 (1989.
9. P. A. Anfinrud, C. Han, and R. M. Hochstrasser, Proc. Natl. Acad. Sci. USA. **86**, 8387 (1989).

Dioxygen Chemistry of Cytochrome Oxidase Revealed by Time-Resolved Resonance Raman Spectroscopy

T. Ogura[1], *S. Takahashi*[1], *K. Shinzawa-Itoh*[2], *S. Yoshikawa*[2], *and T. Kitagawa*[1]

[1]Institute for Molecular Science, Graduate University for Advanced Studies, Myodaiji, Okazaki 444, Japan

[2]Faculty of Science, Himeji Institute of Technology, Kanaji, Kamigori-cho, Ako-gun, Hyogo 678-12, Japan

Cytochrome oxidase has two heme A groups and reduces dioxygen to water. This reaction is coupled with proton translocation through the membrane [1,2]. Decisive structural information of the reaction intermediates is needed to fully understand the reaction mechanism.

Resonance Raman scattering from cytochrome oxidase upon violet laser excitation selectively reveals vibrational spectra of the catalytic site (heme A groups). In some successful cases, the Raman bands associated with the heme-bound oxygen molecule or oxygen atom are observed and give rich structural information of the reaction intermediates, allowing us to discuss the reaction mechanism. Therefore, in this study, time-resolved resonance Raman spectra were observed for reaction intermediates by using the "Artificial Cardiovascular System" [3] and a dual-beam apparatus [4]. The delay time (Δt) was varied between 0.1 and 5.4 ms.

In the spectrum with $\Delta t = 0.1$ ms, there is a band at 573 cm^{-1} for $^{16}O_2$ which is downshifted to 544 cm^{-1} for $^{18}O_2$. This band was previously assigned to the Fe^{2+}-O_2 stretching mode [5-7]. In the spectrum with $\Delta t = 0.5$ ms, there are two new isotope sensitive bands at 788 and 356 cm^{-1} for $^{16}O_2$ which shift down to 749 and 341 cm^{-1} for $^{18}O_2$, respectively. In D_2O the 788 cm^{-1} band upshifts to 802 cm^{-1} but the 356 cm^{-1} shows no frequency shift. We and other groups previously assigned the 788 cm^{-1} band to the $Fe^{4+}=O$ stretching mode [8-10], but we here suggest an alternative assignment of this band to the O-O stretching mode (ν_{OO}) of the hydroperoxy intermediate. The temporal behavior of the two bands at 788 and 356 cm^{-1} are very alike throughout the time range studied, which implies that these two bands arise from the same intermediate. Although both the $Fe^{4+}=O$ stretching and bending vibrations have isotope-sensitivity, the latter mode is expected to be Raman inactive. Accordingly the $Fe^{4+}=O$ structure seems less likely. The ν_{OO} for D_2O_2 is higher by 4 cm^{-1} than that for H_2O_2 [11]. Moreover, the ν_{OO} of oxyhemerythrin, for which the hydroperoxy structure has been proposed, shows a 4 cm^{-1} upshift in D_2O [12]. Since the upshift is caused by vibrational coupling between the O-O stretching and the O-D bending modes [11], the larger upshift in D_2O (14 cm^{-1}) than for hemerythrin is not unreasonable. If the frequency difference between the O-O stretching and the O-D bending modes of the hydroperoxy complex of cytochrome oxidase is much smaller than that of hemerythrin, a larger upshift would be expected.

Springer Proceedings in Physics, Vol. 68 Time Resolved Vibrational Spectroscopy V
Ed.: H. Takahashi

Taking all these things into consideration, the most reasonable assignments of the 788 and 356 cm^{-1} bands at the present stage are to ascribe them to the O-O and the Fe-O stretching modes of the hydroperoxy intermediate.

In the spectrum with Δt_d = 2.2 ms, there is a band at 450 cm^{-1} for $^{16}O_2$ which shows a downshift to 425 cm^{-1} for $^{18}O_2$. In D_2O this band appears at 443 and 415 cm^{-1} for $^{16}O_2$ and $^{18}O_2$, respectively. Accordingly, this band can be assigned to the Fe-OH stretching mode of the ferric-hydroxide intermediate which confirms previous observations [10]. Although we did not observe the ferryloxo Raman bands, we are not ready to exclude the potential involvement of the intermediate in cytochrome oxidase catalysis. However, the present results may demand reconsideration of the reaction mechanism of this enzyme.

References

[1] Wikstrom, M., Krab, K. and Saraste, M., in Cytochrome Oxidase - A Synthesis, Academic Press, New York (1981)

[2] Chan, S. I. and Li, P. M., Biochemistry, 29, 1-12 (1990)

[3] Ogura, T., Yoshikawa, S. and Kitagawa, T., Biochemistry, 28, 8022-8027 (1989)

[4] Ogura, T., Maeda, A., Nakagawa, M. and Kitagawa, T. in Primary Processes in Photobiology (Kobayashi, T. ed.), Springer Proceedings in Physics, 20, 233-241 (1987)

[5] Ogura, T., Takahashi, S., Shinzawa-Itoh, K., Yoshikawa, S. and Kitagawa, T., J. Am. Chem. Soc., 112, 5630-5631 (1990)

[6] Varotsis, C., Woodruff, W. H. and Babcock, G. T., J. Am. Chem. Soc. 112, 1297 (1990)

[7] Han, S., Ching, Y. and Rousseau, D. L., Proc. Natl. Acad. Sci. U.S.A. 87, 2491-2495 (1990)

[8] Ogura, T., Takahashi, S., Shinzawa-Itoh, K., Yoshikawa, S. and Kitagawa, T., J. Biol. Chem., 265, 14721-14723 (1990)

[9] Varotsis, C. and Babcock, G. T., Biochemistry 29, 7357-7362 (1990)

[10] Han, S., Ching, Y. and Rousseau, D. L., Nature (London) 348, 89-90 (1990)

[11] Giguere, P. A. and Srinivasan, T. K. K., J. Raman Spectrosc. 2, 125-132 (1974)

[12] Shiemke, A. K., Loehr, T. M. and Sanders-Loehr, J., J Am. Chem. Soc. 106, 4951-4956 (1984)

Observation of the O-O Stretching Resonance Raman Band for Cytochrome P-450$_{cam}$ Under Catalytic Conditions

T. Egawa[1], *T. Ogura*[1], *R. Makino*[2], *Y. Ishimura*[2], *and T. Kitagawa*[1]

[1]Institute for Molecular Science, Okazaki 444, Japan
[2]Keio University, Shijuku-ku, Tokyo 160, Japan

1. Introduction

Cytochrome P-450 (P-450) is a generic name of heme proteins which have the catalytic activity for monooxygenation of a variety of organic compounds. In their catalytic reaction one oxygen atom of the heme-bound dioxygen is incorporated into a substrate while the other oxygen atom is converted into water. The question to be answered is whether the dioxygen bound to P-450 differs from those of other heme proteins. The O-O stretching (ν_{OO}) frequency is expected to reflect most sensitively the nature of the O-O bond. Bangcharoenpaurpong et al. [1] succeeded in observing the ν_{OO} RR band for noncatalytic O_2 adduct of frozen P-450$_{cam}$ with no electron donor present at -60° C and it was recently confirmed for a solution at -20° C by Nishimura et al. [2]. The observed ν_{OO} frequency was, unexpectedly, very close to that of oxy-myoglobin (oxyMb) in which the heme bound dioxygen is nonreactive. We thought it important to detect the ν_{OO} RR band for oxy P-450$_{cam}$ under catalytic conditions, that is, at room temperature in the presence of electron donor (reduced putidaredoxin; Pd^r) and substrate (D-camphor). Since the reaction becomes faster under such conditions (It is reported that Pd^r and P-450$_{cam}$ associate to form a bimolecular complex and the life time of oxy P-450$_{cam}$ is made significantly shorter by formation of the complex [3]), it is hard to detect oxy P-450$_{cam}$ with an ordinary technique. Accordingly, we applied a home made mixed-flow transient Raman apparatus to observe the ν_{OO} RR band successfully.

2. Experimental Procedures

Both P-450$_{cam}$ and putidaredoxin were reduced enzymatically under anaerobic condition. The mixture of reduced P-450$_{cam}$ and Pd^r was rapidly mixed with O_2 saturated water in the mixed-flow transient Raman apparatus. While the RR spectrum was measured for the flowing sample, the visible absorption spectrum between 500 and 650 nm for the mixed solution in the scattering volume was monitored by using the device for Raman/absorption simultaneous measurement [4]. Raman scattering was excited by the 420 nm line and detected by an OMA-3 system.

Springer Proceedings in Physics, Vol. 68 Time Resolved Vibrational Spectroscopy V
Ed.: H. Takahashi

3. Results and Discussion

We obtained RR spectra at 580 ms after mixing. The RR band at 1141 cm^{-1} was assigned to ν_{O-O} by the isotopic shift using $^{18}O_2$. It is noted that the ν_{OO} frequency observed under the catalytic conditions is almost the same as that observed under the noncatalytic conditions at -60° C (1140 cm^{-1}) [1] and that at room temperature (1139 cm^{-1}) observed for the present preparation.

The product generated after electron transfer from Pd^{r} to oxy P-450_{cam} would be either the Fe^{3+}-O_2^{2-} type dioxygen complex or the Fe^{5+}-O^{2-} type complex. The latter would actually be $Fe^{4+}=O^{2-}$ porphyrin π cation radical. We tried to find an intermediate which is thought to occur following oxy P-450_{cam}. By changing the flow rate, we obtained RR spectra of intermediates at 450 ms, 1,000 ms and 2,300 ms after mixing. The ν_{OO} RR band decreased in intensity for longer delay time and the simultaneously observed absorption spectra also showed the decrease of oxy P-450_{cam}. At 2,300 ms after mixing, the intensity of the ν_{OO} band fell into about half of that at 450 ms, but no other oxygen-isotope sensitive RR band was observed in 700 to 1,500 cm^{-1} frequency region.

References

1) O. Bangcharoenpaurpong, A. K. Rizos, P. M. Champion, D. Jollie, and S. G. Sligar, J. Biol. Chem. 261, 8089 (1986).
2) Y. Nishimura, R. Makino, M. Imai, H. Shimada, H. Koga, T. Horiuchi, and Y. Ishimura, Proc. XIIth Intl. Conf. Raman Spectrosc. (J. R. Durig, and J. F. Sullivan Eds.), pp. 702, John Wiley, (1990).
3) T. C. Pederson, R. H. Austin, and I. C. Gunsalus, in Microsomes and Drug Oxidation (Ullrich et al. Ed.) PP. 275, Pergamon Press, (1976).
4) T. Ogura and T. Kitagawa, Rev. Sci. Instrum. 59, 1316 (1988).

Rapid Mixing Resonance Raman Spectroscopy of Peroxidase Intermediates

V. Palaniappan, M.M. Fitzgerald, A.M. Sullivan, J.R. Shifflett, and J. Terner

Department of Chemistry, Virginia Commonwealth University, Richmond, VA 23284-2006, USA

Peroxidases are heme enzymes which catalyze reactions of hydrogen peroxide and certain other oxidants. The physiological mechanisms of these enzymes involve two oxidized intermediates known as compounds I and II. Compound I is formed by a two-electron oxidation of the resting enzyme and contains an Fe(IV) porphyrin π-radical cation. A one-electron reduction of compound I forms the compound II intermediate which contains an Fe(IV) heme.

Over the past few years, work in our laboratory has centered on horseradish peroxidase compound II and other similar ferryl forms (cytochrome c peroxidase compound ES, ferryl myoglobin, etc.) [1-7]. However, recent efforts have extended our investigations to the compound I type intermediates which contain the porphyrin π-radical cation. We have recently reported resonance Raman spectra of horseradish peroxidase compound I [8]. The purpose of this communication is to provide a brief synopsis of some of the results we have obtained since publication of that report [8].

Chloroperoxidase is a heme enzyme of substantial interest. It contains a thiolate ligated heme in common with the cytochrome P-450 enzymes, rather than a histidine ligated heme as found in horseradish peroxidase, yeast cytochrome c peroxidase, and other well known peroxidases. The intermediates of chloroperoxidase, compounds I and II, are among the least stable (under experimental conditions) of the peroxidase intermediates. It is however possible to stabilize them somewhat at room temperature by avoiding the use of H_2O_2 as a an oxidant. H_2O_2 is commonly used as an oxidant for peroxidases, however chloroperoxidase has catalase activity. Thus H_2O_2 will function also as a substrate, causing rapid turnover of the enzyme.

For generation of the chloroperoxidase intermediates we utilized a Ballou four-jet mixer fed by two 100 ml syringes, which produced a continuous jet of newly formed intermediate. Compound I (Fig. 1b) was formed by mixing the enzyme with an equal volume of peracetic acid in 15-fold excess. Compound II (Fig. 1c) was formed by reacting the peracetic acid with enzyme premixed with a 100-fold excess of ascorbic acid. Under such conditions the intermediates have a lifetime of less than one second at room temperature. The ferrous enzyme (Fig. 1d) was formed by the reduction of the resting (ferric) enzyme (Fig. 1a) with excess sodium dithionite under N_2 in a sealed quartz spinning cell.

The observed resonance Raman frequencies of the chloroperoxidase intermediates and derivatives are similar to those observed for horseradish peroxidase, however there are a number of reproducible differences in frequency, especially with respect to the compound I intermediates.

The in-plane skeletal modes in the resonance Raman spectrum of chloroperoxidase compound II are shown in Figure 1c. The bands are assigned as follows: ν_{10} at 1638 cm^{-1}, ν_{37} at 1606 cm^{-1}, ν_2 at 1582 cm^{-1}, ν_{38} or ν_{11} at 1554 cm^{-1}, ν_3 at 1512 cm^{-1}, ν_{28} at 1476 cm^{-1}, vinyl $=CH_2$ wags at 1345 and 1434 cm^{-1}, ν_{20} or ν_{29} at 1394 cm^{-1} and ν_4 at 1373 cm^{-1}. These assignments are very close in frequency to those previously reported for horseradish peroxidase compound II.

In our previous report [1], we assigned certain high frequency porphyrin skeletal modes of horseradish peroxidase compound I as follows: ν_{10} at 1636 cm^{-1}, ν_{37} at 1614 cm^{-1}, and ν_2 at 1606 cm^{-1}. For chloroperoxidase compound

Springer Proceedings in Physics, Vol. 68 Time Resolved Vibrational Spectroscopy V
Ed.: H. Takahashi

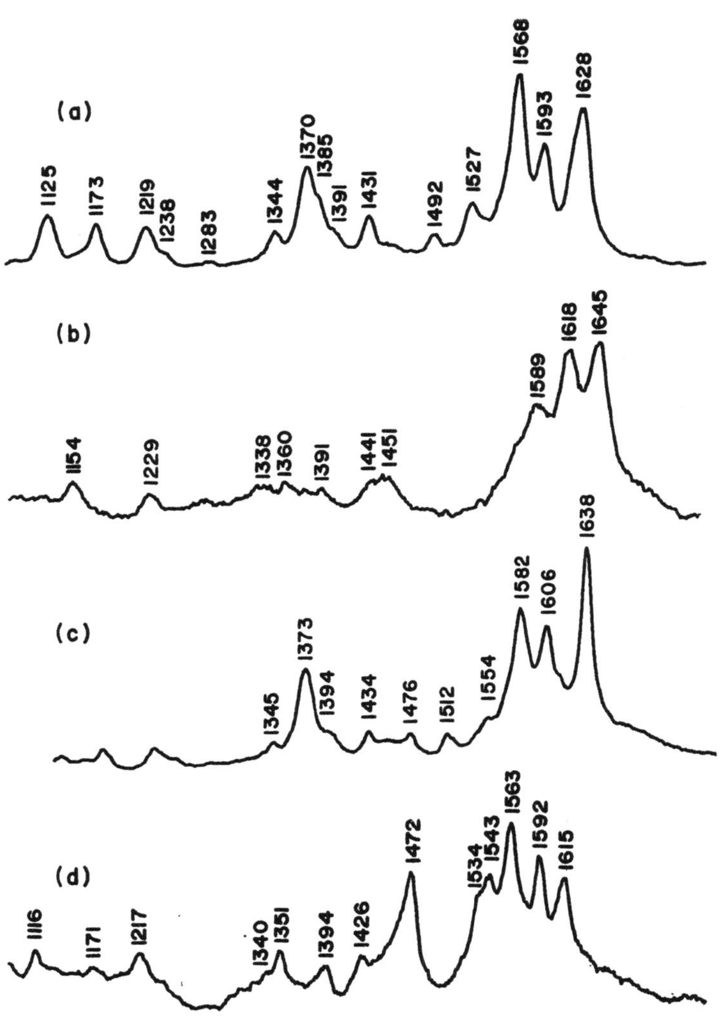

Figure 1. Resonance Raman spectra of chloroperoxidase (3638 Å excitation) at pH 3.8, 0.005 M sodium acetate (a) resting enzyme in a flowing jet stream (b) compound I in a flowing jet produced by mixing the resting enzyme with a 15-fold excess of peracetic acid (c) compound II in a flowing jet produced by mixing a 15-fold excess of peracetic acid with the resting enzyme premixed with a 100-fold excess of ascorbic acid (d) the ferrous enzyme produced by reduction of the resting enzyme with excess sodium dithionite under N_2 in a spinning quartz Raman cell.

I similar bands are found at 1645, 1618, and 1589 cm^{-1}. The band at 1645 cm^{-1} is depolarized, while the bands at 1618 and 1589 cm^{-1} are polarized. We therefore assign the band at 1645 cm^{-1} to ν_{10}, and tentatively assign the bands at 1618 and 1589 cm^{-1} as ν_{37} and ν_2 respectively. Experiments are in progress using additional excitation wavelengths to substantiate these assignments. Though these assignments are preliminary, the compound I intermediates of both peroxidases exhibit resonance Raman spectra attributable to six-coordinate low-spin hemes with frequencies shifted from their

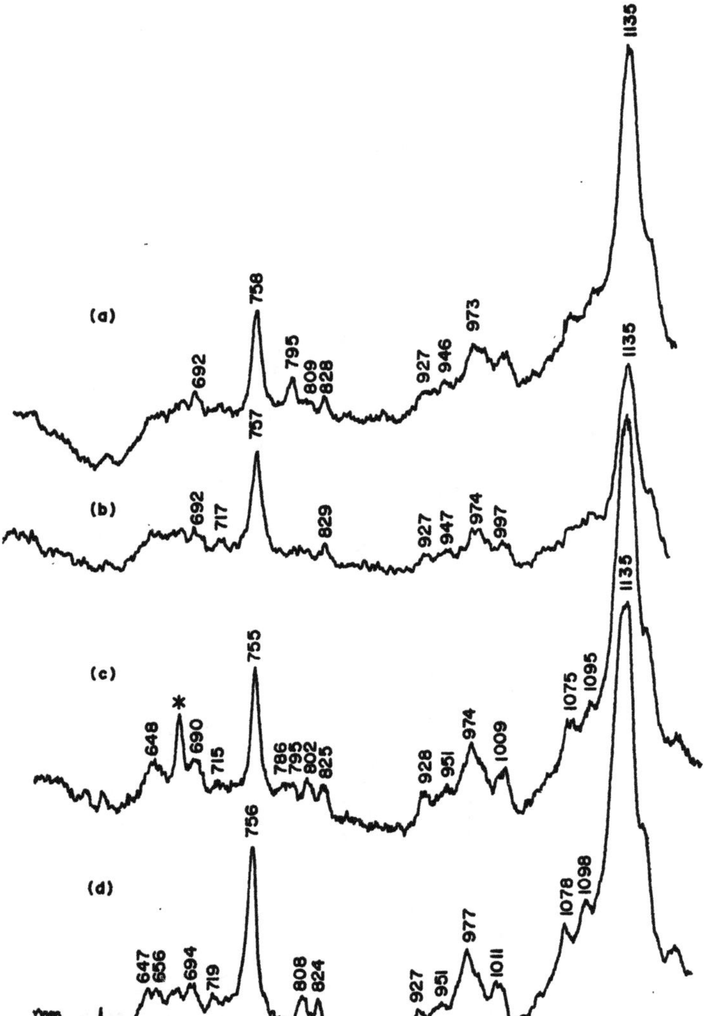

Figure 2. Resonance Raman spectra of horseradish peroxidase compound I formed by mixing the enzyme with equal volumes of H_2O_2 in a two-fold excess (3638 Å excitation). (a) 0.01 M sodium phosphate, pH 6.8; (b) same conditions as "a" except with $H_2{}^{18}O_2$; (c) 0.01 M sodium acetate pH 4.4; (d) same conditions as "c" except with $H_2{}^{18}O_2$. The band marked "*" is a plasma line.

respective compound II values. Nevertheless there are significant frequency differences between the two, especially with regards to ν_{10} and our tentatively assigned ν_2 for chloroperoxidase compound I.

There are a number of possible rationalizations for the frequency differences in the resonance Raman spectra of the compound I intermediates of these two peroxidases. The porphyrin π-radical cation symmetry state assignments have historically been A_1u for chloroperoxidase and A_2u for horseradish peroxidase, supposedly resulting from the differences between thiolate and histidine ligation which themselves are expected to have substantial effects due to differing influences on d_π-porphyrin π^* backbonding. It is also possible that the frequency differences are due to differing interactions of the porphyrin π-radical cation with charged groups in the surrounding heme pocket. Horseradish peroxidase is active at pH 7, however chloroperoxidase must be kept below pH 7 since it will irreversibly convert to an inactive form at pH 7 and above. The data of Figure 1 was therefore obtained near pH 4. Most interestingly, when the pH is lowered from 7 to 4 for horseradish peroxidase compound I, we observed an increase in ν_{10} from 1636 to 1645 cm^{-1}, apparently mimicking the ν_{10} frequency of chloroperoxidase compound I. Concomitant with a raising of the ν_{10} frequency of horseradish peroxidase compound I, we observed a lowering of the Fe(IV)=O frequency. An Fe(IV)=O frequency was previously reported by Oertling and Babcock at 791 cm^{-1} [9], and attributed to compound I. As shown in Figure 2, we observe the band at 795 near pH 7 (Fig. 2a), however a new component at 786 cm^{-1} appears at pH 5 and below (Fig. 2c). Upon ^{18}O-substitution these bands shift to a position that overlaps with a strong band at 756 cm^{-1} (Figs. 2b and d).

Support is acknowledged from grant GM34443 of the U.S. National Institutes of Health.

1. J. Terner and D.E. Reed (1984) Biochim. Biophys. Acta 789, 80-86.
2. J. Terner, A.J. Sitter, and C.M. Reczek (1985) Biochim. Biophys. Acta 828, 73-80.
3. A.J. Sitter, C.M. Reczek, and J. Terner (1985) Biochim. Biophys. Acta 828, 229-235.
4. A.J. Sitter, C.M. Reczek, and J. Terner (1985) J. Biol. Chem. 260, 7515-7522.
5. A.J. Sitter, C.M. Reczek and J. Terner (1986) J. Biol. Chem. 261, 8638-8642.
6. A.J. Sitter, J.R. Shifflett and J. Terner (1988) J. Biol. Chem. 263, 13032-13038.
7. C.M. Reczek, A.J. Sitter and J. Terner (1989) J. Mol. Struct. 214, 27-41.
8. V. Palaniappan and J. Terner (1989) J. Biol. Chem. 264, 16046-16053.
9. W.A. Oertling and G.T. Babcock (1988) Biochemistry 27, 3331-3338.

Picosecond Time-Resolved Vibrational Spectroscopy

G.H. Atkinson

Department of Chemistry and Optical Science Center,
University of Arizona, Tucson, AZ 85721, USA

Introduction

Vibrational spectra of photochemical and photophysical intermediates formed during the initial picoseconds of photo-induced reactions now routinely provide important new insights into the relationship between transient molecular structure and chemical reactivity. In many chemical and biochemical systems, reactivity can be viewed directly in terms of the picosecond structural changes that are reflected in time-resolved vibrational spectroscopy. Among these changes are single and double carbon-carbon bond isomerization, alterations in three-dimensional conformations, differences in electronic interactions which affect delocalized electron densities and conjugation, and bond dissociation.

In addition, time-resolved vibrational data can be used directly to measure reaction dynamics on time scales of a few picoseconds. These dynamical changes are seen in the shifts of frequency positions and changes in the relative intensities of specific vibrational features. The combination of structural and dynamical information found in picosecond time-resolved vibrational data has been used to significantly expand the detailed understanding of the molecular mechanisms which underlie reactivity in even complex biochemical systems. In some cases, new reaction pathways and intermediates have been revealed by these measurements. When time-resolved vibrational data are measured together with transient absorption and fluorescence spectroscopy under experimental conditions which permit all three to be quantitatively related, an even more complete and detailed view of the molecular properties controlling chemical and biochemical reactivity can be obtained.

In this paper, attention is given to picosecond time-resolved resonance Raman (PTR^3) spectroscopy of room temperature photosynthetic and photo-transductive reactions in membrane proteins found in bacteria. A variety of photochemical intermediates and photophysical processes can be observed through the corresponding changes in vibrational degrees of freedom and

Springer Proceedings in Physics, Vol. 68 Time Resolved Vibrational Spectroscopy V
Ed.: H. Takahashi

populations. Both *in vivo* and *in vitro* samples are examined and special emphasis is placed on the roles of vibrationally-excited ground state populations and on the structural changes appearing in excited electronic states involved in energy transfer. Specifically, PTR^3 data are described which have been obtained during photosynthetic and photo-transduction reactions in transmembrane proteins containing retinal and carotenoid chromophores.

Bacteriorhodopsin (BR): Vibrationally-Excited Populations

PTR^3 spectroscopy has been used extensively to characterize the initial photochemical intermediates in the BR photocycle which involve a variety of isomeric and conformational forms of the retinal chromophore [1-3]. The presence of photophysical intermediates such as vibrationally-excited ground state populations in BR-570 (BR') also has recently been found by PTR^3 antiStokes scattering [4,5]. The significance of BR' to the BR photocycle remains to be firmly established, but at least two important characteristics are evident from PTR^3 antiStokes experiments.

First, for pulsed excitation >1 ps, substantial BR' populations can be created since (i) the lifetime of excited electronic state BR-570 (BR^*) optically populated is ≈0.5 ps and (ii) the vibrational relaxation time of BR' to thermally-equilibrated BR-570 is ≈7 ps [4]. For excitation of 5 ps, for example, this means that the early part of the pulse is absorbed by vibrationally "cold" levels while the latter part of the pulse largely excites BR' levels. The two types of excitation should populate different Franck-Condon states in BR^*. The role of different BR^* states on the BR photocycle has yet to be determined.

Second, there appears to be a direct relationship between BR' and J-625, the ground-state species characterized by picosecond transient absorption as the initial ground-state photochemical intermediate in the BR photocycle. An analysis of the frequency positions of PTR^3 bands appearing in both the antiStokes and Stokes spectra shows that the retinal chromophore remains in an all-*trans* configuration [5]. The vibrational anharmonicity of levels comprising BR' causes the associated PTR^3 scattering to be displaced to lower frequencies relative to that from the corresponding vibrationally "colder" levels [5]. A kinetic model derived from picosecond transient absorption and fluorescence data [6,7] as well as from the excitation power dependencies of PTR^3 band intensities is used to obtain the vibrational resonance Raman (RR) spectra of the BR' levels [5].

As expected, the RR spectra recorded from antiStokes and Stokes spectra are the same except with respect to the relative intensities of some RR bands, an observation reflecting differences in the RR cross sections [4,5]. Most importantly, these RR spectra from BR' are essentially identical to that recorded for J-625 [5]. By contrast, there are several clear differences between the BR' RR spectra and the corresponding K-590 RR spectrum [3] which has been assigned to have a 13-*cis* retinal configuration [3]. These comparisons lead to the conclusion that J-625 can be correctly characterized as a vibrationally-excited species which has primarily an all-*trans* retinal configuration. This suggests that retinal isomerization around the $C_{13}=C_{14}$ bond does not occur in BR^*, but rather on the ground-state surface. The J-625 intermediate appears to be a distorted (e.g., vibrationally-excited) retinal which itself undergoes isomerization to K-590 with the involvement of the protein environment [5].

It has long been thought that retinal-protein interactions are important elements in the BR photocycle, but the involvement of the protein environment in the initial isomerization of all-*trans* to 13-*cis* has not been proposed previously. There is evidence from picosecond time-resolved fluorescence (PTRF) spectroscopy, however, that retinal-protein interactions are present from the earliest stages of the BR photocycle [6-9]. For example, PTRF data recorded from artificial BR pigments prepared by replacing all-*trans* retinal by 13-demethylretinal and 11,13 phenylretinal show distinct time and intensity dependencies over the initial 5 ns of the photocycle [8-10]. Since the fluorescence originates in the K^* states of each (PTA data for both artificial pigments show that "J" and "K" intermediates are formed [9]), these changes reflect different nonradiative/radiative competition within K^* which can be associated with difference retinal-protein interactions. In the case of native BR, the PTRF signal decreases significantly over the initial 5 ns of the photocycle suggesting that the protein environment is being altered. From PTRF results of this type, it is evident that retinal-protein interactions play a significant role throughout the initial picosecond time interval of the BR photocycle and that involvement by the protein in the all-*trans* to 13-*cis* isomerization is feasible.

Carotenoids: Excited Electronic States

Carotenoids are widely found to function biochemically in the transfer of excited electronic state energy to acceptors such as chlorophylls with efficiencies approaching 100%. In many carotenoids, the 2^1A_g excited electronic state (lifetime 6-8 ps) is found to be the donor controlling energy transfer even though the optical transition populates the ^1B_u state (lifetime ≈0.25 ps). The ^3B_g also performs a biochemically important function by protecting the system from O_2 quenching. PTR^3 scattering from the 2^1A_g state has been used to determine the specific structural properties which facilitate its involvement in such extremely high efficiency energy transfer [11-17]. These PTR^3 data also monitor the appearance of the ^3B_g state and the population of vibrationally excited ^1A_g levels during electronic state relaxation. Taken together, PTR^3 spectra provide a real time mapping of the energy transfer mechanism occurring within carotenoids.

The PTR^3 spectrum from 2^1A_g β-carotene gives an example which is emulated throughout a large number of synthetic and natural carotenoids. The C-C and C=C bond bands from the 2^1A_g state are shifted by ≈250 cm^{-1} to higher frequencies that the comparable bands form the ^1A_g state [11]. In addition, more than one RR bands of each type appears for 2^1A_g β-carotene (three C-C bands and two C=C bands) [11]. These changes relative to the ^1A_g spectra indicate that either (i) the vibronic coupling between the 2^1A_g and S_n state with which it is in optical resonance generates new relative RR band intensities or (ii) the polyene structure of 2^1A_g β-carotene receives increased electron density and thereby has neither a linear structure nor a completely delocalized electronic distribution. It remains to be determined how these distinct properties are connected (if at all) to the high efficiency energy transfer.

These same general spectral properties are found in the 2^1A_g states of a wide range of carotenoids [13,14]. The frequency shifts of the C=C bands as a function of π-electron conjugation and polyene structure as well as solvent effects are found to be described quantitatively by a vibronic coupling model involving only the ^1A_g, $_1B_u$, and 2^1A_g states [13,14].

The dynamical relaxation mechanism among ^1A_g, 2^1A_g, and ^3B_g states can be seen directly in PTR^3 spectra. Ground state ^1A_g depletion occurs simultaneously with optical pumping, followed by the population of the 2^1A_g state which decays within 6-8 ps [14-17]. The ^1A_g state population does not recover over the same time interval since relaxation from 2^1A_g populates vibrationally-excited ^1A_g levels first. RR bands from the vibrationally-excited population appear displaced to lower frequencies due to vibrational anharmonicities. Vibrational relaxation within ^1A_g is seen to occur over an ≈10 ps interval [15-17]. The ^3B_g state is formed with low yield over the initial 50 ps and only when the carotenoid is located in a pigment-protein complex (chromatophore). The spatial organization within of the chromatophore facilitates fission reactions which increase the efficiency with which the ^3A_g states is formed. No ^3B_g states are observed when the carotenoid is removed from the chromatophore and examined by PTR[3] spectroscopy in organic solvents [16]. The formation and relaxation of vibrationally-excited ^1A_g populations, however, are clearly observed for carotenoids in these same organic solvents [16].

References

[1]G.H. Atkinson, T. L. Brack, D. Blanchard, and G. Rumbles *Chem. Phys.* 131 1-15 (1989).

[2]G.H. Atkinson in *Photobiology*, edited by . Riklis, (Plenum Publishers, New York, 1989).

[3]T.L. Brack and G.H. Atkinson *J. Molec. Structure* 214 289-303 (1989).

[4]T.L. Brack and G.H. Atkinson *J. Phys. Chem.* 95 2352-56 (1991).

[5]T.L. Brack and G.H. Atkinson *Chem. Phys.* (in press)

[6]G.H. Atkinson, D. Blanchard, H. Lemaire, T.L. Brack, and H. Hayashi *Biophys. J.* 55 263-274 (1989).

[7]D. Blanchard, D. Gilmore, T.L. Brack, H. Lemaire, D. Hughes, and G.H. Atkinson *Chem. Phys.* (in press).

[8]G.H. Atkinson and T.L.Brack *J. Luminescence* 48-49 410-14 (1991).

[9]T.L. Brack, W. Gartner, M. Sheves, M. Ottolenghi, and G.H. Atkinson *Biophys. J.* (submitted)

[10]T. Noguchi, S. Kolaczkowski, W. Gartner, and G.H. Atkinson *J. Phys. Chem.* 94 4920-4926 (1990).

[11]T. Noguchi, S. Kolaczkowski, C. Arbour, S. Aramaki, G. H. Atkinson, H. Hayashi, and M. Tasumi, *Photochem. Photobiol.* 50 603-609 (1989).

[12]H. Hayashi, S. V. Kolaczkowski, T. Noguchi, D. Blanchard, and G.H. Atkinson *J. Am. Chem. Soc.* 112 4664-4670 (1990).

[13]T. Noguchi, H. Hayashi, M. Tasumi, and G.H. Atkinson *J. Phys. Chem.* 95 3167-72 (1991).

[14]T. Noguchi, H. Hayashi, M. Tasumi, and G.H. Atkinson *Chem. Phys. Lett.* 75 163-169 (1990).

[15]H. Hayashi, T. Noguchi, M. Tasumi, and G.H. Atkinson *Biophys. J.* (in press).

[16]H. Hayashi, M. Tasumi, and G.H. Atkinson *Chem. Phys. Lett.* 175 163-69 (1990).

[17]H. Hayashi, T. Brack, T. Noguchi, M. Tasumi, and G.H. Atkinson *J. Phys. Chem.* (in press)

Picosecond Time-Resolved Resonance Raman Anti-Stokes Spectroscopy: Vibrationally-Excited Populations in the Bacteriorhodopsin Photocycle

T.L. Brack and G.H. Atkinson

Department of Chemistry and Optical Science Center,
University of Arizona, Tucson, AZ 85721, USA

1 Introduction

Picosecond transient absorption (PTA) and picosecond time-resolved fluorescence (PTRF) results suggest that vibrationally-excited ^1A_g retinal is formed following 5 ps optical initiation of the bacteriorhodopsin (BR) photocycle [1]. This interpretation has been confirmed experimentally by the observation of antiStokes resonance Raman (RR) scattering arising from vibrationally "hot" ^1A_g populations in Br-570 (BR') [2]. These picosecond time-resolved RR (PTR^3) antiStokes data demonstrate that the BR' levels are populated by non-radiative decay from the excited electronic state of BR (BR*) produced by 5 ps optical excitation. PTR^3 measurements using a pump-probe experimental arrangement also have shown that the BR' levels relax with a rate of ≈ 7 ps. [2]. Shifts of the positions of band maxima to lower ferquency displacements on both antiStokes and Stokes RR bands provide an estimate of the ^1A_g vibrational anharmonicities. A detailed kinetic model of the excitation process in the BR phtocycle derived from PTA and PTRF data together with the intensity dependencies of the antiStokes RR bands can be used to establish a clearer understanding of the role of vibrational relaxation and energy redistribution in both the optical excitation process and the BR photocycle.

2 Results and Discussion

AntiStokes RR spectra of BR are obtained using excitation wavelengths in the 563 nm-624 nm region. The relative intensities and positions of the band maxima are essentially the same in all of these spectra indicating that RR scattering arises from the same molecular population which is in strong resonance with the excitation. The band maxima positions are all shifted 2-6 cm^{-1} to lower frequencies relative to the corresponding Stokes RR bands. These shifts are due to vibrational anharmonicity and show that this antiStokes scattering arises primarily from vibrational levels with $v''=1$ or higher.
As the excitation energy from a single picosecond laser pulse is raised, the intensity of the antiStokes RR signal increases almost quadratically with excitation power. This dependence derives from the dual role of the excitation in both creating BR' by pumping BR* and generating RR scattering [3]. Broadening on the low frequency side of the major RR bands also is observed at

the higher excitation powers. This phenomenon is attributable to the increased amount of RR scattering from vibrationally-excited BR' levels which are energetically distinct because of vibrational anharmonicity in the ^1A_g state. These vibrationally "hotter" populations are created by the larger excitation powers which increase the BR* population and the associated nonradiative decay to BR' [3]. Similar broadening is observed in the corresponding Stokes RR bands for increased excitation power [3]. This RR scattering arises from the same BR' population, but is observed only at high excitation powers because of the more intense Stokes scattering from the vibrationally equilibrated ^1A_g population.

The observed functional dependence of antiStokes and Stokes RR band intensities on single laser excitation power is reproduced when the transient populations of all the relevant BR species, including BR' and vibrationally-excited K-590 (K'), are considered. A kinetic model based on PTA and PTRF data acquired for the BR photocycle under the same experimental conditions as those used for PTR^3 spectroscopy [4]. When the contribution of the thermalized ^1A_g BR population is removed from the higher power Stokes RR spectrum, the resulting difference RR spectrum is identical to that of the antiStokes spectrum. Such agreement confirms that RR scattering observed in the broadened antiStokes and Stokes data at higher excitation powers originate from the same vibrationally-excited ^1A_g levels. This result is significant with respect to the BR photocycle since Stokes RR scattering of this type has been assigned to the ground states of the photochemical intermediates K-590 and J-625 [4-6].

These results also suggest that at higher excitation powers, optical pumping involves significant contributions from vibrationally-excited ^1A_g populations.

1. D. Blanchard, D.A. Gilmore, T.L. Brack, H. Lemaire, D. Hughes and G.H. Atkinson, Chem. Phys. (in press).
2. G.H. Atkinson, D. Blanchard, H. Lemaire, T.L. Brack and H. Hayashi, Biophys. J., 55, (1989) 263.
3. T.L. Brack and G.H. Atkinson, J. Phys. Chem. 95, (1991) 2351.
4. G.H. Atkinson and T.L. Brack, Chem. Phys., 131, (1989) 1.
5. R.van den Berg, Du-Jeon-Jang, H. C. Bitting, and M.A. El-Sayed, Biophys. J., 58, (1990) 135.
6. D. Stern and R. Mathies, in M Stockberger and A. Laubereau (Eds.), Time-Resolved Vibrational Spectroscopy, Academic press, New York, 1985, p.250.

Picosecond Time-Resolved Spectroscopy of Artificial Bacteriorhodopsin Pigments

T.L. Brack[1], *W. Gärtner*[2], *M. Sheves*[3], *M. Ottolenghi*[4], *and G.H. Atkinson*[1]

[1]Department of Chemistry and Optical Sciences Center, University of Arizona, Tucson, AZ 85721, USA

[2]Institut für Biologie, University of Freiburg, W-7800 Freiburg, Fed. Rep. of Germany

[3]Department of Organic Chemsitry, The Weizmann Institute of Science, Rehovot, Israel

[4]Department of Physical Chemistry, The Hebrew University of Jerusalem, Jerusalem, Israel

1. Introduction

Artificial bacteriorhodopsin (BR) pigments prepared by reconstituting the apoprotein opsin with chemically modified retinals have been shown to have a variety of distinct chemical and spectroscopic properties [1,2]. Initially investigated as models of the biophysical system operating in bacteria, these artificial BR pigments have also become recognized as new materials for optical processing applications. The influence of different retinal chromophores on the chemical, dynamical, and spectroscopic properties of the respective artificial BR pigment can provide new information on the structural changes undergone by the all-*trans* retinal in the native pigment as well as the associated chromophore-protein interactions.

Recent picosecond transient absorption (PTA), picosecond time-resolved fluorescence (PTRF), and resonance Raman (PTR^3) studies have revealed that the differences between the spectroscopic properties of the native and artificial BR pigments extend into the picosecond time regime [2,3]. The picosecond time-resolved laser spectroscopy of artificial BR pigments prepared with 13-demethyl retinal (removal of the C_{13} methyl group) and a retinal containing a six-membered ring at the C_{12} to C_{14} positions is described in this paper.

2. Results and Discussion

PTA data indicate that both 13-demethyl and native BR behave similarly with respect to producing two red-shifted intermediate within the initial 10 ps of their respective photocycles [3,4]. These intermediates are associated with the J-625 and K-590 species identified in the native BR photocycle [5].

Different kinetics over the initial 10 ps are observed in the PTRF data when the native and artificial BR pigments are compared [3,4]. The 13-demethyl pigment exhibits a significantly higher (≈ 3) fluorescence yield when measured >13 ns after the initial 5 ps excitation. This slowly increasing fluorescence signal is in contrast to the PTRF signal from native BR which has a rapid (8 ps) rise (5 ps excitation [6]) which reaches a plateau remaining constant for several ns.[4].

The enhanced fluorescence in the native BR is assigned to K-590 [5]. The increased fluorescence signal in 13-dm BR also is associated with the formation of a "K-like", isomerized intermediate.

In native BR, a slow decrease in PTRF signal is found with pump-probe delays > 2 ns. The observed increase in fluorescence for 13-dm BR over this time interval suggests significantly different nanosecond dynamics. In both cases, changes in the PTRF signals are attributed to changes in the non-radiative relaxation pathways affecting the excited electronic states of the respective K-like intermediates. Differences in chromophore-protein interactions are considered likely sources of such differences.

The artificial pigment reconstituted with the retinal containing a six-membered ring at the C_{12}-C_{14} positions has PTA properties which are analogous to those in native BR ("J" formed in <3 ps and "K" formed in ≈8 ps [5]), but exhibits significantly different PTRF behavior. There is a slow (300 ps) recovery of the fluorescence signal following the initial 5 ps excitation and no increase in the fluorescence signal above that from the ground-state level which can be attributed to the formation of a "K-like" intermediate. PTR^3 data for this artificial pigment are currently being recorded.

Although more detailed studies are required to fully characterize these artificial BR pigments, it is clear that the contrasting fluorescence properties observed to date suggest that the influence of protein environment on the excited-state electronic properties of the retinal chromophore are significantly changed as the chemical-modifications of the retinal are altered. PTRF spectroscopy. therefore, provides an opportunity to examine such chromophore=protein interactions.

1. M. Ottolenghi and M. Sheves, J. Mem. Biol., **112**, (1989).
2. T. Noguchi, S. Kolaczkowski, W. Gärnter and G.H. Atkinson, J. Phys. Chem., **94**, (1990) 4920.
3. T.L. Brack, M. Sheves, M. Ottolenghi and G.H. Atkinson, J. Chem. Phys. (submitted).
4. T.L. Brack, W. Gärtner and G.H. Atkinson, Chem. Phys. Lett. (submitted).
5. D. Blanchard, D.A. Gilmore, T.L. Brack, H. Lemaire, D. Hughes and G.H. Atkinson, Chem. Phys. (in press).

Picosecond Time-Resolved Resonance Raman Spectroscopy of Bacteriorhodopsin: Structure and Kinetics of the J, K, and KL Intermediates

S.J. Doig, P.J. Reid, and R.A. Mathies

Chemistry Department, University of California, Berkeley, CA 94720, USA

1. Introduction

Bacteriorhodopsin (BR) is an intrinsic membrane protein that functions as a light-driven proton pump [1]. Light absorption by its all-*trans* retinal protonated Schiff base prosthetic group initiates a *trans* → 13-*cis* isomerization. The first ground-state photoproduct, called J, forms in only 500 fs [2]. J decays to the K intermediate in 3 ps [3] which forms L in about 1 μs. The structural changes of the chromophore that accompany the J to K transition are not well defined. Also, although some studies have suggested the presence of an additional intermediate between K and L called "KL" [4, 5], this transition has not been characterized.

The structure and kinetics of the early intermediates can be examined using time-resolved resonance Raman spectroscopy. Single-pulse Raman experiments with 40-ps resolution have observed the spectrum of a red-shifted photoproduct with fingerprint vibrations characteristic of a 13-*cis* chromophore [6, 7]. A study with 8-ps pulses exhibited a fingerprint that was thought to arise from a mixture of J and K [8]. Recently, El-Sayed and coworkers [9] recorded a Raman spectrum using 800-fs pulses which was assigned to J. Two-color, time-resolved Raman experiments have recently been used to obtain the spectrum of K [8]. We now extend the use of the two-color, pump-probe technique by employing an apparatus with sufficient time resolution to resolve the J to K transition and to determine the role of vibrational relaxation in this transition.

2. Materials and Methods

A schematic of our laser system is presented in Fig. 1. A mode-locked argon ion laser was used to synchronously pump a dye laser to produce 3-ps pulses with an energy of ~200 pJ. These pulses were selectively amplified using a four-stage prism dye amplifier pumped by a 50 Hz Nd:YAG laser (Quantel YG 581C). The amplified pulse at 589 nm had an energy of 1 mJ, ASE of <1%, and an autocorrelation trace of 2.8 ps FWHM. The pump pulse was created by focusing 90% of the amplified pulse into a 10 cm cell of water to create a continuum. Light at 550 nm was selected with a 10 nm bandpass filter and then amplified with a two-stage prism amplifier pumped by the YAG laser. The amplified 550 nm pulse had an energy of 600 μJ, an autocorrelation width of 2.5 ps, and a cross-correlation width with the probe of 2.8 ps.

Purple membrane fragments containing bacteriorhodopsin were isolated from an overproducing strain of *Halobacterium halobium.* The light-adapted BR suspension was recirculated through a vertically mounted square capillary. The pump and probe beams were combined at a beamsplitter and then focused in the flow cell with cylindrical lenses to produce an illuminated spot about 1 x 0.5 mm. The pump pulse energy was 14 μJ at 550 nm which produced about 20% photolysis. The low-photoalteration probe pulse had an energy of 1.4 μJ. Raman scattering was collected with a camera lens and focused into a Spex 500M single spectrograph equipped with a PAR 1421 OMA detector.

3. Results

Three sets of spectra were recorded at each time delay: pump-only, pump-and-probe, and probe-only. First, the pump-only background was subtracted from the pump-and-probe spectrum to reveal the photolysis spectrum. The photoproduct spectrum was obtained by

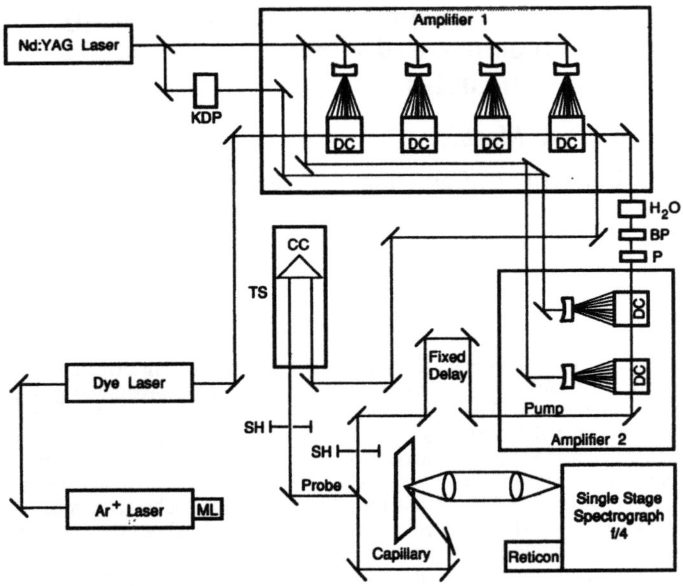

Fig. 1. Schematic of the picosecond Raman apparatus. A single YAG laser operating at 50 Hz is used to pump two amplifiers to provide tuneable pump and probe pulses [11].

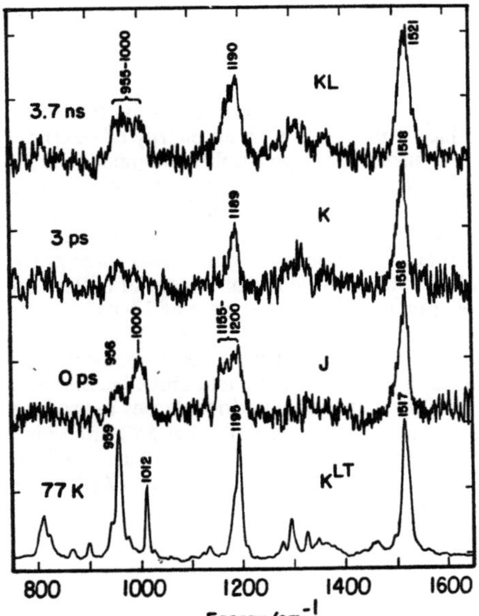

Fig. 2. Resonance Raman spectra of the photoproduct of bacteriorhodopsin for time delays of 0 ps, 3 ps, and 3.7 ns [11]. Spectra were obtained using probe and pump wavelengths of 589 and 550 nm. A spectrum of K trapped at low temperature is shown at the bottom [12].

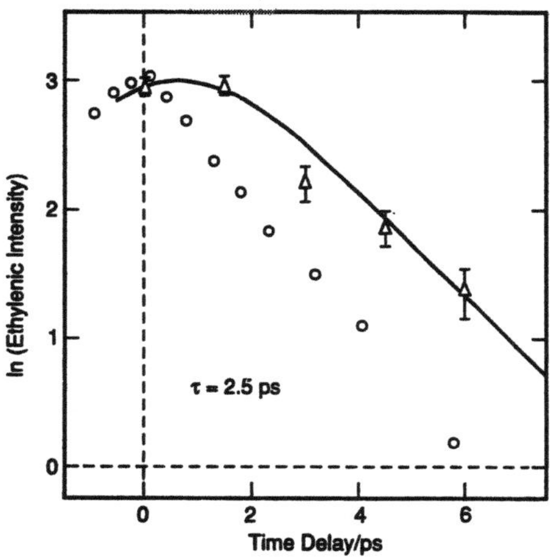

Fig. 3. Natural logarithm of the anti-Stokes ethylenic intensity versus time (Δ), The fit (solid line) was obtained by convolving a Gaussian fit to the experimental (O) cross-correlation trace with an exponential, $J(t) = J(t_0)e^{-kt}$. The best fit gave a (1/e) cooling time of 2.5 ± 1 ps.

removing the scattering from BR by subtracting a fraction of the probe-only spectrum. The subtraction parameter was determined by subtracting the BR spectrum until negative features were observed at the locations of prominent BR lines (1201 and 1527 cm^{-1}).

Time-resolved Raman spectra of the bacteriorhodopsin photoproduct are presented in Fig. 2 along with a spectrum of the photoproduct trapped at low temperature which we denote K^{LT}. The spectra at 0 ps, 3 ps and 3.7 ns have been selected from the kinetic data because they represent the spectra of the J, K, and KL intermediates, respectively. For J and K the ethylenic line is at 1517-1518 cm^{-1} and it shifts up to 1521-1522 cm^{-1} in KL. The fingerprint region of the J spectrum is complex with a broad band of lines extending from 1155 to 1200 cm^{-1}. By 3 ps the fingerprint has collapsed to a single broad line centered at 1189 cm^{-1} which is characteristic of a 13-*cis* chromophore [10], and this feature is essentially unchanged in KL. The hydrogen out-of-plane (HOOP) wagging intensity of J at 956 and 1000 cm^{-1} is initially very strong but it collapses by the time K is formed. There is an increase in the HOOP intensity at ~100 ps which is evident when the integrated area is plotted versus time [11].

To look for evidence of vibrational cooling, anti-Stokes resonance Raman spectra were recorded using temporally distinct 588 nm pump and probe pulses. Pump-and-probe, pump-only and probe-only spectra were recorded and then the pump-only and probe-only contributions were directly subtracted to yield the photolysis spectra. These spectra are composed of scattering from both vibrationally-excited photoproduct and BR. However, the 1521-1523 cm^{-1} ethylenic frequency observed in the anti-Stokes data suggests that the spectrum is dominated by scattering from the photoproduct. Figure 3 presents a plot of the natural logarithm of the ethylenic intensity versus time. The convolution of the system response with a first-order decay gave a satisfactory fit with a decay time of 2.5 ps.

4. Discussion

These time-resolved data provide new information about the structure of the chromophore in J. The intense HOOP modes indicate that the chromophore in J is conformationally distorted. The anti-Stokes scattering intensity directly shows that the chromophore is vibrationally excited. The breadth and complexity of the lines in the fingerprint region of the Stokes spectrum of J is probably due to the anharmonicity of these modes. The spectrum of the low-temperature trapped photoproduct K^{LT} shows that conformational distortion (as indicated by HOOP intensity at 959 cm^{-1}) is not sufficient to cause the distortion of the fingerprint observed in J. Thus, the J intermediate contains a twisted, thermally hot chromophore.

These Raman data show that the transition from J to K is due to cooling and a relaxation of conformational distortion. The close correspondence between the vibrational cooling time (2.5 ps) and the time for the J to K transition, 3 ps [3], argues that cooling is the origin of this transition. Since the K spectrum exhibits the fingerprint pattern expected for a 13-*cis* chromophore, we conclude that isomerization is complete on a 3 ps time scale. The lack of any shift of the ethylenic line between J and K suggests that there is little change in the electronic structure of the chromophore or of the interaction between the Schiff base and the protein.

The long-time kinetics exhibit a transition around 100 ps that is assigned to the K→KL transition. The ethylenic line shifts from 1518 to 1522 cm^{-1} and the HOOP intensity increases. We take these observations as evidence for an additional transition from K to a species denoted as KL on a 100-ps time scale. This transition has been previously suggested in time-resolved absorption studies [4, 5].

These data allow us to develop a relatively complete picture of the photochemical dynamics in bacteriorhodopsin. The rapid isomerization in a restrictive protein binding pocket produces a highly twisted and vibrationally hot photoproduct. Vibrational cooling and conformational relaxation of the chromophore occurs in 3 ps to produce the 13-*cis* photoproduct, K. This suggests that vibrational cooling and conformational relaxation on this time scale will be a common feature of the picosecond dynamics in photochemical reactions. At much longer times, the chromophore-protein complex undergoes a relaxation (K→KL) in which the Schiff base forms a stronger hydrogen-bond with its protein counterion and conformational distortion is reintroduced in the chromophore as a result of this protein conformational change. This transition provides an indication that the presence of the isomerized chromophore in the protein binding site drives further conformational changes of the protein. Isomerization-driven protein conformational changes are now thought to be a key feature of the proton-pumping mechanism in BR [1, 13].

References

1. R.A. Mathies, S.W. Lin, J.B. Ames, W.T. Pollard: Annu. Rev. Biophys. Biophys. Chem. **20**, 491 (1991)
2. R.A. Mathies, C.H. Brito Cruz, W.T. Pollard, C.V. Shank: Science **240**, 777 (1988)
3. H.J. Polland, M.A. Franz, W. Zinth, W. Kaiser, E. Kolling, D. Oesterhelt: Biophys. J. **49**, 651 (1986)
4. S.J. Milder, D.S. Kliger: Biophys. J. **53**, 465 (1988)
5. Y. Shichida, S. Matuoka, Y. Hidaka, T. Yoshizawa: Biochim. Biophys. Acta **723**, 240 (1983)
6. C. Hsieh, M.A. El-Sayed, M. Nicol, M. Nagumo, J. Lee: Photochem. Photobiol. **38**, 83 (1983)
7. D. Stern, R. Mathies: In *Time-Resolved Vibrational Spectroscopy*, ed. M. Stockburger, A. Laubereau (Springer-Verlag, Berlin, New York 1985) Vol. 4, p. 250
8. G.H. Atkinson, T.L. Brack, D. Blanchard, G. Rumbles: Chem. Phys. **131**, 1 (1989)
9. R. van den Berg, D. Jang, H.C. Bitting, M.A. El-Sayed: Biophys. J. **58**, 135 (1990)
10. S.O. Smith, J.A. Pardoen, J. Lugtenburg, R.A. Mathies: J. Phys. Chem. **91**, 804 (1987)
11. S.J. Doig, P.J. Reid, R.A. Mathies: J. Phys. Chem. (in press)
12. M. Braiman, R. A. Mathies: Proc. Natl. Acad. Sci. USA **79**, 403 (1982)
13. S.P.A. Fodor, J.B. Ames, R. Gebhard, E.M.M. van den Berg, W. Stoeckenius, J. Lugtenburg, R.A. Mathies: Biochem. **27**, 7097 (1988)

Structural Variety of Bacteriorhodopsin

*I. Harada, T. Yamagishi, K. Uchida, S. Hashimoto, H. Takeuchi, and F. Tokunaga**

Pharmaceutical Institute and Faculty of Science*, Tohoku University, Aobayama, Sendai 980, Japan

1. Introduction

Bacteriorhodopsin (bR), the only protein contained in the purple membrane of *Halobacterium halobium*, functions as a light-driven proton pump. Such function must result from a series of concerted changes in conformation, microenvironments, and deprotonation/protonation of both the visible chromophore (retinylidene Schiff base) and the protein (opsin) during a photocycle. The aim of our ultraviolet resonance Raman (UVRR) study on bR is to obtain information on the dynamic structure of opsin during the photocycle. UVRR spectroscopy is a powerful technique for examining the structure of proteins [1-3]. Aromatic sidechains and mainchains in proteins absorb deep ultraviolet light and their vibrational scattering can be selectively enhanced by choosing appropriate excitation wavelengths. It is applicable to bR films as well as aqueous suspensions and possible structural differences between them can be detected. This technique has recently been applied to bR suspensions, the reported spectra being partly controversial [4-6]. In the present study (1) suspension UVRR spectra of bR from different batches were examined and two types of bR, containing and not containing a tyrosinate were identified; (2) film UVRR spectra were compared with those of the aqueous suspensions and structural differences in the mainchain and tyrosine and tryptophan sidechains were observed; and (3) a 240-nm spectrum of a photocycle intermediate K was obtained and a slight change from bR568 in the tryptophan scattering was noticed.

2. Results and Discussion

2.1. Bacteriorhodopsins Containing and not Containing a Tyrosinate

UVRR spectra of light-adapted bR (LA) and dark-adapted bR (DA) reported by Harada et al. indicated the presence of a tyrosinate (a 1603-cm^{-1} peak with 253-nm excitation) in bR568 [bR568(+), bR568 containing a tyrosinate] and the possible absence in bR548 [4]. On the other hand, Ames et al. denied the presence of tyrosinate, because their sample did not show the 1603-cm^{-1} peak [bR(-)]. In order to find out the origin of this discrepancy we examined the 253-nm Raman spectra of bR samples from several batches, which had been incubated under similar conditions, and noticed that there were two types of bR, one contained a tyrosinate at pH 8 [bR(+)] and the other did not [bR(-)] (Fig. 1). These two bR's could not be discriminated by visible absorption, because the absorption spectra were practically the same. bR(-) did not give the tyrosinate band even after the following treatments: pH changes up to 11; sonication for 5 min at low temp; drying followed by suspension formation; addition of SDS, Triton X-100, or methanol; and 3-mW UV-laser irradiation. Conservation of the N- and C-terminals was examined on bR(+) and bR(-) samples by an enzyme immuno assay (ELISA) method and the results indicated that the presence or absence of a tyrosinate did not correlate with the conservation of the terminals. Our tentative conclusion is that the sequence of the peptide chain buried in the membrane interior is

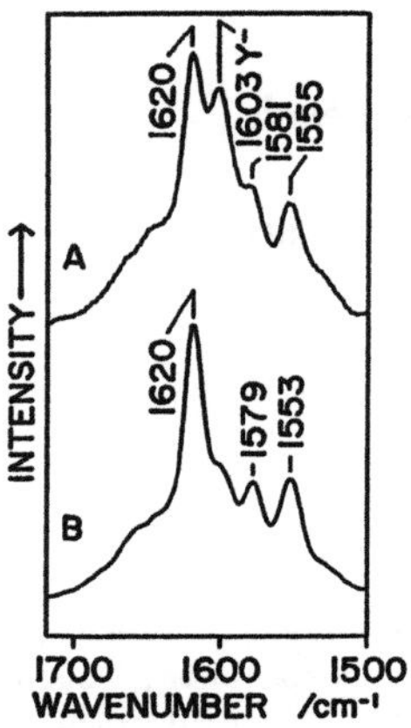

Figure 1. 253-nm Raman spectra of LA in aqueous suspension (pH 8). Sample A contains a tyrosinate (the 1603-cm^{-1} peak) while sample B does not.

different between the two bR's, which causes a change in the local structure around a sidechain group consisting of Tyr-57, Tyr-185, and Asp-212 [7].

As reported earlier [4], the tryptophan scattering with 240-nm excitation (e.g. the intensity ratio of the 1360/1340 cm^{-1} doublet) was different between DA and LA of bR(+). For bR(-), however, the intensity change of the doublet between DA and LA was smaller than that for bR(+).

2.2. LA(-) Aqueous Suspension and Hydrated Film

It has been known that the protein structure in bR film varies depending on the water content [8]. We have taken the UVRR spectra of hydrated films (H_2O, ca 30%) and compared them with those of bR suspensions. In parts A and B of Figure 2 are shown the 240-nm spectra of LA(-) in suspension (A) and in the film (B). Apparently, the spectra are different from each other, especially for the 1360/1340-cm^{-1} tryptophan doublet, the 1620-cm^{-1} scattering, and the amide I band.

The tryptophan doublet is known to arise from Fermi resonance between a skeletal stretch fundamental and one or both of certain out-of-plane combinations [1,9]. The intensity ratio I(1360)/I(1340) is sensitive to the hydrophobicity around the indole ring. With 240-nm as well as visible excitation, the ratio is high when the ring is surrounded by aliphatic groups and it is low when it is exposed to the solvent water [4]. The difference in the shape of tryptophan doublet between the suspension and the film indicates that the average hydrophobicity is different between the two.

The 1617-cm^{-1} peak is very strong in the LA film (B). Since this band is due to an overlap of scattering from tryptophan and tyrosine, it is necessary to separate the individual contributions, in order to see the origin of such an intensity increase. We prepared a bR sample in which the tyrosine phenol rings were deuterated. In parts C and D are shown the spectra of the tyrosine-deuterated LA(-) in suspension (C) and in film (D). Clearly, the down-shifted tyrosine scattering is much stronger from the film than from the suspension, while the tryptophan scattering is not much different. The broadening of the tyrosine band is due to the presence of imperfectly deuterated phenol rings. In the case of N-acetyl-L-tyrosine ethyl ester, scattering from tyrosine is about twice stronger in chloroform than in water. On such basis the environments of some tyrosine sidechains are more hydrophobic in the film than in the suspension.

The amide I band around 1650 cm^{-1} is very weak and partly overlapped by the strong 1620-cm^{-1} scattering in undeuterated bR (parts A and B in Figure 2). In order to see the amide I band clearly we deuterated both the trypto-

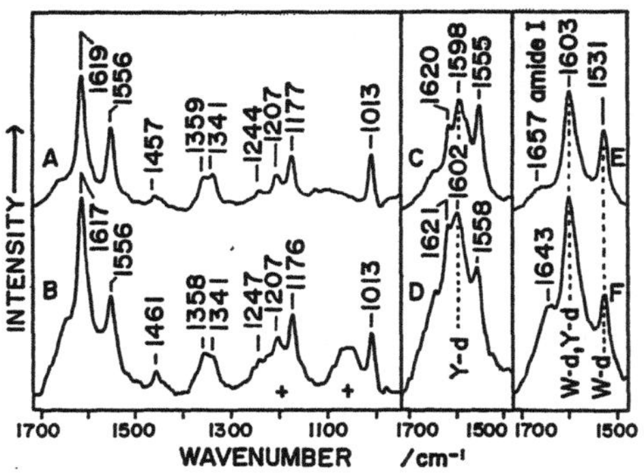

Figure 2. 240-nm Raman spectra of aqueous LA(-) suspensions (A, C, and E) and LA(-) films (B, D, and E). A and B, undeuterated bR; C and D, Tyr-deuterated bR; and E and F, Trp- and Tyr-deuterated bR. †, quartz.

phan indole and tyrosine phenol rings. Although weak, the amide I bands are well observed in the spectra of such bR samples (parts E and F). The band is much sharper and the peak frequency lower in the film (F) than in the suspension (E). This observation indicates that the mainchain α-helices are uniform and tightly wound in the film.

2.3. K Intermediate

UVRR spectra of the K intermediate were obtained by pumping at 532 nm (pulse width 5 ns, 30 Hz, 30 mW) and probing at 240 nm (0.3-1.2 mW) with a delay time of 1 ns. Under this condition most of the photocycle intermediates are expected to take the K state with a lifetime of 2 μs. The amount of remnant bR568(+) at the condition was estimated to be less than 20% from the 532-nm Raman intensity at 1008 cm^{-1} relative to an external intensity standard at 934 cm^{-1} of perchlorate anion in a divided rotating cell. Internal intensity standard was not used, because the UVRR spectra were found to be affected by addition of a large amount of salt such as potassium sulfate. In Figure 3 are compared the 240-nm Raman spectra of bR(+) in the DA, LA,

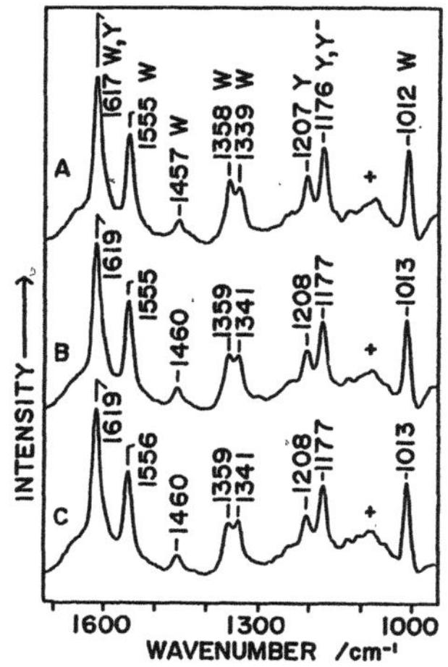

Figure 3. 240-nm Raman spectra of bR(+) in the DA (A), LA (B), and K (C) states. †, quartz.

and K states. Spectral differences between DA and LA were discussed earlier [4]. The spectra of LA and K are mostly identical with the exception of the shape of the tryptophan doublet. Accordingly, a slight change in the environments of tryptophan sidechains takes place on going from LA to K.

3. Conclusion

We have found two types of bacteriorhodopsin, one contains a tyrosinate at pH 8 and the other does not. The bR structure was very different between in suspension and in hydrated films, because of low contents of bulk liquid water in the films. The average environment of the tryptophan sidechains is different between the K intermediate and LA(+).

References

1. I. Harada and H. Takeuchi, in *Spectroscopy of Biological Systems*, ed. by R. J. H. Clark and R. E. Hester (Wiley, Chichester, 1986), Chap. 3.
2. T. G. Spiro and C. A. Grigon, *J. Mol. Struct.* **173**, 79 (1988).
3. S. A. Asher, *Ann. Rev. Phys. Chem.* **39**, 537 (1988).
4. I. Harada, T. Yamagishi, K. Uchida, and H. Takeuchi, *J. Am. Chem. Soc.* **112**, 2443 (1990).
5. A. V. Feofanov, R. G. Efremov, and I. R. Nabiev, in *XII International Conference on Raman Spectroscopy*, ed. by J. R. Durig and J. F. Sullivan (Wiley, Chichester, 1990), pp. 544-545.
6. J. B. Ames, S. R. Bolton, M. M. Netto, and R. A. Mathies, *J. Am. Chem. Soc.* **112**, 9007 (1990).
7. R. Henderson, J. M. Baldwin, T. A. Ceska, F. Zemlin, E. Beckmann, and K. H. Downing, *J. Mol. Biol.* **213**, 899 (1990).
8. M. S. Braiman, P. L. Ahl, and K. J. Rothschild, *Proc. Natl. Acad. Sci. USA* **84**, 5221 (1987).
9. I. Harada, T. Miura, and H. Takeuchi, *Spectrochim. Acta* **42A**, 307 (1986).

Time-Resolved Resonance Raman Studies on Bacteriorhodopsin – Implications for Chromophore/Protein Interaction

R. Lohrmann and M. Stockburger

Max-Planck-Institut für biophysikalische Chemie,
Am Fassberg, W-3400 Göttingen, Fed. Rep. of Germany

Bacteriorhodopsin (bR) is a retinal–binding protein in the purple membrane of Halobacteria, where it acts as a light–driven proton pump. On excitation, bR's retinylidene chromophore, BR–570, runs through a cyclic reaction with several intermediate states (K, L, M, N, O) [1]. However, little is known about the chromophore–protein interaction at the various stages of the cycle.

It is well established that the reaction step from BR–570 to K, which occurs within 5 ps, involves *trans–cis* isomerization around the 13,14 double bond of the retinal chain [2]. We believe that the subsequent dark reaction from K to L, which under physiological conditions occurs within ~2 μs, is of important mechanistic relevance. To substantiate this idea we have studied the RR spectra of K and L in time–resolved experiments. Important conclusions on structural changes of the chromophore and the protein environment could be inferred from the vibrational pattern.

Results and Discussion

The RR spectra of a diluted aqueous suspension of purple membranes were obtained using the spinning cell technique and 514–nm excitation. When appropriate pump conditions are used and the dwell time, Δt, of a sample element in the laser beam is ~2 μs, a nearly 1:1 mixture of K and L is established in the beam [3]. In order to obtain the pure spectra of K well–defined subtraction criteria are required. Here we used the fact that the C=N stretching bands of BR and K which lie between 1610 and 1640 cm^{-1} are well resolved.

In Fig. 1 the spectrum of K obtained in this way is compared with the spectra of the parent, BR, and its direct product L [4]. It must be noted that the K spectrum in Fig. 1 represents this intermediate 1–2 μs after photoexcitation, just before it is converted to L. Distinctive differences in the spectra are observed in the region of the C=C and C=N stretching bands between 1500 and 1650 cm^{-1} which is depicted on an expanded scale in Fig. 1. Whereas BR shows a single strong C=C stretch at 1528 cm^{-1} the intermediates K and L are characterized by two strong bands. The low–frequency band was assigned to a normal mode, $[C=C]_+$, in which the various C=C stretches of the retinal chain move in phase, whereas the high frequency band, $[C=C]_-$, refers to an out–of–phase motion [3].

Spectra as in Fig. 1 were also obtained in suspensions of D_2O (not shown). The results are summarized in Table 1. For an understanding of the bands between 1620 and 1650 cm^{-1} which refer to the protonated Schiff base (–C=NH) by which retinal is linked to the protein, one has to recall that the respective normal modes, besides the C=N stretch, also involve a N–H bending component [5]. Since its frequency (~1350 cm^{-1}) strongly depends

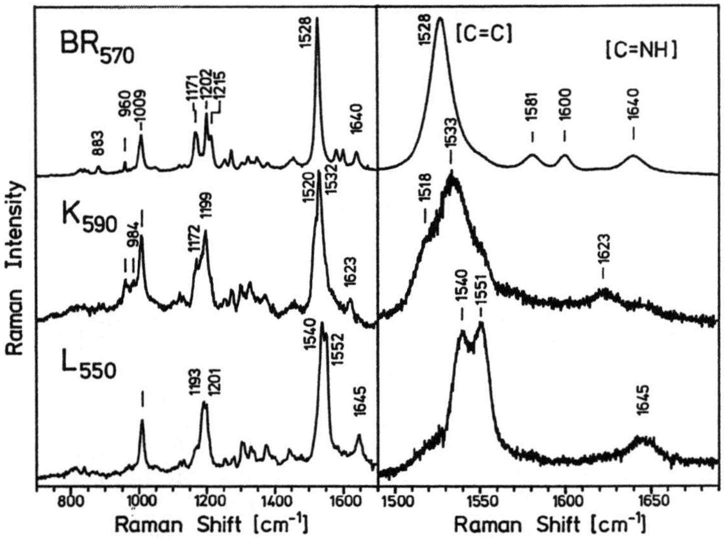

Fig. 1. RR spectra of BR–570 and the intermediates K–570 and L–550 as described in the text

on environmental interactions such effects are also reflected in the [C=NH] normal mode. For instance strong hydrogen bonding or solvation implies a frequency up–shift of the N–H bend and consequently of the [C=NH] normal mode [6]. In D_2O the proton at the nitrogen of the Schiff base group is exchanged for a deuteron. Since the frequency of the N–D bend is low (~970 cm^{-1}) there is no efficient coupling with the C=N stretch. The corresponding [C=ND] normal mode therefore essentially reflects the pure C=N stretching co ponent.

The data in Table 1 show an inverse linear correlation between λ(max) and $\nu[C{=}C]_+$ which had been reported for a variety of retinal chromophores. In the literature it was proposed that such 'color' changes are correlated with variations of the distance between the positively charged Schiff base group and a negative counterion [7]. This model would require that ν[C=ND] varies in the same way as $\nu[C{=}C]_+$. However, as is seen from Table 1 such a relation holds for the transition from BR to K but not for the relaxation step from K to L. This implies that the large changes in λ(max) and $\nu[C{=}C]_+$ during this step cannot be explained by the one–counterion model. It is

Table 1. Characteristic wavelengths and band frequencies of the chromophores BR, K and L

	λ(max) [nm]	$\nu[C{=}C]_+$	$\nu[C{=}C]_-$	ν[C=ND] [cm^{-1}]	ν[C=NH]
K	590	1518	1533	1610	1623
BR	570	1528	–	1622	1640
L	550	1540	1551	1618	1648

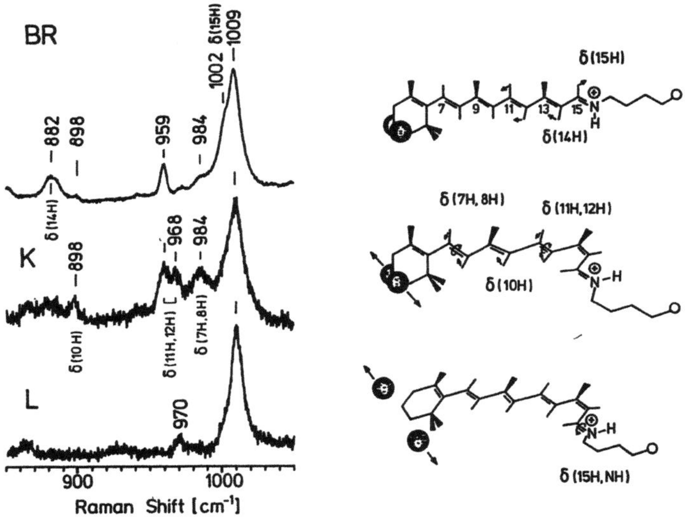

Fig. 2. RR spectra of BR, K, L in the HOOP region

rather suggested that they are induced by conformational changes of the protein which influence the retinal moiety [3].

The frequency up–shift between ν[C=NH] and ν[C=ND] has been correlated with the strength of hydrogen bonding [6]. The observed changes in Table 1 indicate that during the transition from BR to K such bonds are partially broken but reestablished in the step from K to L.

In Fig. 2 RR spectra are shown in the region of the hydrogen out–of–plane bending modes (HOOP–modes). Such modes reflect in–phase motions of two H atoms adjacent to C=C double bonds or motions of single H atoms as indicated on the right–hand side of Fig. 2. HOOP modes are only RR active if the retinal chain is distorted. The high intensity of the HOOP bands in the K spectrum thus shows that in this intermediate state the protein holds the chromophore in a distorted position. The fact that the HOOP bands disappear in the spectra of L indicates that the torsional strain is released.

A simple model is sketched on the right–hand side of Fig. 2. In BR the retinal moiety is kept in a fixed position by surrounding aromatic side groups which is modelled by the two balls adjacent to the β–ionone ring. In the transition from K to L the two balls are kicked away, i.e. the torsional energy which is stored in K is transferred to the protein backbone which then may undergo functionally important conformational changes. This idea is supported by time–resolved infrared experiments which have demonstrated that protonation changes of internal aspartic acid residues already occur at the stage of L [8].

References

1. W. Stoeckenius, R.A. Bogomolni: Annu.Rev.Biochem. 51, 587 (1982)
2. M. Braiman, R. Mathies: Proc.Natl.Acad.Sci.USA 79, 403 (1982)
3. R Lohrmann, I. Grieger, M. Stockburger: J.Phys.Chem. 95, 1993 (1991)

4. R. Diller, M. Stockburger: Biochemistry 27, 7641 (1988)
5. B. Aton, A.G. Doukas, D. Naura, R.H. Callender, U. Dinar, B. Honig: Biophys.J. 29, 70 (1980)
6. T. Baasov, N. Friedman, M. Sheves: Biochemistry 26, 3210 (1987)
7. M.F. Grossjean, P. Tavan, K. Schulten: J.Phys.Chem. 94, 8059 (1990)
8. M. Engelhard, K. Gerwert, B. Hess, W. Kreutz, F. Siebert: Biochemistry 24, 400 (1985)

Chromophore and Protein Reactions of Bacteriorhodopsin Studied by Sub-Microsecond Time-Resolved Step-Scan FTIR Spectroscopy

K. Noelker[1], *O. Weidlich*[1,2], *and F. Siebert*[1,2]

[1]Institut für Biophysik and Strahlenbiologie der Albert-Ludwigs-Universität, Albertstraße 23, W-7800 Freiburg, Fed. Rep. of Germany
[2]Max-Planck-Institut für Biophysik, Kennedyallee 70, W-6000 Frankfurt am Main 70, Fed. Rep. of Germany

Time-resolved IR difference spectroscopy is an ideal tool to study molecular processes in photobiological systems, since it is sensitive to both chromophore and protein alterations. Recently, a new powerful method, time-resolved FT-IR absorption spectroscopy using a step-scan interferometer, has been described [1]. The time-resolution has now been increased to ca. 200 ns, and the accuracy for the slower processes has been increased by the use of a quasi-logarithmic transient-recorder. In this report, this method is applied to the study of the BR->K (KL) transition and to the molecular discrimination of the M and N intermediates.
In Fig. 1, the time-resoved (500 ns after the flash) and the static (temperature 80 K) BR->K difference spectra are compared. The overall agreement is surprisingly good. It is noteworthy that the same spectral features in the amide-I range are observed in both spectra, indicating comparable protein structural changes at 80 K vs. 500 ns after the flash at room temperature. However, the small band at 1555 cm^{-1}, present in the low-temperature spectrum, is missing in the time-resolved studies. The main difference, however, can be seen in the HOOP spectral range below 1000 cm^{-1}: whereas in the low-temperature spectrum a HOOP mode at 955 cm^{-1} (11,12-HOOP) dominates, in the room-temperature spectrum a mode at 980 cm^{-1}, which is absent in the former spectrum, exhibits the largest intensity. A similar behaviour is also observed in resonance Raman investigations [2-5]. High intensity HOOP bands, can be correlated, both in infrared [6] and resonance Raman spectra, with twists of the retinal near the corresponding HC=CH group. Although the mode at 980 has not yet been assigned, it is tempting to indentify it with the 15-HOOP which is usually located between 1010 and 980 cm^{-1} [7]. If this is correct, the relative intensities would signify that in the K (KL) intermediate at room temperature, the twist of the retinal is more localized near the Schiff base, whereas in the low-temperature K the twists extend at least to the middle part of the chromophore. A twist of the 14,15 single bond has been proposed as a cause for the pK reduction of the Schiff base, leading to its deprotonation. Also shown in Fig. 1 is the subtraction

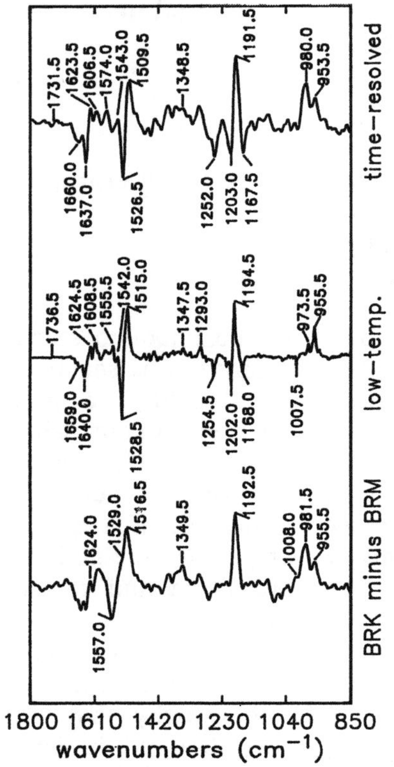

Figure 1 Comparison of time-resolved (upper) und static (middle) BR->K difference spectra. Lower trace:difference between BR->K and BR->M spectra, revealing the K bands.

of the BR->M spectrum from the BR->K spectrum. The former spectrum was obtained in the same measurement as the latter. Since the M bands have very low intensity below 1550 cm^{-1}, the overlap with the negative BR bands is removed by the subtraction and the K bands are revealed. It is noteworthy that the ethylenic shows a shoulder at 1529 cm^{-1}.

Whereas in M the Schiff base is deprotonated, in N it has already taken up a proton. This is a crucial step, since the uptake must occur from the other side than that to which the proton is ejected. From studies of mutants lacking asp96 [8,9] it was concluded that asp96 is the direct proton donor for the reprotonation of the Schiff base. However, the recent structural model of BR places asp96 rather far from the Schiff base. Therefore, we applied our step-scan technique to tackle this problem. Due to the quasi-logarithmic transient-recorder, high-precision BR->M and BR->M<=>N spectra can be recorded. Representative results obtained at

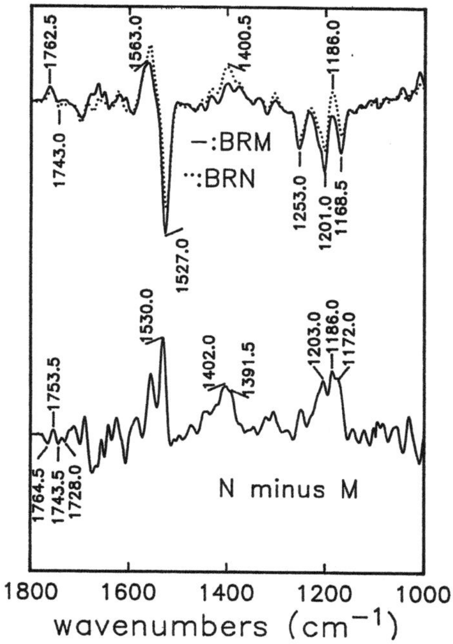

Fig. 2 Time-resolved BR->M and BR->M<=>N spectra (upper traces) and their subtraction (lower trace).

pH 8.5 are shown in Fig. 2. The upper part shows the BR->M and BR->M<=>N difference spectra, at 185 μs and 6 ms after the flash, respectively, the lower part represents the subtraction, cancelling the BR bands. It thus represents the molecular changes involved in the M->N transition. The subtraction reveals clearly the N bands. A difference band is located around 1750 cm^{-1}, the spectral region where protonated carboxyl groups show up. The negative band at 1740 cm^{-1}, due to asp96, has been interpreted as deprotonation of this group [10]. To arrive to such a conclusion, one has, however, to demonstrate that the positive band at 1754 cm^{-1} is not due to asp96. From specific labeling of asp212 on the one hand and asp85, asp96 and asp115 on the other hand we know that the positive band is not due to asp212 and that the negative band at 1765 cm^{-1} is due to asp85 (K. Fahmy, H. Sigrist, M. Engelhard and F. Siebert, in preparation). The data can be interpreted in three ways: 1. asp85, which accepts the proton from the Schiff base in M, gives it already away in N and asp96 stays protonated but changes its environment, the corresponding band being shifted from 1742 to 1754 cm^{-1}. 2. asp85 stays protonated but changes its environment, the corresponding band being shifted from 1764 to 1755 cm^{-1} and asp96 becomes deprotonated. 3. Both groups stay protonated and only the environment is changed. The small size of the bands argue rather for the last possibility.

[1] W. Uhmann, A. Becker, Ch. Taran and F. Siebert, Appl. Spectrosc. **45**, 390 (1991).

[2] J. Terner, C.-L. Hsieh, A.R. Burns and M.A. El-Sayed, Proc.Natl. Acad.Sci. USA **76**, 3046 (1979).

[3] D. Stern and R.A. Mathies, in Time-Resoved Vibrational Spectroscopy (Springer Proceedings in Physics 4), A. Laubereau and M. Stockburger, Eds. (Springer-Verlag, Berlin, 1985), p. 250.

[4] C. Pande, J.K. Lanyi and R.H. Callender, Biophys.J. **55**, 425 (1989).

[5] R. Lohrmann, I. Grieger and M. Stockburger, J.Phys.Chem. **95**, 1993 (1991).

[6] K. Fahmy, M.F. Großjean, F. Siebert and P. Tavan, J.Mol.Struct. **214**, 257 (1989).

[7] M. Stockburger, T. Alshuth, D. Oesterhelt and W. Gärtner, in Spectroscopy of Biological Systems, J.H. Clark and R.E. Hester, Eds. (Wiley & Sons, Chichester, 1986), p. 483.

[8] K. Gerwert, B. Hess, J. Soppa and D. Oesterhelt, Proc. Natl. Acad.Sci. USA **86**, 4943 (1989).

[9] L.J. Stern, P.L. Ahl, T. Marti, T. Mogi, M. Dunach, S. Berkowitz, K.J. Rothschild and H.G. Khorana, Biochemistry **28**, 10035 (1989).

[10] K. Gerwert, G. Souvignier and B. Hess, Proc.Natl.Acad.Sci. **87**, 9774 (1990).

Time-Resolved FTIR Studies on Bacteriorhodopsin

K. Gerwert

Max-Planck-Institut für Ernährungsphysiologie,
Rheinlanddamm 201, W-4600 Dortmund 1, Fed. Rep. of Germany

By two different time-resolved FTIR techniques the light driven proton pump bacteriorhodopsin is investigated:

i) the rapid scan technique yielding a few ms time-resolution [1]

ii) the stroboscope technique yielding a few μs time-resolution [2];

The idea of the rapid scan mode is the following: first a reference spectrum is taken of the protein ground state, than the protein is activated (e.g. by a laserflash) and during the proteinactivity (e.g. during the photocycle) interferograms are taken in a much shorter time than the half-life of the reaction. The interferograms are Fourier Transformed to obtain the spectra. Than difference spectra are calculated between the ground state spectrum and the spectra taken after protein activation. Thereby, only absorbance changes of those molecular groups undergoing reactions are selected from the background absorbance.

In the experimental set up used today the scanner velocity can not be improved significantly due to experimental problems e.g. the acceleration at the turning points is several g. Without principle alterations the time-resolution for the rapid-scan technique in such set-ups is restricted to the ms time domain.

Without changes in the experimental set up, but only in the software the stroboscope method yields a time-resolution of a few microseconds.

The idea is the following: If a time T is needed to take a complete interferogram of length X only T/N time is consumed to take a X/N segment of the interferogram. Recording in succeeding experiments all X/N different interferogram segments and combine this segments to a new interferogram, the combined interferogram has a T/N better time-resolution. This will be illustrated in the following: In Fig. 1 the protein activity is indicated by an absorbance change in the visible (VIS). The interferograms of length X are divided into ten different (N=10) segments. The start of the interferograms relative to the investigated process is shifted from experiment to experiment as indicated in Fig. 1.

Then the segments are rearranged (A^2 B^3 ... K^{11}) into a new interferogram which has a 10 times increased time-resolution. Because out of one interferogram several segments can be used (e.g. A^1 B^2 C^3 ... K^{10}) to form new interferograms with better time-resolution the multiplex advantage is not completely gone.

With the stroboscope technique also the L to M transition in the BR photocycle can be investigated at almost room temperature. A 3-D plot of the difference spectra describing the L to M to N to O to BR reaction is given in Fig. 2.

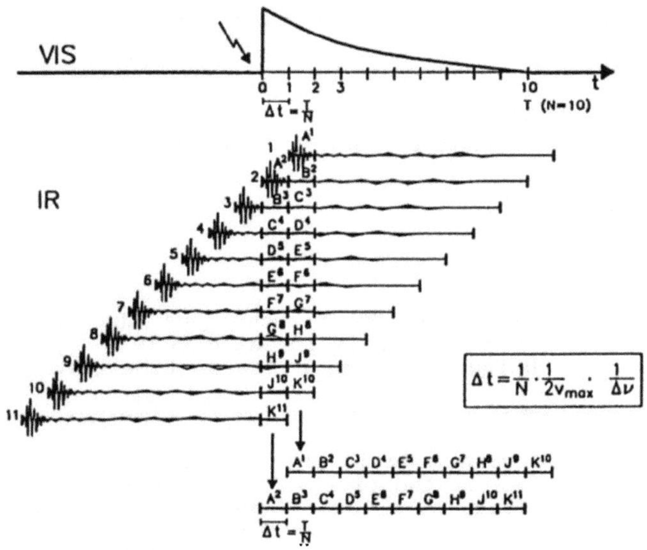

Fig. 1: Stroboscope technique

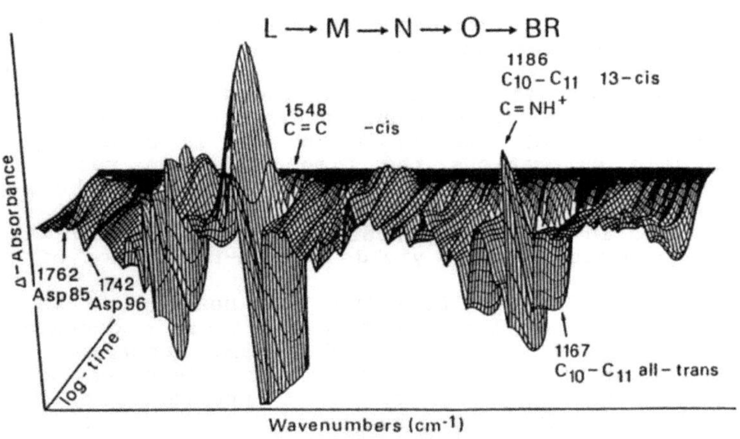

Fig. 2: 3 D Plot of IR-absorbance changes between 1800 cm^{-1} and 1000 cm^{-1}

Negative band showing disappearance of BR bands positive bands correspond to the respective intermediates.

In Fig. 3 the absorbance change at 1189 cm^{-1} in dependence of time on a log-time-scale out of Fig. 2 is shown. It represents the deprotonation of the Schiff Base in the L to M transition and reprotonation and reisomerization in the M to BR reaction. In Fig. 4 the absorbance change at 1762 cm^{-1} is shown indicating transient protonation and deprotonation of asp 85 [4]. Transient protonation of asp 85 can be described by the same rate constants as deprotonation of the Schiff Base (compare Figs. 3,4). Mutation of this residue inhibits this

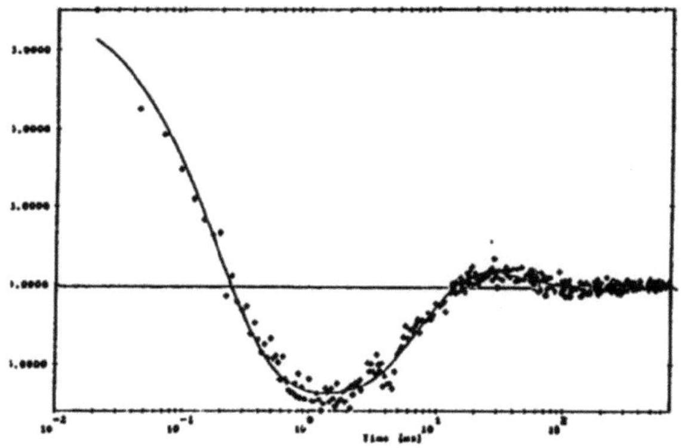

Fig. 3: 1189 cm^{-1}

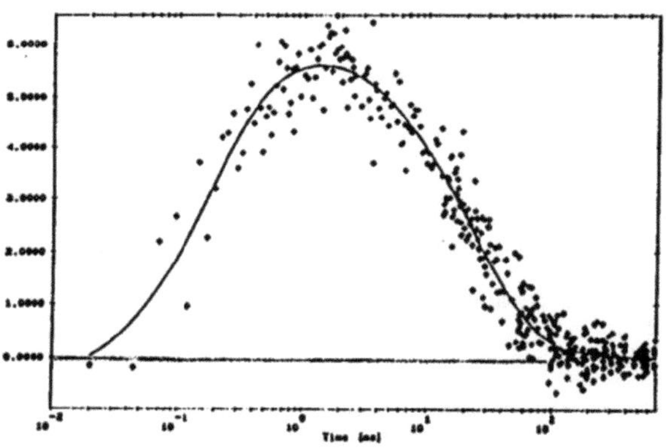

Fig. 4: 1762 cm^{-1}

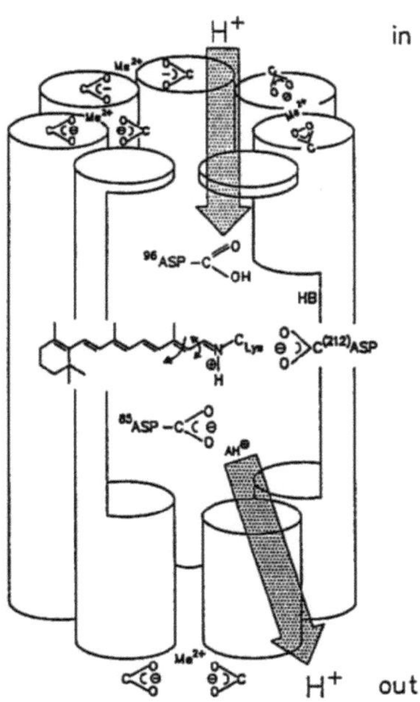

Fig. 5: Proton pump model of bacteriorhodopsin

step [4,5]. It was concluded that asp 85 is the internal proton acceptor of the Schiff Base proton [3-6].

The results are summarized in Fig. 5 showing asp 85 as katalytic proton binding side on the proton release pathway and asp 96 on the proton uptake pathway (3,6).

ACKNOWLEDGEMENT

I gratefully acknowledge a Heisenberg fellowship of the DFG and thank Regina Eichas-Nell and Georg Souvignier for their help.

REFERNCES

1) Gerwert, K.
Ber. Bunsenges. Phys. Chem. 92, 978-982 (1988)

2) Souvignier, G. and Gerwert, K.
submitted

3) Gerwert, K., Hess, B., Soppa, J. and Oesterhelt D.
Proc. Natl. Acad. Sci. USA 86, 4943-4947 (1989)

4) Holz, M., Drachev, L.A., Mogi, T., Otto, H., Kaulen, A.D., Heyn, M.P., Skulachev, V.P. and Khorana, H.G., Proc. Natl. Acad. Sci. USA 86, 2167-2171 (1989)

5) Butt, H.J., Fendler, K., Bamberg, E., Tittor, J. and Oesterhelt, D. (1989)
EMBO J. 8, 1657-1663

6) K. Gerwert, G. Souvignier, and B. Hess
Proc. Natl. Acad. Sci. USA, 87, 9774-9778 (1990)

FTIR-TRS Studies of the Influence of pH on BR Photocycle

Xinyao Hu, Yanying Pi, Suqin Sun, and Weiqin Lu

Department of Chemistry, Tsinghua University, Beijing 100084, P.R. of China

The purple membrane of Halobacteria halobium is one of the new photo–electric converting materials. Bacteriorhodopsin(BR), the integral protein in the purple membrane, functions as a light–driven proton pump. FTIR–TRS difference spectroscopy is a useful technique for measuring the photoreactive proteins derived from Bacteriorhodopsin in the photocycle. This paper studies the influence of pH (at room temperature) on BR photocycle (M412 transient–state).

All spectra were obtained on a Nicolet 170 sx FTIR spectrometer equipped with a liquid nitrogen cooled Hg / Cd / Te detector and linked with our own–built automatic flash photolysis apparatus system.

The protonation state in proton release and inflow has relevance to decay of M412 photoproduct. pH influence protonation state that changes during a transfomation of protonation. pH values play an important role in M412 decay kinetics. Transient FTIR difference spectra between M412 and BR555 are shown in Figs. 1–4.

In Figs. 1,2 and 3, transient time is 2ms to 44ms. In Fig.4, transient time is 0.2ms to 2ms (for pH7 only). Absorbance at 1560 cm^{-1} peak is plotted vs. T(ms) for several pH values as shown in Fig.5 and Fig.6 (for

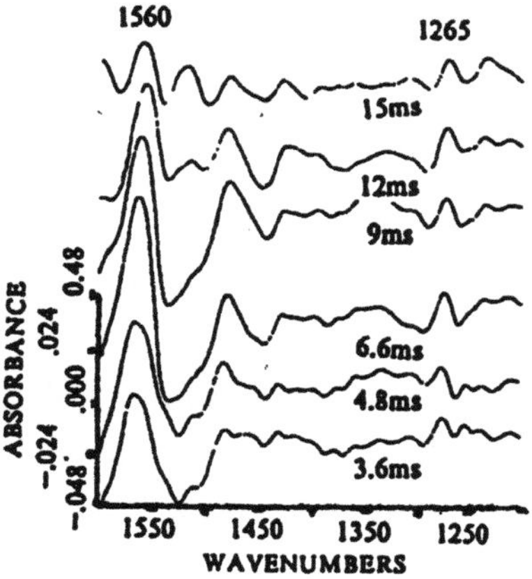

Fig.1 FTIR difference spectra in different photo–transient–state conditions (2ms–44ms) for M412 to BR555 at PH7

Springer Proceedings in Physics, Vol. 68 Time Resolved Vibrational Spectroscopy V
Ed.: H. Takahashi

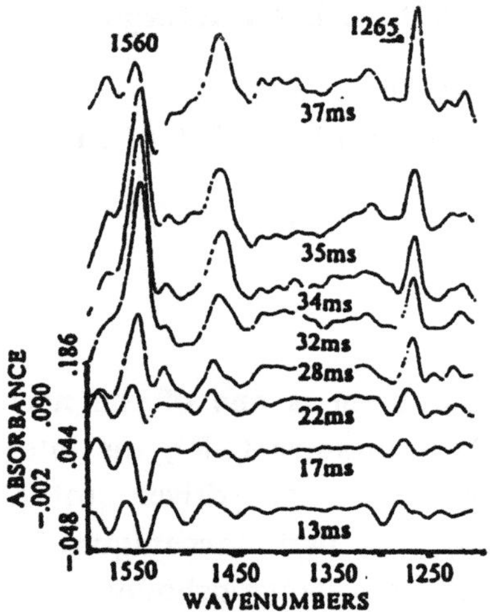

Fig.2 FTIR difference spectra in different photo-transient-state conditions (2ms-44ms) for M412 to BR555 at pH9.9

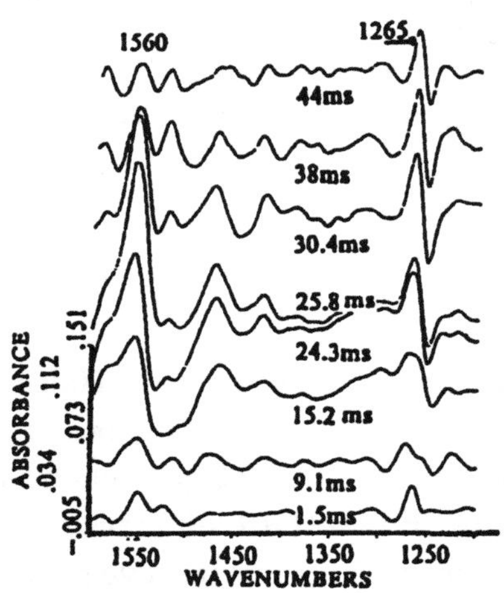

Fig.3 FTIR difference spectra in different photo-transient-state conditions (2ms-44ms) for M412 to BR555 at pH5.5

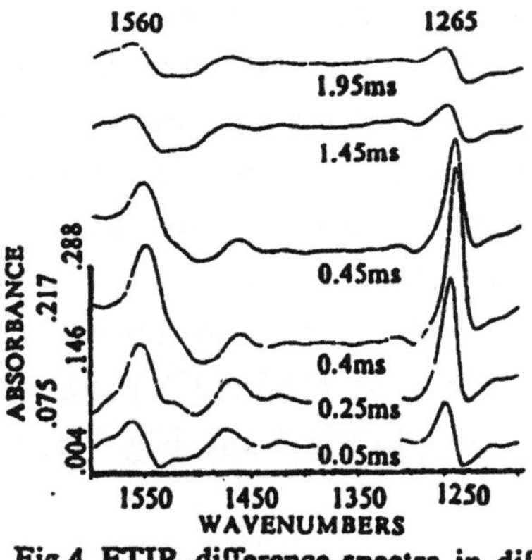

Fig.4 FTIR difference spectra in different photo-transient-state conditions (0.2ms-2ms) for M412 to BR555 at PH7

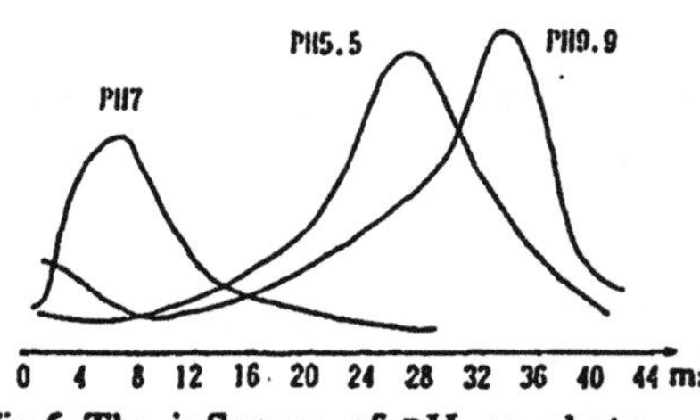

Fig.5 The influence of pH on photo-transient-state (2ms-44ms) for M412 to BR555

pH7 only). Under different pH values M412 kinetic behavior is different.

(1) For different pH palues, the peak of the maximum M412 photoproduct appears at different time intervals. For example, the maximum M412 photoproduct yields were observed at 8ms, 28ms, 36ms for pH7, pH5.5 and pH9.9 respectively (see Fig.5).

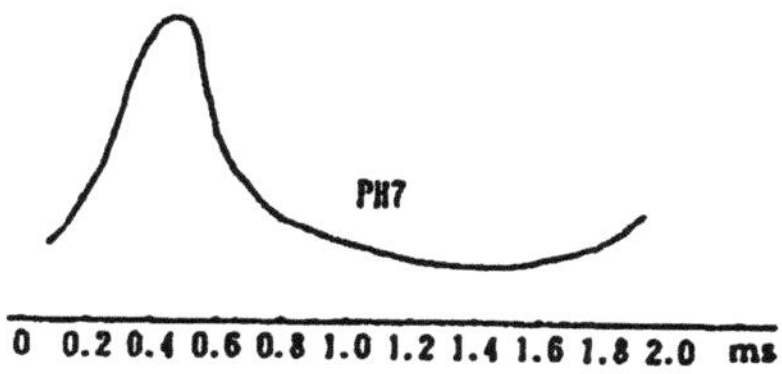

Fig.6 photo-transient-state (0.2ms-2ms) for M412 to BR555 at pH7

(2) At pH7, maximum M412 photoproduct yield appears twice. Besides the 8ms one mentioned alove, the other peak was observed at 400 μs (see Fig.6). At pH5.5, there is similar situation.

(3) Besides the peak at 1560 cm^{-1}, the similar phenomenon for different pH values can be seen at 1265 cm^{-1}, which is in the range of the tyrosinate c—o stretch vibration (see Figs. 2 and 3).

Such spectra show the changes in protonation between BR and M412. This technique and result provide important experimental tests of proposed proton–translocation mechanism for BR photocycle.

REFERENCE

[1] Mantele M.; Siebert F.; Kreutz W.; Methods Enzymol. 88, 729 (1982).

[2] Mark S. Braiman; Patrick L. Ahl; Kenneath J. Rothschild. Proc. Natl. Acad. Sci. USA, Vol.84. 5221(1987).

[3] Klaus Gerwert, Ber. Bunsenges. Phys. Chem. 92, 978 (1988).

[4] Paul Roepe, Patrick L.Ahl, S. K. Das Gupta, Judith Herzfeld. K.J.Rothschild Biochemisty, 26, 6696(1987).

Low Temperature FTIR Studies on the Photointermediates L, M, and N of Bacteriorhodopsin

A. Maeda, J.-M. Pfefferlé, J. Sasaki, Y. Shichida, and T. Yoshizawa

Department of Biophysics, Faculty of Science,
Kyoto University, Kyoto 606, Japan

Abstract. Light-dependent changes around the protonated Schiff base of bacteriorhodopsin were studied by low temperature FTIR spectroscopy on intermediates **K**, **L**, **M** and **N**. A special effort was done for the recording of the spectrum of **N**. Upon formation of **L**, a strong hydrogen bonding occurs between the Schiff base and the aspartic acid residues. It causes the twisting of the chromophore. Such a twisting is lost during the stage of **M** and the strong interaction with the aspartic acid residues reforms upon formation of **N**.

Bacteriorhodopsin (bR) is a protein localized in the purple membrane of Halobacterium halobium and transports protons across the membrane upon absorption of light by the retinylidene chromophore. bR is active only when the chromophore is all-*trans* (**BR**). Light energy is first used for the isomerization of the chromophore to the 13-*cis* form. The photocycle has been described by a series of intermediates;

BR -> **K** <-> **L** <-> **M1** -> **M2** <-> **N** <-> **O** -> **BR.** [1]

The conversion of light energy to proton gradient is thought to be mediated through the interaction of the Schiff base of the chromophore with the protein part. The process for the **L** to **M** conversion, which is accompanied by the deprotonation of the Schiff base, is related to the proton release [2]. Formation of **N** is coupled with the reprotonation of the Schiff base and the decay of **N** is accomplished with the uptake of proton from the opposite side [3]. To obtain an insight of the energy conversion process, the structure of the intermediates involved in these processes, **L**, **M** and **N**, must be characterized. Low temperature FTIR spectroscopy, which enables accurate recordings of the FTIR spectrum of intermediates by stabilizing them, is one of the most useful techniques for this purpose.

RESULTS

(1) To obtain a full account for the process, the recording of the spectrum of **N** is necessary. It is stable at alkaline pH [4]. By use of the highly hydrated bR film at pH 10, the difference FTIR spectrum of **N** to **BR** was obtained by the photoreaction at 274K [5], where **N** lives longer than **M**. A similar spectrum was also obtained for **N** at pH 7. A difference was found only in the frequencies of the C=C stretching vibrational band; 1544 cm^{-1} at pH 10 and 1532 cm^{-1} at pH 7.

(2) The C=O stretching vibrational bands of protonated aspartic acids are useful for the analysis of the protonated

Springer Proceedings in Physics, Vol. 68 Time Resolved Vibrational Spectroscopy V
Ed.: H. Takahashi

state of aspartic acid residues. The deprotonation of Asp-96 is shown by a negative band at 1742 cm^{-1} and the protonation of Asp-85 by a positive band at 1755 cm^{-1} [5]. The difference spectrum of **M** to **BR** at pH 7 always contains a contribution of **N** as seen by the presence of these bands. However, that recorded at 230 K for bR at pH 10 is free from the spectrum of **N** and does not contain any signals due to Asp-96. Thus, the deprotonation of Asp-96 is caused by the protonated Schiff base in **L** and **N**, probably through an electrical effect on Asp-96.

(3) A comparison of the spectrum of **K** and **L** reveals a striking difference between the frequencies of the N-H in-plane bending vibration-containing bands, and also the intensities of the C_{15}-H in-plane bending vibration-containing bands. Results can be interpreted as that **L** has much stronger H-bonding with the protein than **K**, and that **L** has a twisting encompassing the C_{12}-H and C_{15}-H region. Thus, the chromophore is distorted in **L**. The distortion must be forced by the strong interaction of the Schiff base with the protein. This will decrease pK_a of the Schiff base, resulting in the deprotonation of the Schiff base. The proton released from the Schiff base is trapped by Asp-85 located in a close proximity of the Schiff base. The proton at Asp-85 remains up to the stage of **N**. Thus, the proton released to the outside should be originated from another site in bR.

(4) A similar comparison for those in-plane bending vibrational modes was also done between **L** and **N**. The twisting of **L** is relieved in **N** as shown by the increased intensity of the C_{15}-H in-plane bending vibrational mode [5], though the interaction with the protein is equally strong. This would be related to the increased pK_a of the Schiff base at the **N** stage, resulting in the reprotonation of the Schiff base.

DISCUSSION

As a summary, a following scheme for the light-induced reaction can be depicted. In the early time range for **K** (pico and nano second ranges), the Schiff base has no special interactions with the protein part. The interactions are settled only upon **L** formation (microsecond range). The interactions involve the twisting of the retinal and the deprotonation of Asp-96. This step is prerequisite for the deprotonation of the Schiff base. The formation of **M** rescues the distortion of the chromophore by disrupting the interaction with the protein. It instead leaves the reversal of the protonated states of Asp-96 and -85, and also specific changes in the peptide bonds.

REFERENCES

[1] G. Varo & J. K. Lanyi: Biochemistry 29, 2241 (1990).
[2] S. Y. Liu: Biophys. J. 57, 943 (1990).
[3] L. A. Drachev, A. D. Kaulen, V. P. Skulachev & V. V. Zorina: FEBS Lett. 226, 139 (1987).
[4] A. Maeda, T. Ogura & T. Kitagawa: Biochemistry 25, 2793 (1986). T. Kouyama, A. -N. Kouyama, A. Ikegami, M. K. Mathew & W. Stoeckenius: Biochemistry 27, 5855 (1988).
[5] J.-M. Pfefferle, A. Maeda, J. Sasaki & T. Yoshizawa: Biochemistry in press (1991).

Excited States of Retinoids, Carotenoids, and Chlorophylls

Y. Koyama

Faculty of Sciences, Kwansei Gakuin University,
Uegahara, Nishinomiya 662, Japan

1. Introduction

The Raman spectra of all-trans-retinal and β-carotene in the T_1 state were first reported in 1981 [1] and 1979 [2], respectively. The S_1 Raman spectrum of β–carotene, on the other hand, was first reported as late as in 1989 [3-5]. The first T_1 Raman spectra of chlorophyll a (Chl) and bacteriochlorophyll a (BChl) were reported in 1989 [6] and 1990 [7], and the S_1 Raman spectrum of BChl has been recorded very recently [8].

First recording of an excited-state Raman spectrum was a breakthrough, which demonstrated the applicability of time-resolved (transient) resonance Raman spectroscopy (hereafter abbreviated as TR^3S) to the field of photochemistry and photobiology, but it cannot be a goal in itself. Information obtained by Raman spectroscopy is useful, but is limited. In most cases, combination of TR^3S with other techniques is necessary to obtain a result of scientific interest. This lecture is not intended to be a comprehensive review; it aims, by using typical examples, to evaluate the contribution of TR^3S in the rather historical field of T_1 retinoids and carotenoids, and to show its potential in the new field of S_1 carotenoids and T_1 and S_1 chlorophylls.

2. Retinoids

T_1 potential of retinoids. Transient Raman spectra of T_1 species which were generated from the all-trans, 7-cis, 9-cis, 11-cis and 13-cis isomers of retinal (C20 aldehyde) showed that all the isomers except for 13-cis generate the same T_1 species (assigned as the "all-trans" T_1) and that the 13-cis isomer generates the "13-cis" T_1 in addition to the above "all-trans" T_1 [9]. (Slightly different results were reported by Wilbrandt et al. [10]). Picosecond time-resolved absorption spectra of the isomeric retinal showed that each isomer generates its own T_1 species immediately after excitation and that each of them subsequently isomerizes toward the "all-trans" T_1. After 5ns, all the isomer except for 13-cis showed the same $T_n \leftarrow T_1$ absorption spectrum, supporting the above result of transient Raman spectroscopy [11,12]. Analyses by HPLC of triplet-sensitized isomerization showed isomerization of each isomer into the all-trans isomer, the quantum yields being 0.8, 0.8, 0.9 and 0.2 for 7-cis, 9-cis, 11-cis and 13-cis, respectively. The above results strongly suggested a potential surface with minima at each cis and the all-trans positions; the all-trans minimum is deeper than the cis minima. The conclusive evidence for the two-fold potential

Springer Proceedings in Physics, Vol. 68 Time Resolved Vibrational Spectroscopy V
Ed.: H. Takahashi

around each C=C bond was the observation that the quantum yield of triplet sensitized isomerization exceeds unity at higher concentrations of cis isomers, indicating that the resultant "all-trans" T_1 functions as a supplementary sensitizer for the ground state cis isomers [13].

Transient Raman [14] and time-resolved absorption [15] spectroscopies as well as analyses by HPLC of triplet-sensitized isomerization [13] of retinylideneacetaldehyde (C22 aldehyde) showed that C22 aldehyde has a similar potential surface.

A discrete change in the T_1 potential was found between β-ionylideneacetaldehyde (C15 aldehyde) and β-ionylidenecrotonaldehyde (C17 aldehyde). Transient Raman spectroscopy and analyses by HPLC of triplet-sensitized isomerization showed that the former has T_1 potential minima with similar depth at each cis and the all-trans positions, and that the latter has a deep minimum at the all-trans position and shallow minima at cis positions as in the case of C20 and C22 aldehydes [13].

Here, TR^3S has taken a role to identify T_1 species with different cis-trans configurations.

Triplet-excited region in retinoids. Comparison of the T_1 Raman spectra of C15, C17, C20 and C22 aldehydes (all-trans isomers) lead to the idea of "the triplet-excited region" with a limited span and a similar magnitude of changes in the bond orders [13,14]. The strongest Raman line to be assigned to the stretching of the central "C=C" bond (*vide infra*) was seen in the 1556-1550 cm^{-1} region for all the aldehydes, indicating a common C=C bond order in the center of the triplet-excited region. Further, upon triplet excitation, the terminal C=O stretching Raman line shifted to the lower frequencies by 54, 77, 64 and 32 cm^{-1} for C15, C17, C20 and C22 aldehydes, respectively. The discrete change in the amount of shift between C20 and C22 aldehydes indicates that the ordinary span of the triplet-excited region is approximately six conjugated double-bond long [13].

Assignments of the T_1 Raman lines for all-trans-retinal were given based on the effects of 10-D, 14-D, 15-D, 7,8-D_2, 11,12-D_2, 9-methyl-D_3 and 13-methyl-D_3 substitutions. For example, the central "C=C" ("C11=C12") stretching shifted only by the 10-D and 11,12-D_2 substitutions, while the terminal "C=O" stretching shifted only by the 15-D substitution. Both observations support the assignments of the above key Raman lines (unpublished results).

Here, T_1 Raman lines have provided a unique piece of information concerning the bond orders. Normal coordinate analyses will provide more quantitative information on the bond orders within and out of "the triplet-excited region" as a set of force constants.

3. Carotenoids

T_1 potential of β–carotene. In order to reveal the reason for the selection of the 15-cis carotenoid in bacterial reaction centers for the photo-protective function, the physical properties of the T_1 species which were generated from different cis-trans isomers of β–carotene were compared. Transient Raman spectroscopy of T_1 species generated from the all-trans and 15-cis isomers showed exactly the same Raman spectrum [16,17], but T_1 species from the 7-cis, 9-cis and 13-cis isomers showed their own unique Raman spectra [17]. Transient absorption spectroscopy supported

the Raman results; the all-trans, 7-cis, 9-cis and 13-cis isomers showed their own $T_n \leftarrow T_1$ absorption and T_1 lifetimes, while the 15-cis isomer showed the same $T_n \leftarrow T_1$ absorption and T_1 lifetime as those of the all-trans isomer [18]. Analyses by HPLC of triplet-sensitized isomerization revealed a unique isomerization property of the 15-cis isomer; it isomerizes exclusively into the all-trans isomer with a quantum yield as high as 0.98 [19]. PPP-SD-CI calculations of the bond-orders of model polyenes, in which the bond lengths were optimized, showed the presence of "the triplet-excited region" in the central part of the polyene chain; a large decrease in the bond order of the central C=C bond is supposed to trigger the above extremely efficient isomerization [19]. All the above results indicate a potential surface with the deepest minimum at the all-trans position and a very shallow minimum at the 15-cis position, which causes very efficient isomerization in the T_1 state. The efficient isomerization in the T_1 state is ascribed to the reason why the 15-cis configuration has been selected for the photo-protective function [20,21]. Transient Raman spectroscopy of reaction centers showed a unique structure of the bound carotenoid [22].

Here also, TR^3S has been powerful to differentiate the configurations of the T_1 species. It also has potential to determine the bond orders within and out of the triplet-excited region.

The S_1 state of carotenoids. It has long been suggested that the optically forbidden $2A_g$ state below the optically allowed B_u state is responsible for energy transfer to chlorophylls. Picosecond Raman spectroscopy of all-trans-β–carotene in benzene solution detected a C=C stretching Raman line of which frequency was as high as 1777 cm^{-1} [3]. By inference based on shorter polyenes, it was ascribed to an a_g type C=C stretching vibration which is vibronically coupled with the $S_0(A_g)$ state. This indicates that the S_1 state probed is actually the $2A_g$ state [3-5]. The $2A_g$ state as the S_1 state was also substantiated for carotenoids bound to photosynthetic membranes [23,24]. The particular Raman line was recorded for a variety of carotenoids in a variety of solvents, and vibronic coupling between the $S_2(B_u)$ and $S_1(2A_g)$ states was also identified for polar carotenoids [25,26].

Here, a C=C stretching Raman line detected by TR^3S evidenced the $2A_g$ state.

4. Chlorophylls

The T_1 and S_1 state structures of BChl. Comparison of the Raman lines of T_1 BChl with those of S_0 BChl showed a shift of the ring-breathing mode (ν_r) to the lower frequencies and a shift of the C_a-N stretching mode to the higher frequencies. The result indicates changes in the bond orders within the macrocycle [7]. Solvent effects on the T_1 Raman spectra showed that the ν_r Raman line can be used as a coordination-marker. However, no discrete shift of the particular Raman line, such as that used to discriminate the penta- and hexa-coordinated states in the S_0 state, was observed [27]. Comparison of the ν_r Raman lines among the S_0, T_1 and S_1 states showed that changes in the bond orders take place upon the $S_1 \leftarrow S_0$ excitation as well, and that the conjugation effect is further enhanced in the S_1 state [8].

Here, TR^3S showed its potential to determine the electronic and molecular structures in the excited states, when it is combined with PPP-SD-CI calculations and normal coordinate analyses of the parent and various isotope-substituted species.

References

[1] R. Wilbrandt and N.-H. Jensen, J. Am. Chem. Soc. **103**, 1036 (1981).
[2] R. F. Dallinger, J. J. Guanci, Jr., W. H. Woodruff, and M. A. J. Rodgers, J. Am. Chem. Soc. **101**, 1354 (1979).
[3] H. Hashimoto and Y. Koyama, Chem. Phys. Lett. **154**, 321 (1989).
[4] T. Noguchi, S. Kolaczkowski, C. Arbour, S. Aramaki, G. H. Atkinson, H. Hayashi, and M. Tasumi, Photochem. Photobiol. **50**, 603 (1989).
[5] H. Hashimoto and Y. Koyama, Chem. Phys. Lett. **163**, 251 (1989).
[6] E. Nishizawa, H. Hashimoto, and Y. Koyama, Chem. Phys. Lett. **164**, 155 (1989).
[7] E. Nishizawa and Y. Koyama, Chem. Phys. Lett. **172**, 317 (1990).
[8] E. Nishizawa, H. Hashimoto, and Y. Koyama, Chem. Phys. Lett. in press.
[9] H. Hamaguchi, H. Okamoto, M. Tasumi, Y. Mukai, and Y. Koyama, Chem. Phys. Lett. **107**, 355 (1984).
[10] R. Wilbrandt, N.-H. Jensen and C. Houée-Levin, Photochem. Photobiol. **41**, 175 (1985).
[11] Y. Hirata, N. Mataga, Y. Mukai, and Y. Koyama, Chem. Phys. Lett. **134**, 166 (1987).
[12] Y. Mukai, Y. Koyama, Y. Hirata, and N. Mataga, J. Phys. Chem. **92**, 4649 (1988).
[13] Y. Mukai, H. Hashimoto, and Y. Koyama, J. Phys. Chem. **94**, 4042 (1990).
[14] H. Hashimoto, Y, Mukai, and Y. Koyama, Chem. Phys. Lett. **152**, 319 (1988).
[15] Y. Hirata, N. Mataga, Y. Mukai, and Y. Koyama, J. Phys. Chem. **91**, 5238 (1987).
[16] R. Wilbrandt and N.-H. Jensen, in *Time-Resolved Vibrational Spectroscopy*, edited by G. H. Atkinson (Academic Press, 1983), p.273.
[17] H. Hashimoto and Y. Koyama, J. Phys. Chem. **92**, 2101 (1988).
[18] H. Hashimoto, Y. Koyama, K. Ichimura, and T. Kobayashi, Chem. Phys. Lett. **162**, 517 (1989).
[19] M. Kuki, Y. Koyama, and H. Nagae, J. Phys. Chem. in press.
[20] Y. Koyama, I. Takatsuka, M. Kanaji, K. Tomimoto, M. Kito, T. Shimamura, J. Yamashita, K. Saiki, and K. Tsukida, Photochem. Photobiol. **51**, 119 (1990).
[21] Y. Koyama, in *Carotenoids: Chemistry and Biology*, edited by N. I. Krinsky, M. M. Mathews-Roth, and R.-F. Taylor (Plenum Press, 1990), p.207.
[22] M. Lutz, L. Chinsky, and P.-Y. Turpin, Photochem. Photobiol. **36**, 503 (1982).
[23] M. Kuki, H. Hashimoto and Y. Koyama, Chem. Phys. Lett. **165**, 417 (1990).
[24] H. Hashimoto and Y. Koyama, Biochim. Biophys. Acta **1017**, 181 (1990).
[25] T. Noguchi, H. Hayashi, M. Tasumi, and G. H. Atkinson, Chem. Phys. Lett. **175**, 163 (1990).
[26] T. Noguchi, H. Hayashi, M. Tasumi, and G. H. Atkinson, J. Phys. Chem. **95**, 3167 (1991).
[27] E. Nishizawa and Y. Koyama, Chem. Phys. Lett. **176**, 390 (1991).

Vibrational Spectroscopy of Excited Electronic States in Carotenoids *in vivo*: Picosecond Time-Resolved Resonance Raman Scattering

H. Hayashi[1,2,3], *T. Noguchi*[1,2], *M. Tasumi*[2], *and G.H. Atkinson*[1]

[1]Department of Chemistry and Optical Science Center, University of Arizona, Tucson, AZ 85721, USA

[2]Department of Chemistry, Faculty of Science, The University of Tokyo, Bunkyo-ku, Tokyo 113, Japan

[3]Present address: National Institute for Basic Biology, Okazaki, Aichi 444, Japan

1. Introduction

Three excited states of carotenoids are involved in photosynthetic systems in two major ways: (i) light-harvesting in which radiation energy is absorbed via the allowed $^1Bu \leftarrow {}^1Ag$ transition and subsequently transferred through the 2^1Ag excited state and (ii) protection from photodamage through the formation of the excited ^3Bu state. The energetic positions and population dynamics of these excited electronic states are of importance to understand the biochemical function of carotenoids. In this study, picosecond time-resolved resonance Raman (PTR^3) spectroscopy is used to measure the vibrational spectra and population dynamics of the excited 2^1Ag and ^3Bu states and the ^1Ag ground state of the *in vivo* and *in vitro* carotenoids of photosynthetic bacteria.

2. Experimental

The *in vivo* sample (chromatophores) was prepared from a photosynthetic bacterium, *Chromatium vinosum*, and suspended in a 50 mM Tris buffer. The *in vitro* sample (spirilloxanthin) was extracted and purified from the same bacteria, and dissolved in benzene. Two dye lasers were synchronously pumped by the second harmonic of a mode-locked Nd:YAG laser to produce 6 ps pulses at a 1 MHz repetition rate. For both the *in vivo* and *in vitro* samples, one dye laser at 565 nm with 5-10 mW average power was used as a pump for excitation of carotenoids, and the other at 590-610 nm with 2-4 mW average power, as a probe laser to generate resonance Raman (RR) scattering.

3. Results and Discussion

Figures 1A and 1B show respectively the PTR^3 spectrum (565 nm pump/601 nm probe) at 0 ps time delay and the probe laser only spectrum observed from carotenoids in chromatophores. The broad band around 1770 cm^{-1} in the PTR^3 spectrum has been assigned to the 2^1Ag state of carotenoids and the intense band at 1063 cm^{-1} has been assigned to coherent Stokes Raman scattering (CSRS) [1]. The 2^1Ag band of *in vitro* carotenoid (spirilloxanthin in benzene, one of the dominant carotenoids in the chromatophores and the one which mostly contributes to the RR spectrum in Figure 1A) is also detected around 1770 cm^{-1}. The similarity of the *in vivo* and *in vitro* 2^1Ag bands shows there to be no significant modifications in the vibronic coupling caused by the *in vivo* environment.

Springer Proceedings in Physics, Vol. 68 Time Resolved Vibrational Spectroscopy V
Ed.: H. Takahashi

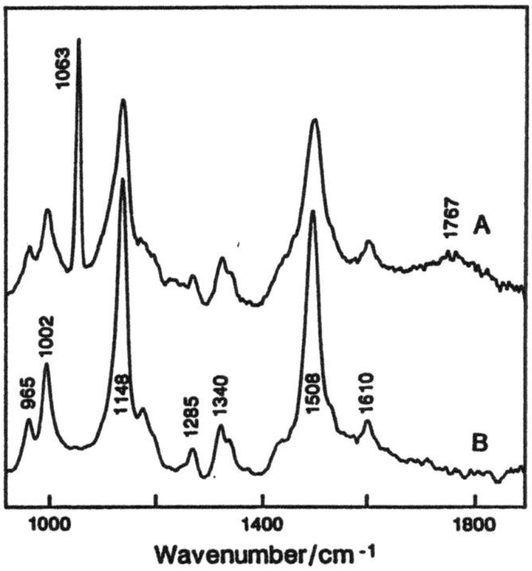

Figure 1. Resonance Raman spectra of carotenoids in chromatophores from *Ch. vinosum.* (A) Recorded with a pump laser (565 nm, 7 mW) and a probe laser (601 nm, 3.5 mW) at 0 ps time delay. (B) Recorded with a probe laser only.

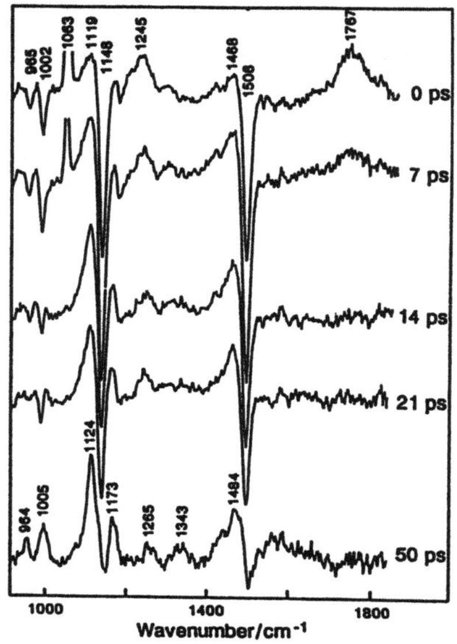

Figure 2. Population dynamics of excited and ground states carotenoids in *Ch. vinosum* chromatophores. The probe laser only spectrum is subtracted from the PTR^3 spectra at the indicated time delays.

The difference between the PTR^3 spectra at time delays between 0 ps and 50 ps and the probe laser only spectrum of *in vivo* carotenoid is shown in Figure 2. As seen from the time-dependence of the positive band around 1770 cm^{-1}, the 2^1Ag state decays with an ≈5 ps rate. The decay rate of the 2^1Ag state of *in vivo* carotenoids is faster than that of *in vitro* sample (≈10 ps for spirilloxanthin in benzene) due to the additional relaxation channels (e.g., energy transfer to bacteriochlorophylls) present in the chromatophore. Positive bands in the PTR^3 spectrum at 50 ps delay are assignable to the ^3Bu state of carotenoids in chromatophores [2]. Formation of the ^3Bu state is observed for *in vivo* carotenoids, but not for *in vitro* spirilloxanthin, indicating that the ^3Bu state is formed by fission processes originating from the spatial organization of pigments within chromatophores. Comparisons of the time-dependent intensities of 2^1Ag and ^1Ag RR bands (negative band in Figure 2) suggest that vibrationally-excited levels in ^1Ag are populated directly by the decay of the 2^1Ag state and that molecules in these vibrationally-excited levels relax into a thermalized distribution in 50 ps.

References

1. H. Hayashi, T. Noguchi, M. Tasumi, G.H. Atkinson: Biophys. J., in press.
2. H. Hayashi, S.V. Kolaczkowski, T. Noguchi, D. Blanchard, G.H. Atkinson: J. Am. Chem. Soc. **112**, 4664 (1990)

Vibrational Relaxation in Carotenoids *in vivo* and *in vitro*: Picosecond Time-Resolved Anti-Stokes Resonance Raman Spectroscopy

H. Hayashi[1,2,3], *T. Brack*[1], *T. Noguchi*[1,2], *M. Tasumi*[2], and *G.H. Atkinson*[1]

[1]Department of Chemistry and Optical Science Center, University of Arizona, Tucson, AZ 85721, USA
[2]Department of Chemistry, Faculty of Science, The University of Tokyo, Bunkyo-ku, Tokyo 113, Japan
[3]Present address: National Institute for Basic Biology, Okazaki, Aichi 444, Japan

1. Introduction

The transient population of excited electronic states in carotenoids is a fundamental part of the photophysical mechanism by which carotenoids contribute to the biochemical function of the pigment-protein complexes in photosynthetic organella. Whereas the roles of electronic state in the overall relaxation mechanism have begun to emerge [1], the contributions and even the presence of vibrationally-excited levels in the relaxation mechanism has remained largely unexplored. In this study, the vibrationally-excited populations in the *in vivo* and *in vitro* carotenoids are examined in real time using picosecond time-resolved anti-Stokes resonance Raman scattering.

2. Experimental

The *in vivo* sample (chromatophores) was prepared from a photosynthetic bacterium, *Chromatium vinosum*, and suspended in a 50 mM Tris buffer. The *in vitro* sample (spirilloxanthin) was extracted from the same bacteria, purified and dissolved in either benzene or toluene. Two dye lasers were synchronously pumped by the second harmonic output of a mode-locked Nd:YAG laser to produce 6 ps pulses at a 1 MHz repetition rate. One dye laser (567 nm, 6-7 mW) was used for excitation of carotenoids, and the other dye laser (576 nm, 2-3 mW) was used to generate the time-resolved anti-Stokes resonance Raman signal.

3. Results and Discussion

Figures 1A and 1B show the anti-Stokes resonance Raman spectra of carotenoids in chromatophores, recorded with pump-probe conditions at 7 ps and 50 ps delay times, respectively. Whereas the intensities of the anti-Stokes bands arising from the pump laser (indicated with closed circles) remain constant, the anti-Stokes bands generated by the probe laser at 7 ps time delay (indicated with frequency displacements) are much larger than the corresponding bands in the 50 ps spectrum and the probe laser only spectrum (Figure 1C). Clearly, with 7 ps pump-probe conditions, a vibrationally excited ^1Ag population is created by electronic state relaxation [2]. The vibrational relaxation of this population is complete by 50 ps.

Springer Proceedings in Physics, Vol. 68 Time Resolved Vibrational Spectroscopy V
Ed.: H. Takahashi

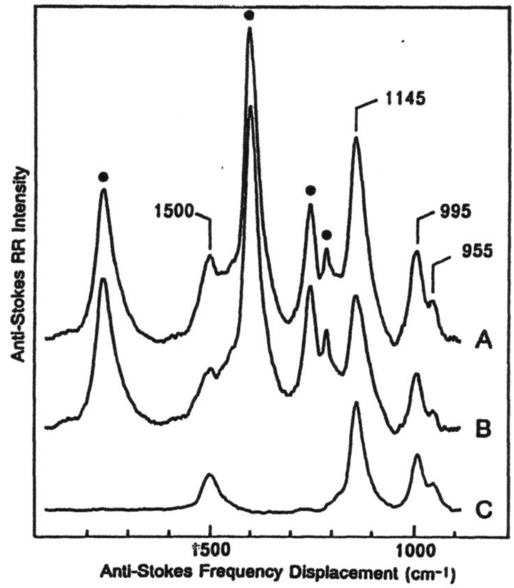

Figure 1. Anti-Stokes resonance Raman spectra of carotenoids in *Ch. vinosum* chromatophores. (A) and (B) Recorded with pump and probe pulses at 7 ps and 50 ps time delay, respectively. (C) Recorded with a probe lasers only. Anti-Stokes bands generated by the pump laser are indicated by filled circles.

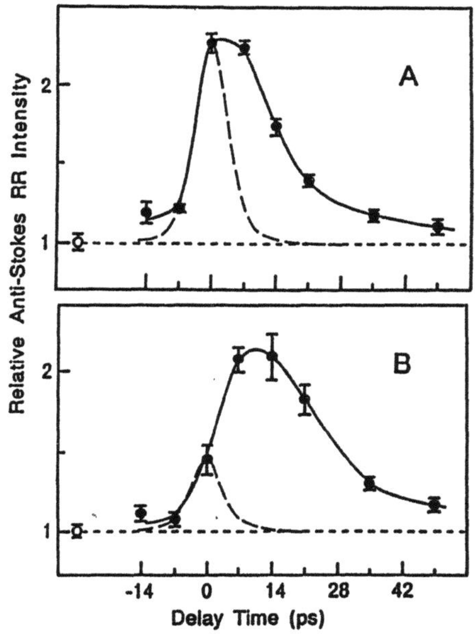

Figure 2. Time dependencies of anti-Stokes Raman band intensities at 1145 cm^{-1} relative to those in the probe laser only spectrum. (A) Carotenoids in chromatophores. (B) Spirilloxanthin in benzene. The cross correlation time is schematically shown as a dashed curve centered at 0 ps.

The time dependencies of the intensities of the anti-Stokes bands at 1145 cm^{-1} (C-C stretching mode) observed for the *in vivo* (A) and *in vitro* (B) samples are shown in Figure 2. The anti-Stokes bands at 1500 cm^{-1} of the *in vivo* and *in vitro* samples exhibit time dependencies similar to those seen in Figures 2A and 2B, respectively. The time-dependent intensities of these bands reflect the population dynamics associated with the initial phases of electronic relaxation as well as vibrational relaxation toward thermal equilibrium in ^1Ag. For *in vitro* samples, anti-Stokes resonance Raman signals reach maxima at ≈10 ps after optical excitation before decaying with a rate of ≈15 ps. For the *in vivo* case, the maximum intensities are found within the cross-correlation time (≈8 ps) defined by the 567 nm pump and 576 nm probe laser pulses (6 ps FWHM), while the decay rate is similar to that of the *in vitro* samples.

These data document for the first time the contribution of vibrationally excited ^1A_g levels to the picosecond relaxation mechanism operating in both *in vivo* and *in vitro* carotenoids.

References

1. H. Hayashi, T. Noguchi, M. Tasumi, G.H. Atkinson: Biophys. J., in press.
2. H. Hayashi, M. Tasumi, G.H. Atkinson: Chem. Phys. Lett. 178, 388 (1991)

Time-Resolved FTIR Difference Spectroscopy of Photosynthetic Bacterial Reaction Centers

D.L. Thibodeau[1], *E. Nabedryk*[1], *R. Hienerwadel*[2], *F. Lenz*[2], *W. Mäntele*[2], *and J. Breton*[1]

[1]SBE/DBCM, CEN Saclay, F-91191 Gif-sur-Yvette Cedex, France
[2]Institut für Biophysik, Universität Freiburg, W-7800 Freiburg, Fed. Rep. of Germany

1. Introduction

Primary photosynthetic reactions feature rapid kinetics with high quantum yield of stable charge separation occurring in basic units called reaction centers (RC). The X-ray structure of bacterial RC [1,2] suggests that the efficiency and the stability of the charge separation rely at least partly on the protein environment to optimize the orientation and the localization of the various redox components involved in electron transport. The differences in the nature and organization of the aminoacids forming the binding pocket of the two quinones, Q_A and Q_B, are thought to be responsible for their different properties in RC. Indeed, the two quinones are unequivalent, Q_A (menaquinone for *Rps. viridis* and ubiquinone for *Rb. sphaeroides*) is more tightly bound to the protein than Q_B (ubiquinone for both RC), and they display different redox properties *i.e.* Q_A is a one-electron acceptor while Q_B can accept two electrons under physiological conditions.

Following photoexcitation, the electron is transferred from the primary donor (a dimer of bacteriochlorophyll), P, to the acceptor quinone, Q_A. When the loosely bound quinone, Q_B, is still present in the RC the electron proceeds from Q_A to Q_B in approx. 100 μs at room temperature. The $P^+Q_A^-$ state in *Rps. viridis* RC has a lifetime in the 1-ms range while the transient state $P^+Q_AQ_B^-$ lasts 100 ms, at room temperature [3]. In *Rb. sphaeroides* RC, the recombination rate of $P^+Q_A^-$ and $P^+Q_AQ_B^-$ are known to be 60–100 ms and a few seconds, respectively [4].

Structural changes which accompany the formation of the states $P^+Q_A^-$ and $P^+Q_AQ_B^-$ in *Rb. sphaeroides* and *Rps. viridis* RC have been previously characterized under steady-state illumination conditions using light-induced FTIR difference spectroscopy [5,6]. Vibrational studies of transient systems with simultaneous temporal and spectral resolution can provide detailed information on rate constants and population distribution of intermediates or products. The most attractive attribute of time-resolved (TR) FTIR difference spectroscopy is that it is possible to record information on transient states simultaneously on all the IR spectral range. TR FTIR difference spectroscopy has been recently implemented for *Rb. sphaeroides* RC [7] with a time-resolution of 25 ms for a spectral resolution of 4 cm^{-1} to yield spectra containing contributions of the quinones and their protein environment following photoexcitation. In this study, we present results obtained from *Rps. viridis* RC.

2. Material and Methods

A cell made from two CaF_2 windows contains the *Rps. viridis* samples with a 5-fold excess of ubiquinone-10. The TR FTIR difference experiments have been performed as in [7] with modifications to account for the difference in the kinetics of *Rps. viridis* RC recombination rate. The Rapid Scan experiment involves the consecutive digitization of a series of 10 groups of n interferograms (n=1 for *Rb. sphaeroides* and n=5 for *Rps. viridis*). After Fourier transform, 10 single-beam spectra are obtained, where each one represents the IR single-beam spectrum of the sample at a given time after the start of the data acquisition. The first single-beam of the series is taken as a reference on which every subsequent single-beam will be

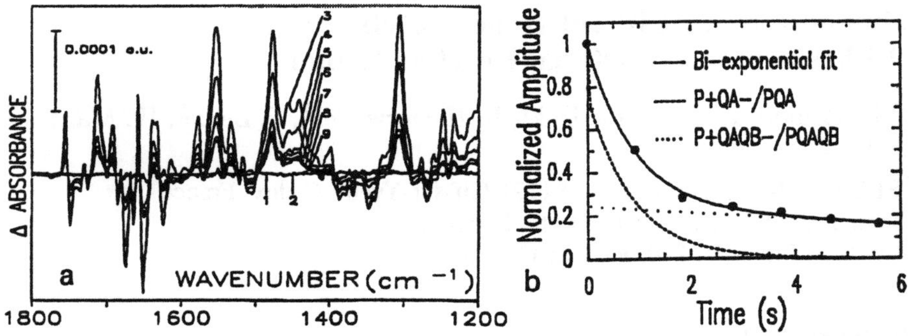

Fig. 1. a) Series of TR FTIR difference spectra (9 600 interferograms) for *Rps. viridis* RC at 170K. The background spectra are denoted 1 and 2 and show the noise level. The spectrum denoted 3 is recorded under illumination. b) Change in amplitude (average at several wavenumbers) as a function of time. At 170K, the initial concentrations of $P^+Q_A^-$ and $P^+Q_AQ_B^-$ are (75± 2)% and (25 ± 2)% respectively, the half-times are estimated to (0.7 ± 0.3) s and (15 ± 7) s.

ratioed to yield 9 successive difference spectra (every 185 ms for *Rb. sphaeroides* and every 925 ms for *Rps. viridis*). The duration of data acquisition of a series for *Rb. sphaeroides* was approx. 2 s while it lasted approx. 10 s for *Rps. viridis*. The actinic event (100 ms for *Rb. sphaeroides* and 1 s for *Rps. viridis*) is delayed from the start of the series of data collection to occur at the beginning of the digitization of the 4th group of interferograms of the series and this corresponds to the spectrum denoted 3 in fig.1a. After the data acquisition, there is a pause of at least 50 s to complete the return to the ground state. The same cycle is repeated several hundreds times to improve the signal-to-noise ratio.

3. Results

Figure 1a represents a series of TR FTIR difference spectra of *Rps. viridis* at 170K each separated by 925 ms. The amplitude of the difference spectra is proportional to the concentration of the $P^+Q_A^-$ and $P^+Q_AQ_B^-$ transient species produced under light-saturating conditions. The amplitude of most features in the spectra recorded during and after illumination seem to decay at the same rate. This is demonstrated in fig.1b where the change in amplitude at several wavenumbers is plotted as a function of time. By analyzing the charge recombination kinetics into fast and slow phases, the relative proportion of photochemically-active $P^+Q_A^-$ and $P^+Q_AQ_B^-$ were determined. The $P^+Q_AQ_B^-/PQ_AQ_B$ spectrum is calculated from the average of the last 4 spectra (see ref. 7 for more detail) from fig.1a; typically the $P^+Q_A^-$ species no longer contributes to the spectra recorded 2.5 s after the end of the illumination (fig.1b). The $P^+Q_A^-/PQ_A$ spectrum is the result of the subtraction of the spectrum recorded under illumination and the calculated $P^+Q_AQ_B^-/PQ_AQ_B$ spectrum. Figures 2a and 3a show the subtraction of the normalized $P^+Q_A^-/PQ_A$ and $P^+Q_AQ_B^-/PQ_AQ_B$ spectra for *Rps. viridis* and *Rb. sphaeroides*, respectively. The resulting spectrum is denoted $Q_A^-Q_B/Q_AQ_B^-$ to stress that the contributions of P and P^+ have been cancelled assuming that the signal of P^+/P is the same whether the acceptor quinone is Q_A or Q_B. It is important to keep in mind that in these $Q_A^-Q_B/Q_AQ_B^-$ spectra the positive bands contain contributions of neutral Q_B or charged Q_A^- while the negative features represent contributions originating from neutral Q_A or charged Q_B^-. In figs. 2 and 3, the dashed line represents band positions that are believed to represent contributions from Q_A^- and Q_A spectral bands while the dotted lines is for the Q_B^- and Q_B bands.

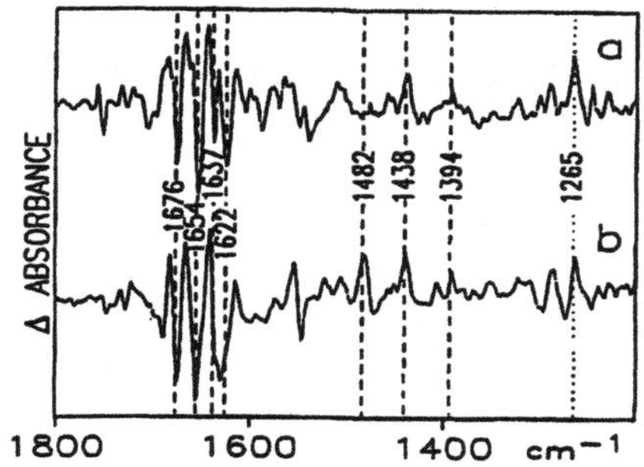

Fig. 2. Comparison of $Q_A^-Q_B/Q_AQ_B^-$ spectra for *Rps. viridis* RC obtained by (a) TR spectroscopy at 170K and (b) steady–state illumination at 283K.

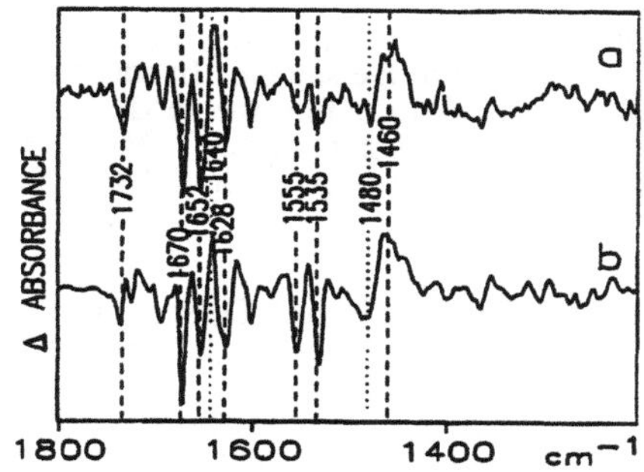

Fig. 3. As in Fig. 2 for *Rb. sphaeroides* RC at (a) 280K for the TR FTIR experiments and (b) 283K for the steady–state experiments.

4. Discussion

The quinone vibrational modes have been difficult to establish in steady–state FTIR spectra [6,8] owing to the masking effect of the large contributions of P and P^+ in the light–induced difference spectra. The distinct decay lifetime of the $P^+Q_A^-$ and $P^+Q_AQ_B^-$ transient species is the key element on which the TR FTIR technique discriminates the contributions of $P^+Q_A^-/PQ_A$ and $P^+Q_AQ_B^-/PQ_AQ_B$. The spectra obtained by steady–state and TR FTIR difference spectroscopy have already shown many similarities [8]. In addition, new information is extracted from TR FTIR difference spectra because the dominant contributions of P^+ and P could be reliably removed.

Recently, light–induced Q_A^-/Q_A and Q_B^-/Q_B FTIR difference spectra of *Rb. sphaeroides* and *Rps. viridis* have been obtained under steady–state illumination conditions and in the presence of reductants and mediators which rapidly rereduce P^+ [9]. For purpose of comparison, the subtraction of these two difference spectra is performed to generate $Q_A^-Q_B/Q_AQ_B^-$ spectra presented in fig.2b for *Rps. viridis* and fig.3b for *Rb. sphaeroides.*

The comparison of spectra obtained by TR FTIR difference spectroscopy to the steady–state spectra is remarkable in most features for *Rps. viridis* (fig.2) and for almost every band in *Rb. sphaeroides* (fig.3). It supports the assumption that the spectral features observed by TR FTIR difference spectroscopy are of the same nature than the contributions obtained by the steady–state method. In particular for *Rb. sphaeroides* RC (fig.3), the 1480–cm^{-1} (Q_B^-) and the ≈1460–cm^{-1} (Q_A^-) bands have been attributed to C–O^- vibrations of the semiquinone anion. In *Rps. viridis* RC, although the two sets of spectra were recorded under different temperature conditions, the TR FTIR difference spectrum (170K) also presents several bands that find their counterpart in the steady–state $Q_A^-Q_B/Q_AQ_B^-$ spectrum (280K). It is worthwhile to note the similarity of $Q_A^-Q_B/Q_AQ_B^-$ spectra in the 1680–1600 cm^{-1}–region for both RC. In particular, the 1670 cm^{-1}, 1652 cm^{-1}, and 1628 cm^{-1} bands in *Rb. sphaeroides* are also represented in *Rps. viridis* at 1676 cm^{-1}, 1654 cm^{-1} and ≈ 1630 cm^{-1}. The temperature difference would explain the splitting of the bands into two components at 1637 cm^{-1} and 1622 cm^{-1} in fig.2a. In the 1500–1390–cm^{-1}region, since the three positive bands of Q_A^- in *Rps. viridis* have been well characterized, the absence of the band at 1482 cm^{-1} in fig.2a is difficult to justify at this point and will require further investigation. Similarly, some minor contributions from RC–bound cytochrome cannot be excluded to explain the differences around 1550 cm^{-1}.

First applied to *Rb. sphaeroides* [7] and here to *Rps. viridis*, TR FTIR difference spectroscopy has proven to be a very valuable method to obtain the contributions of only Q_A, Q_B^-, Q_A^-, Q_B and their aminoacid partners with a minimum of perturbation to the sample. In addition, TR FTIR difference spectroscopy studies are advantageous to define the mode frequencies of interest for kinetic IR studies at individual wavelengths (see W. Mäntele *et al.*, these proceedings) which can follow electron transfer steps at shorter time resolutions.

1. H. Michel, O. Epp, J. Deisenhofer: EMBO J. **5**, 2445 (1986)
2. J.P. Allen, G. Feher, T.O. Yeates, H. Komiya, D.C. Rees: Proc. Natl. Acad. Sci. USA **85**, 8487 (1988)
3. R.J. Shopes, C.A. Wraight: Biochim. Biophys. Acta **806**, 348–356 (1985)
4. R.K. Clayton, H.F. Yau: Biophys. J. **12**, 867 (1972)
   J.R. Codgell, D.C. Brune, R.K. Clayton: FEBS Lett. **45**, 344 (1974)
5. W. Mäntele, A. Wollenweber, E. Nabedryk, J. Breton: Proc. Natl. Acad. Sci. USA **85**, 8468 (1988)
6. E. Nabedryk, K.A. Bagley, D.L. Thibodeau, M. Bauscher, W. Mäntele, J. Breton: FEBS Lett. **266**, 59 (1990) and *refs. therein*
7. D.L. Thibodeau, E. Nabedryk, R. Hienerwadel, F. Lenz, W.G. Mäntele, J. Breton: Biochim. Biophys. Acta **1020**, 253 (1990)
8. D.L. Thibodeau, J. Breton, C. Berthomieu, K.A. Bagley, W. Mäntele, E. Nabedryk: *Bacterial Reaction Centers, Structure and Dynamics* (Springer Series in Biophysics, Berlin, 1990) vol. **6**, pp. 87–98 and *refs. therein*
9. J. Breton, D.L. Thibodeau, C. Berthomieu, W.G. Mäntele, A. Verméglio, E. Nabedryk: FEBS Letters **278**, 257 (1991)
   J. Breton, C. Berthomieu, D.L. Thibodeau, E. Nabedryk (in preparation) (1991)

Time-Resolved Vibrational Analysis of Quinone Binding and Interaction in the Electron Transfer of Bacterial Photosynthetic Reaction Centers

R. Hienerwadel[1], *D. Thibodeau*[2], *F. Lenz*[2], *J. Breton*[2], *E. Nabedryk*[2], *W. Kreutz*[1], *and W. Mäntele*[1,*]

[1]Institut für Biophysik, Albertstraße 23,
W-7800 Freiburg, Fed. Rep. of Germany
(* author for correspondence)
[2]DBCM, CEN Saclay, F-91191 Gif-sur-Yvette Cedex, France

1. INTRODUCTION

The understanding of the sequence of events, the mechanisms of reactions and the identity of reaction partners in the primary electron transfer of photosynthesis mainly arises from time-resolved spectroscopic techniques (for a review, see [1]). These primary steps of energy conversion take place in reaction centers (RC) which contain well-organized pigments and redox components. In purple photosynthetic bacteria, electron transport upon photoexcitation proceeds within 2-3 picoseconds from the primary electron donor "P" (a $BChl_2$) to the monomeric BPheo "H" and from H to the primary acceptor quinone Q_A within approx. 200 psec. In the presence of the secondary quinone Q_B, electron transfer from Q_A to Q_B proceeds within approx. 150 μsec. Q_A and Q_B are the first electron acceptors stable in the msec-to-second range. If the redox energy stored in the $P^+Q_B^-$ state is not used by secondary redox reactions, charge recombination with a half-time of 0.5-1 sec observed. In the absence of Q_B, charge recombination from Q_A proceeds within ca. 70 msec.

The efficiency, speed and specificity of this electron transfer rely on the arrangement of the pigments in the protein matrix, which has been shown by the structures derived from X-ray crystallography [2,3]. The protein matrix, in addition, seems to provide specific bonding and interactions for the redox cofactors, thus modulating their spectral and redox properties. The X-ray structures provide an averaged static picture of the quiescent state of the RC and suggest specific interactions of the pigments and the quinones with amino acid side chains in their binding pockets. Information on the dynamic nature of this interaction, however, and the involvement of the protein has to be collected by spectroscopic techniques.

Infrared (IR) spectroscopy can provide such information, since vibrational modes are highly sensitive to bond distances and angles. In previous work, molecular changes concomitant with primary electron transfer and radical formation in RC have been characterized using static Fourier-transform IR (FTIR) spectroscopy [4-7]. In these investigations, difference spectra between stable or quasi-stationary states of the RC were obtained, with a level of sensitivity high enough to detect single molecular bonds in large pigment-protein complexes. Using rapid-scan FTIR spectroscopy, charge recombination from the $P^+Q_A^-$ and the $P^+Q_B^-$ states [8] has been investigated.

In previous reports [9,10], we have described an IR photometer based on tunable IR diode lasers which allows to record very small ($\approx 10^{-4}$) flash-induced IR absorbance changes in the time range from below μsec to sec. Almost continuous tuning of the lasers gives access to the main vibrational modes of the pigments, quinones and the protein in the 1400 to 1800 cm^{-1} range. We use here this new technique for the identification of vibrational modes of Q_A and Q_B and of amino acid residues in their protein sites involved in the primary electron transfer in bacterial RC.

2. MATERIALS AND METHODS

Rb. sphaeroides R-26 RC were prepared and reconstituted with different amounts of Q_B as in [8]. Preparation of IR samples is described in detail in [11]. Water and protein content were adjusted to give at least 10% transmission at the amide I/water absorbance bands around 1650 cm^{-1}. The kinetic photometer used for the detection of small light-induced IR transmittance changes, its resolution, tunability and sensitivity have been described [9,10]. A second measuring beam using a near-IR LED source at 960 nm (P^+ absorbance) and a silicon diode detector were used to determine the degree of occupation of the Q_B site in the IR samples. The time constants of charge recombination were analyzed by fitting to a sum of two exponentials. By correlation with the kinetic parameters of the IR signals, attribution of modes to Q_A or Q_B was achieved [11].

3. RESULTS AND DISCUSSION

Examples of kinetic signals at discrete wavenumbers in the 1700-1450 cm^{-1} range and at 960 nm are shown in fig. 1. The kinetic analysis of the decay at 960 nm (data not shown) indicates about 95 % of functional Q_B. A rapid, unresolved ($t_{1/2} < 0.5$ µsec) and a slow ($t_{1/2} \approx 150$ µsec) kinetic component are superimposed for most of the IR signals. The ca. 150 µsec component shows only 10-15 % of the amplitude of the strongest step signal at 1682 cm^{-1}.

The rapid, unresolved signals can be attributed to the processes of immediate charge separation between the primary electron donor P and the primary electron acceptor Q_A ($PQ_AQ_B \rightarrow P^+Q_A^-Q_B$), while the ca. 150 µsec component is due to the transfer of the electron from the primary to the secondary quinone acceptor ($P^+Q_A^-Q_B \rightarrow P^+Q_AQ_B^-$). Signals at 1486 cm^{-1} for RC samples with different content of Q_B (data not shown) clearly indicate that the amplitude of the step signal is not affected, while the amplitude of the 150 µsec component is proportional to the amount of Q_B. The presence of a 150 µsec component, together with the analysis of the decay half-time, can thus be taken to assign a vibrational mode to Q_A or Q_B (or to modes in their respective protein environment) [11]. Indeed, the wavelength dependance of the total signal amplitude closely matches static light-induced FTIR difference spectra obtained by charge separation (P^+Q^-/PQ) [4,6,9], while those constructed from the slow component compare those of quinone reduction [7,8,12].

Electron transfer to Q_A and further to Q_B should give rise to strong difference bands of the C=O ($\approx$ 1670-1650 cm^{-1}) and the C=C (ca. 1610 cm^{-1}) modes [13]. Indeed, the signal at 1670 cm^{-1} shows a rapid decrease of the absorbance, which is restored with the kinetics of the $Q_A \rightarrow Q_B$ electron transfer. The same time course is observed for the signal at 1632 cm^{-1}. The complementary signal is observed at 1467 cm^{-1}. It appears reasonable to attribute these signals to modes of Q_A (1670,1632 cm^{-1}) and $Q_A^{-\cdot}$ (1467 cm^{-1}) and/or its binding pocket. According to model quinone spectra [13] the band at 1467 cm^{-1} can be assigned to the C-C/C-O modes of $Q_A^{-\cdot}$. In this line, the band at 1668 cm^{-1} might be assigned to the C=O mode of the neutral Q_A, with its vibrational frequency characteristic for a non-protic environment. Preliminary experiments with ^{13}C isotopic labelling of Q_A, however, did not show the expected shift for this band [8]. For a final assignment, further studies using isotope-labelled and chemically-modified quinones at the Q_A site are in progress.

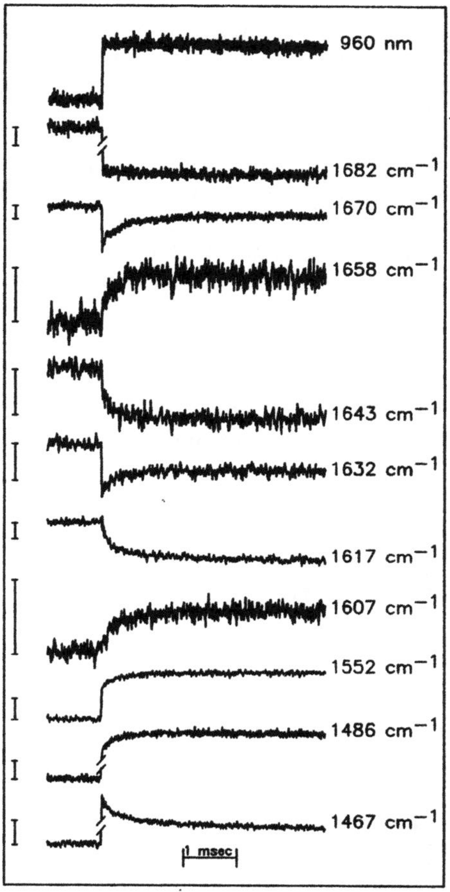

Fig. 1: Flash-induced kinetic IR absorbance changes from *Rb. sphaeroides* RC at 10° C. Signals are shown at the same time scale but at different absorbance scales (bar length corresponds to $2*10^{-4}$ AU). For the signals at 1467, 1486 and 1682 cm^{-1}, the amplitude of the rapid rise ($1.1*10^{-3}$ AU) is further compressed with respect to the slower rise.
Excitation: 595 nm, 5 mJ, 15 nsec.
Up to 300 flashes were averaged.

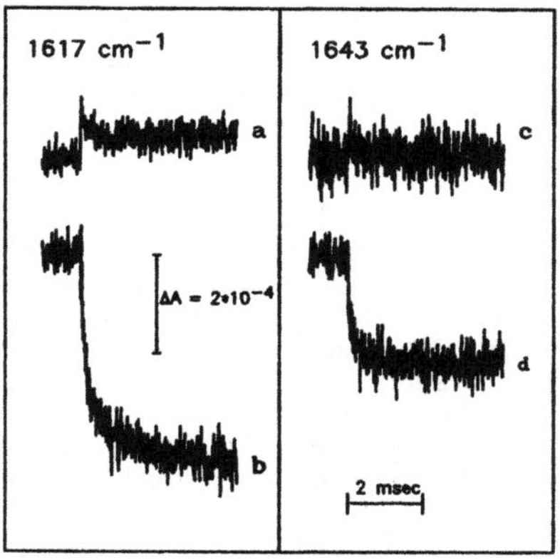

Fig. 2 IR signals at 1617 cm^{-1} and 1643 cm^{-1} without (2a,c) and with (2b,d) the secondary electron acceptor Q_B present. Excitation as in fig. 1. Note the different kinetics in 2b and 2d

The signals at 1658 cm^{-1}, 1643 cm^{-1}, 1617 cm^{-1} and 1607 cm^{-1} exhibit 100-200 μsec components without any noticeable step signal. In the absence of Q_B, no signals are observed at these wavelengths (see fig. 2c). On this basis, one would ascribe these signals to modes of $Q_B/Q_B^{-\cdot}$ and amino acid residues in its environment. The signal at 1643 cm^{-1} appears a reasonable candidate for the C=O vibration of Q_B, showing an absorbance decrease upon formation of $Q_B^{-\cdot}$. The vibration frequency appears to be rather low for the C=O as compared to isolated quinones [13]. However, H-bonding of the quinone to HIS 190 and SER 223 [14] might account for the lowering. Nevertheless, contributions of modes from amino acid side chain groups or peptide carbonyls are equally posssible in this frequency region as suggested in [15]. The signal at 1658 cm^{-1} corresponds to an absorbance increase upon Q_B reduction; it is thus an unlikely candidate for the Q_B C=O. Assignment to the amide I mode of one of the amino acids forming the Q_B pocket appears more likely. Nabedryk et al. have proposed a conformational change of the protein backbone near Q_A upon Q_A reduction [15]. However, no signal characteristic for Q_A is observed at 1658 and 1643 cm^{-1} (fig 1). Whether this is due to very narrow band structures or to the different experimental conditions needs to be settled.

Among these signals due to $Q_B^{-\cdot}$ formation, the signal at 1617 cm^{-1} is exceptional (Fig 2b), in that it cannot be approximated with a single 100-200 μsec component. An additional kinetic component with a half time of 1-2 msec is detected. No chromophore signal exhibits this reaction half-time upon RC single flash turnover. However, a non-chromophoric acid-base group has been characterized that undergoes a pK shift and protonation upon $Q_B^{-\cdot}$ formation [16,17]. The 1617 cm^{-1} signal could be due to the strong antisymmetric C-O vibration of a carboxylate group, presumably GLU L212 and/or ASP L213; in this view the sign of the signal corresponds to a partial protonation. We currently investigate the frequency region below 1450 cm^{-1} for the disappearance of the symmetric C-O vibration of the carboxylate and the 1730-1750 cm^{-1} region for the appearance of the carboxyl C=O group.

ACKNOWLEDGEMENTS: We are indebted to the laser technology group of the Fraunhofer-Inst. für Physikalische Meßtechnik (Freiburg) for producing and supplying the lasers. Support by grants from the DFG to WM is gratefully acknowledged.

1 W.W. Parson, B. Ke (1982), in *"Photosynthesis: Energy conversion by Plants and Bacteria"* (Govindjee, ed.) Academic Press, p.331.
2 H. Michel, O. Epp, J. Deisenhofer (1986), The EMBO Journal 5, 2445-2451.
3 J.P. Allen, G. Feher, T.O. Yeates, H. Komiya, D.C. Rees (1987) Proc. Natl. Acad. Sci. USA **84**, 5730-5734
4 W. Mäntele, E. Nabedryk, B.A. Tavitian, W. Kreutz, J. Breton (1985), FEBS Lett. **187**, 227-232
5 E. Nabedryk, W. Mäntele, B.A. Tavitian, J. Breton (1986), Photochem. Photobiol. **43**, 461-465
6 W. Mäntele, A. Wollenweber, E. Nabedryk, J. Breton (1988), Proc. Natl. Acad. Sci. USA **85**, 8468-8472
7 W. Mäntele, M. Leonhard, M. Bauscher, E. Nabedryk, J. Breton, D.A. Moss (1990) in: "Reaction Centers of Photosynthetic Bacteria" (M.E. Michel-Beyerle, ed.) Springer Series In Biophysics **6**, 31-44
8 D.L. Thibodeau, E. Nabedryk, R. Hienerwadel, F. Lenz, W. Mäntele, J. Breton (1990), Biochim. Biophys. Acta **1020**, 253-259
9 R. Hienerwadel, W. Kreutz, W. Mäntele (1989) in: *"Spectroscopy of Biological Molecules -State of the Art"* (A. Bertoluzza, C. Fagnano and P. Monti, eds.) Societa Editrice Esculapio, Bologna, pp. 315-316
10 W. Mäntele, R. Hienerwadel, F. Lenz, W.J. Riedel, R. Grisar, M. Tacke (1990), Spectroscopy International **2**, 29-35
11 R. Hienerwadel, D.L.Thibodeau, F. Lenz, E. Nabedryk, J. Breton, W. Kreutz, and W. Mäntele (1991) manuscript in preparation
12 J. Breton, D.L. Thibodeau, C. Berthomieu, W. Mäntele, A. Vermeglio, E. Nabedryk (1991) FEBS Lett. **278**, 257-260
13 M. Bauscher, E. Nabedryk, K.A. Bagley, J. Breton, W. Mäntele (1990) FEBS Lett. **261**, 191-195
14 M.L. Paddock, S.H. Rongey, G. Feher, M.Y. Okamura (1989) Proc. Natl.Acad. Sci. USA **86**, 6602-6606
15 E. Nabedryk, K.A. Bagley, D.L. Thibodeau, M. Bauscher, W. Mäntele, J. Breton (1990) FEBS Lett **266**, 59-62
16 C.A. Wraight (1979) Biochim. Biophys. Acta **548**, 309-327
17 P. Brzezinski, M.L. Paddock, S. H. Rongey, M. Y. Okamura, G. Feher (1991) Biophys. J. **59**, 143a

Part II

Metal Complex and Organometallic Systems

Fast and Ultrafast Processes in Organometallic Chemistry

J.J. Turner

Department of Chemistry, University of Nottingham,
Nottingham NG7 2RD, UK

Organometallic Photochemistry has become increasingly important because of its role in synthesis and homogeneous catalysis, and its relevance to electron transfer, biology and solar conversion. A typical photo-reaction in hydrocarbon solution is:

$$Mo(CO)_6 \ + \ PPh_3 + h\nu \rightarrow Mo(CO)_5PPh_3 \ + \ CO,$$

which proceeds via the square pyramidal intermediate $Mo(CO)_5$. Such simple schemes may however disguise some extremely complex chemistry, involving more than one excited state, several dissociation pathways, and production of different intermediates, which may be electronically or vibrationally excited or be in equilibrium with each other. The timescales of these processes can vary, or be made to vary, dramatically. For instance, the intermediate $Fe(CO)_4$ (photolytically generated from $Fe(CO)_5$) has essentially an infinite lifetime in a 10K matrix [1]. On the other hand dissociation of CO from excited $Cr(CO)_6$ takes place in a few hundred femtoseconds [2].

Vibrational spectroscopy has played a key role in unravelling some of these complex processes and we shall largely concentrate on its applications to (i) Fragments and their kinetics, (ii) Excited states, and (iii) Fast exchange processes.

1. Organometallic Fragments/Intermediates and their Kinetics

TR^3 has been widely applied to biological molecules but hardly at all to organometallic species [3,4], partly because parents and intermediates are both so photosensitive. On the other hand the $\nu(CO)$ IR bands of the CO groups in organometallic species are very intense, structurally very revealing and have been employed in a vast array of studies. Low-temperature IR data can provide the basis for monitoring fast reactions by time-resolved IR (TRIR) at ambient temperature.

Low-temperature methods Photolysis of a parent metal carbonyl in a matrix of solid Ar at 10K (or hydrocarbon glass at 77K) can produce fragments via CO loss. The energy factored CO force field, combined with the use of ^{13}CO isotopes and intensity measurements, allows the determination of the structures of fragments, including bond angles, quite accurately [6].

The matrix technique is appropriate for very unstable, coordinatively unsaturated, species; more stable species can be examined in cold solutions. Given their inertness and spectral transparency, liquid noble gases have proved to be the solvents of choice [7]. For example, $Cr(CO)_5(\eta^2\text{-}H_2)$ can be generated in liquid Xe doped with H_2 and, as well as the intense $\nu(CO)$ bands, the extremely weak $\nu(H\text{-}H)$ band is detected. This species has a lifetime of hours at -50°C - and can thus be studied by conventional FTIR spectroscopy [8] - whereas fast TRIR is required at room temperature [9].

Time-resolved IR on the μsec timescale There is now a bewildering array of TRIR techniques but all applications to metal carbonyls [10] have employed flash generation of transient combined with IR spectral changes monitored at a particular frequency (glowbar + monochromator; CO laser; diode laser); the experiment is repeated at different IR wavelengths to build up appropriate time-resolved spectra.

There have been many solution TRIR studies on organometallic intermediates [5,11]. In the gas phase [12], depending on the metal and on the wavelength of the photolysis laser, metal carbonyls can lose one or more CO groups. The bands in the gas phase are broad because of unrelaxed vibrational structure and matrix data has been crucial in understanding the spectra of the fragments. The correspondence between the IR spectra of "naked" gas and matrix isolated species demonstrates that the matrix does not perturb the structure of the fragments.

Time-resolved IR on the ps/fs timescale Any technique which relies on IR detection is limited by the available detectors to a time resolution of the order of 100 ns. This is entirely adequate for most "chemical" purposes but not necessarily for the detection of primary photochemical events. There have been many demonstrations of the advantages of "upconversion" with consequent detection in the UV/visible and hence improvement in time resolution. Hochstrasser and colleagues [13] have been able to detect changes in the ν(CO) IR spectrum of $[CpFe(CO)_2]_2$ ($Cp=\eta^5\text{-}C_5H_5$) down to 100 femtoseconds; from about 100 ps onwards the spectral changes are similar to those observed in the earlier μsec TRIR work [14] which proved that the parent dissociates into either two $CpFe(CO)_2$ radicals or the triply bridged $CpFe(\mu\text{-}CO)_3FeCp$ plus CO. At shorter timescales, although it is clear that there are ν(CO) bands which cannot be assigned to any of the μsec intermediates, the interpretation is complicated by band profile changes associated with vibrational relaxation [15]. Spears and colleagues have probed the photodissociation of $Cr(CO)_6$ on the ps timescale [16]. Again the picture is complicated by vibrational relaxation, as revealed by both anti-Stokes TR^3 [4] and UV/visible fs/ps studies [2,17,18] and there is some argument about the solvation dynamics. Unfortunately none of the proposed very short-lived species have been detected in matrices so the spectra are not well characterised. Moreover, the breadth of the bands and the low resolution of the techniques will make isotopic studies less revealing. However it is clear that ps/fs IR will provide valuable clues about the early stages of photodissociation.

2. Excited States

Until recently most information on the dynamics [19] and structures [20] of the excited states of transition metal complexes came from electronic spectroscopy. However based on the Lee-Heller approach [21], the distortions in the excited state of, for example, $W(CO)_5$(pyridine) can be determined from the intensities of the resonance Raman spectrum (RR) of the ground state molecule [22]. The RR spectra of a number of metal carbonyl systems in their excited states have been measured directly. For instance TR^3 of the bpy bands of the lowest excited state of $ClRe(CO)_3(bpy)$ shows that electron transfer has occurred from metal to bpy ligand [23]. TR^3 does have its limitations, and we have recently applied TRIR to the excited state of $ClRe(CO)_3(4,4'\text{-}bpy)_2$ [24]; the ν(CO) bands, measured 250 ns after the laser flash, are shifted some 60

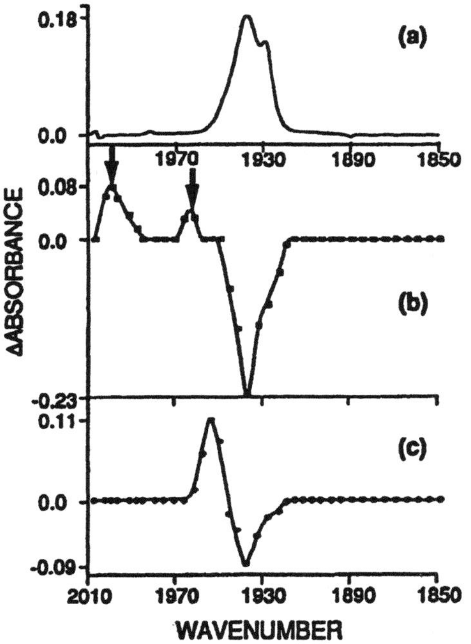

(a) FTIR spectrum of $W(CO)_5(4\text{-CNpyr})$ in MCH; weak band at 2072 cm^{-1}. TRIR spectra at (b) 150 ns and (c) 5 μs after dye laser flash (510 nm, 10 ns). Arrowed bands in (b) due to excited state. Positive band in (c) due to $W(CO)_5$..MCH. Reproduced, with permission, from ref. [27].

cm^{-1} to high frequency of the ground state - thus confirming that charge transfer occurs from Re to bpy ligand - and the decay kinetics match lifetimes previously measured via UV emission [25]. This shift in ν(CO) bands can also be detected in the TR^3 spectrum of the MLCT excited state of $W(CO)_4$(diimine) species [26]. More recently it has been possible to follow the quite complex photochemistry of $W(CO)_5(4CN\text{-pyr})$ from charge transfer excited state to dissociation to recombination [27]; see figure.

3. Fast Exchange Processes

The application of NMR spectroscopy to the study of dynamic processes is well known [28]. In principle similar band broadening and coalescence might be expected to occur in vibrational spectroscopy, provided the dynamic processes are much faster. In practice it is very difficult to separate these spectral effects from other band broadening mechanisms [29]. Grevels and colleagues [30] observed a striking effect in the ν(CO) IR spectra of some diene iron complexes, (eg (η^4-norbornadiene)$Fe(CO)_3$-1). On raising the temperature, the pattern changed from three IR bands to two, consistent with very fast exchange among the CO groups. To test if this effect is "dynamic", we examined the spectrum of 1, 42% enriched with ^{13}CO, from -146°C in liquid krypton to 85°C in supercritical xenon. It proved possible to analyse, very accurately, the energy factored CO force field at the low temperature and to predict, again very accurately, the high temperature collapsed spectrum by averaging the force constants [31]. We believe that this presents an opportunity to monitor exchange dynamics on the picosecond timescale which could complement ultra fast laser techniques.

References

[1] M. Poliakoff and J. J. Turner, J. Chem. Soc. Dalton Trans. 2276 (1974).
[2] A. G. Joly and K. A. Nelson, J. Phys. Chem. 93, 2876 (1989).
[3] S. E. J. Bell, K. C. Gordon and J. J. McGarvey, J. Am. Chem. Soc. 110, 3107 (1988).
[4] S-C. Yu, X. Xu, R. Lingle and J. B. Hopkins, J. Am. Chem. Soc. 112, 3668 (1990).
[5] M. Poliakoff and E. Weitz, Adv. Organomet. Chem. 25, 277 (1986).
[6] J. K. Burdett, M. Poliakoff, J. J. Turner and H. Dubost, Adv. IR and Raman Spect. 2, 1 (1977).
[7] J. J. Turner, M. Poliakoff, S. M. Howdle, S. A. Jackson and J. G. McLaughlin, Faraday Disc. Chem. Soc. 86, 271 (1988), and references therein.
[8] R. K. Upmacis, M. Poliakoff and J. J. Turner, J. Am. Chem. Soc. 108, 3645 (1986).
[9] S. P. Church, F-W. Grevels, H. Hermann and K. Schaffner, J. Chem. Soc. Chem. Commun. 30 (1985).
[10] There are groups working with metal carbonyls in Mulheim (Germany), Nottingham (UK), Ottawa (Canada) and in the USA at Northwestern (2), Davis, Berkeley, Santa Barbara, North Texas, Purdue and Pennsylvania.
[11] F-W. Grevels, W. Klotzbucher and K. Schnaffner (in preparation).
[12] Much of the gas phase work has been reviewed in E. Weitz, J. Phys. Chem. 91, 3945 (1987).
[13] P. A. Anfinrud, C-H. Han, T. Lian and R. M. Hochstrasser, J. Phys. Chem. 95, 574 (1991) and references therein.
[14] B. D. Moore, M. B. Simpson, M. Poliakoff and J. J. Turner, J. Chem. Soc. Chem. Commun. 972 (1984); A. J. Dixon, M. A. Healy, M. Poliakoff and J. J. Turner, J. Chem. Soc. Chem. Commun. 994 (1986).
[15] E. J. Heilwell, R. R. Cavanagh and J. C. Stephenson, Chem. Phys. Letters, 134, 181 (1987); J. Chem. Phys. 89, 230 (1988).
[16] L. Wang, X. Zhu and K. G. Spears, J. Am. Chem. Soc. 110, 8695 (1988); J. Phys. Chem. 93, 2 (1989).
[17] X. Xie and J. D. Simon, J. Am. Chem. Soc. 112, 1130 (1990), and references quoted therein.
[18] M. Lee and C. B. Harris, J. Am. Chem. Soc. 111, 8963 (1989).
[19] A. J. Lees, Chem. Rev. 87, 711 (1987).
[20] eg H. Yersin, H. Otto, J. I. Zink and G. Gliemann, J. Am. Chem. Soc. 102, 952 (1980).
[21] E. J. Heller, R. L. Sundberg and D. Tannor, J. Phys. Chem. 86, 1822 (1982).
[22] K-S. K. Shin and J. I. Zink, Inorg. Chem. 28, 4358 (1989).
[23] W. K. Smothers and M. S. Wrighton, J. Am. Chem. Soc. 105, 1067 (1983).
[24] P. Glyn, M. W. George, P. M. Hodges and J. J. Turner, J. Chem. Soc. Chem. Commun. 1655 (1989).
[25] P. J. Giordano and M. S. Wrighton, J. Am. Chem. Soc. 101, 2888 (1979).
[26] J-H. Perng and J. I. Zink, Inorg. Chem. 29, 1158 (1990).
[27] P. Glyn, F. P. A. Johnson, M. W. George, A. J. Lees and J. J. Turner, Inorg. Chem. in press.
[28] J.Sandstrom, *Dynamic NMR Spectroscopy* (Academic Press, London, 1982).
[29] K. A. Wood and H. L. Strauss, J. Phys. Chem. 94, 5677 (1990).
[30] F-W. Grevels, J. Jacke, W. E. Klotzbucher, C. Kruger, K. Seevogel and Y-H. Tsay, Angew. Chem. Int. Ed. Engl. 26, 885 (1987); F-W. Grevels, J. Jacke and K. Seevogel, J. Mol. Struct. 174, 107 (1988).
[31] J. J. Turner, F-W. Grevels, S. M. Howdle, J. Jacke, M. T. Haward and W. E. Klotzbucher, submitted for publication.

Femtosecond Infrared Spectroscopy: The Condensed Phase Photochemistry of Metal Carbonyl Dimers

T. Lian, B. Locke, M. Iannone, and R.M. Hochstrasser

Department of Chemistry, University of Pennsylvania, Philadelphia, PA 19104, USA

Introduction: The photochemistry of cyclopentadienyl transition metal carbonyl dimers has been extensively studied by various IR techniques. By low temperature matrix isolation and microsecond time resolved infra-red spectroscopy, the photoreaction pathways of these compounds have been generally agreed to be:

$$[CpM(CO)_n]_2 \longrightarrow CpM(CO)_n + CpM(CO)_n$$

$$\longrightarrow Cp_2M_2(CO)_{2n-1} + CO$$

Recently, picosecond (3) and femtosecond (4) IR spectroscopy were used to study the photochemistry of $[CpFe(CO)_2]_2$ in room temperature solution. It was found that the primary events of this photoreaction happen on the few ps time scale and involve many intermediate species. In this work, femtosecond IR spectroscopy is used to study the primary events of the photochemistry of $[CpFe(CO)_2]_2$ (Cp=C_5H_5 and C_5Me_5) in cyclohexane and DMSO, and $[CpNiCO]_2$ in hexane and cyclohexane. Also presented here is the study of the ultrafast motion of these molecules in solution.

Experimental Section: The details of this femtosecond IR spectrometer have been published elsewhere (4). Briefly, The molecules are pumped by 300 fs optical pulses at 580 nm to initiate photoreactions. The change of IR absorbance induced by the pump is probed by a c.w. CO laser gated by 300 fs optical pulses to achieve the subpicosecond time resolution. The frequency resolution of the system is determined by the probe c.w. CO laser which has lines spaced by 4 cm^{-1}.

Photochemistry of $[CpFe(CO)_2]_2$: $[C_5H_5Fe(CO)_2]_2$ molecules exist in room temperature solution as both trans and cis isomers. The ratio of trans to cis isomers depends on the polarity of the solvent. Polar solvents favor cis isomers. $[C_5Me_5Fe(CO)_2]_2$ molecules exist as trans isomers only in room temperature solution.

The photochemistry of $[CpFe(CO)_2]_2$ in cyclohexane (trans: cis ratio, 1.8:1) has been studied by femtosecond IR spectroscopy (4). It was found that 580 nm photolysis proceeds through an excited electronic state having a lifetime of about 1 ps followed by the breakage of one or both bridging bonds. About 50% of the intermediates reform the broken bridging bonds with a time constant of ca. 5 ps to form a metastable doubly bridged dimer that then relaxes with a time constant of ca. 15 ps .

Based on the results a model for the photochemistry of this molecule was proposed (4). The intermediate species are not unambiguously identified because the IR difference spectra are very complicated consisting of four CO groups from every species and from both cis and trans isomers.

To separate the dynamical processes associated with the cis and the trans isomers, we studied the photoreaction of $[C_5Me_5Fe(CO)_2]_2$ in cyclohexane (trans only) and $[C_5H_5Fe(CO)_2]_2$ in DMSO (cis dominant). Out results in the bridging CO region show that the dynamical processes after 580 nm photolysis associated with trans and cis isomers are very similar, and are similar to that of $[C_5H_5Fe(CO)_2]_2$ in cyclohexane (trans and cis mixture), suggesting that trans and cis may have similar photochemistry at t<1ns.

Photochemistry of $[CpNiCO]_2$: The IR spectrum of $[CpNiCO]_2$ in hexane after 580 nm photolysis is shown in Figure 1, while Figure 2 shows the kinetics of $[CpNiCO]_2$ dissociation in cyclohexane solution. The recovery of the bleach signal fits to a single exponential with a 7 ps time constant.

Our results show that no significant amount of CO is lost. The photodissociation proceeds through what appears to be an excited state with a lifetime of about 0.7 ps and 80% of the photoproducts geminately recombine with a time constant of about 7 ps to form hot dimers. The hot dimers cool down with a time constant of about 20 ps. There is evidence from the kinetics at 1845 cm^{-1} (see Fig. 1) that the 7 ps geminate recombination time is determined by the need for vibrational energy to be removed from its reaction coordinates. This absorption signals the CO stretch of molecules not fully equilibrated along the geminate rebinding coordinate.

From our results, it is suggested that photochemistry of $[CpNiCO]_2$ in room temperature solution is similar to that of $[CpFe(CO)_2]_2$ as proposed in ref (4). The intermediate species have not been investigated yet, and they will be investigated in our future experiments. This molecule, having only two CO groups, should have simpler and distinguishable IR spectra corresponding to each intermediate species and, hence, can be used as a model system to study the photodissociation of metal carbonyl dimers in solution.

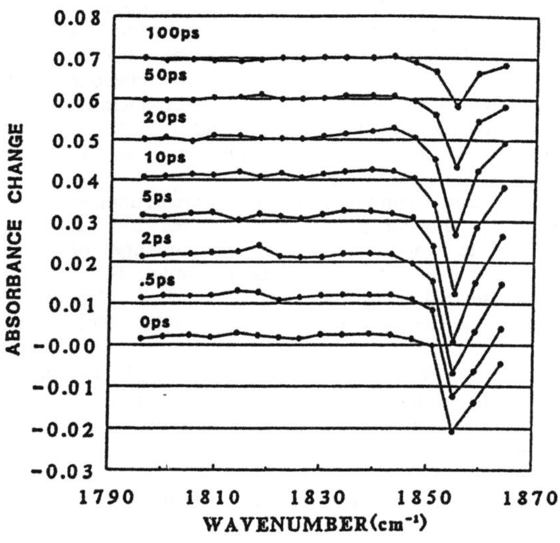

Figure 1.
Transient IR spectra of $[CpNiCO]_2$ at various times after photolysis.

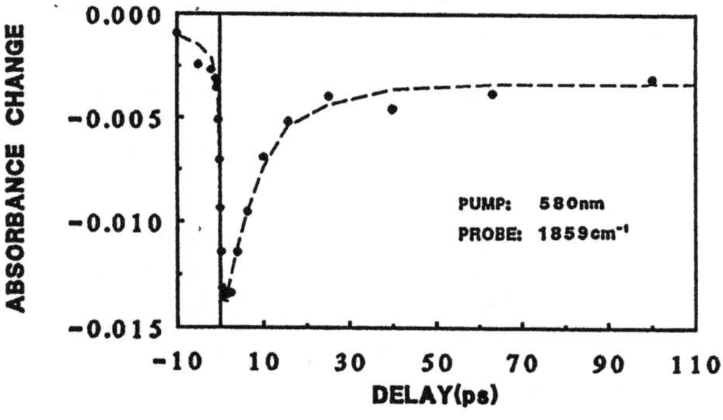

Figure 2.
Geminate rebinding of [CpNiCO] radicals in n-hexane at 300 K.

Ultrafast Motion: The IR anisotropy decay of $[CpFe(CO)_2]_2$ and $[CpNiCO]_2$ in different solvents was measured by pumping at 580 nm and probing at the bridging CO stretching bands. The results are summarized in the following Table.

Table 1. Comparison of predicted and measured orientational relaxation times

Solution	[1/6D]/ps	T/ps
$[C_5H_5Fe(CO)_2]_2$ in cyclohexane	60	25
$[C_5H_5Fe(CO)_2]_2$ in DMSO	150	42
$[C_5Me_5Fe(CO)_2]_2$ in cyclohexane	110	55
$[C_5H_5NiCO]_2$ in cyclohexane	60	15

Our results show that the measured orientational relaxation time is faster than the time constant predicted from the hydrodynamic model with stick boundary conditions (5). It is known that the cyclopentadienyl ring is nearly free to rotate about the axis normal to and through the center of the ring. It is suggested here that the rotational relaxation of these metal carbonyl compounds in solution has some degree of slip due to this property of the cyclopentadienyl groups.

Two Photon Absorption: Since the excited states suggested by the present and previous (4) work have very short lifetimes, comparable with the pulse auto-correlations, the possibility that the signals represent vibrational resonances in an IR-optical two photon absorption cannot be excluded at this time. Further experimental work is needed to establish this point.

Conclusion: Femtosecond IR spectroscopy has made it possible to study the primary processes of the photochemistry of metal carbonyl dimers which happen on the 10 ps time scale. The rapid reforming of the initial dimers after photo-dissociation is due to the solvent "cage effect", and can be used to study the solvent effects on condensed phase chemical reactions. Presumably the solvent attack on the coordinatively unsaturated nickel competes unfavorably. The motion of these cyclopentadienyl metal dimers in solution has some degree of slip due to the "slippery" property of the cyclopentadienyl ring.

Acknowledgements: This research was supported by NSF and NIH.

References:

1. T. J. Meyer and J. V. Caspar, Chem. Rev. 85, 187 (1985).
2. J. J. Turner, M. Poliakoff and M. A. Healy, Pure & Appl. Chem. 1, 787 (1989).
3. J. N. Moore, P. A. Hansen and R. M. Hochstrasser, J. Am. Chem. Soc. 111, 4563 (1989).
4. P. A. Anfinrud, C. Han, T. Lian and R. M. Hochstrasser, J. Phys. Chem. 95, 574 (1991).
5. R. M. Hochstrasser, et al., in preparation.

Excited States of Metalloporphyrin Photosensitizers

R.E. Hester

Department of Chemistry, University of York, York YO1 5DD, UK

1. Introduction

The porphyrins and metalloporphyrins lie at the heart of a multitude of systems and processes, finding applications in fields as diverse as photodynamic therapy and optical waveguide gas sensors. Resonance Raman (RR) spectroscopy has been used to good effect to study the structure and bonding in porphyrins and to report on changes in these properties caused, for example, by binding as prosthetic groups in active sites of proteins or by optical absorption by porphyrin photosensitizers. The structures and properties of these molecules may be significantly changed in electronically excited states, but the commonly very short lifetimes of such states dictate the use of sophisticated pulsed laser, time-resolved RR methods for their characterization.

Time-resolved resonance Raman spectroscopy (TR3S) has previously been used to monitor structural changes associated with ligand-field excitation of nickel porphyrins [1] and for following the structural relaxation in haem proteins which follows their photoexcitation and subsequent rapid dissociation of axially coordinated ligands [2]. Earlier work on zinc tetraphenylporphryin (TPP) [3,4] has been facilitated by the relatively long lifetime of its transient photoexcited state (> 1 ms) and its high quantum yield for formation (*ca.*. 0.8) [5]. By contrast, the lifetimes of electronically excited iron porphyrins are extremely short. Only for the complex Fe(II)(TPP)Cl has any phosphorescence been detected [6]. Flash photolysis studies have shown this to arise from an excited state with only a 30 ps lifetime and a 1% quantum yield for formation [7,8]. Nonetheless, equipped only with two nanosecond lasers of independently tunable wavelength, we have been able to measure the strongest features in the TR3 spectrum of this state [9].

We also have used two-colour TR3S to probe porphyrin excited states in complexes (exciplexes) with DNA and model polynucleotide systems. This relates significantly to photodynamic therapy, in that some porphyrins are known to become anti-tumour active when exposed to light. Specifically, we have studied the water-soluble metalloporphyrin tetrakis(1-methylpyridinium-4-yl) porphinato copper(II), Cu(T4MPyP), complexed with the synthetic model nucleic acid poly (dA-dT), under irradiation in its B(Soret) absorption band. Our Q-band probed TR3S data are indicative of a 20-ps lifetime exciplex involving the A-T sites of the nucleic acid [10].

2. Photoexcited iron(III) tetraphenylporphyrin

Fig. 1 shows RR spectra of Fe(TPP)Cl obtained using single-wavelength irradiation at 408 nm, which is close to the Soret absorption maximum. At low laser irradiance (10 μJ/pulse, 100 μm beam diameter) the ground-state spectrum of Fe(TPP)Cl was obtained (Fig. 1a). Increasing laser energy at the sample (to 2

Springer Proceedings in Physics, Vol. 68 Time Resolved Vibrational Spectroscopy V
Ed.: H. Takahashi

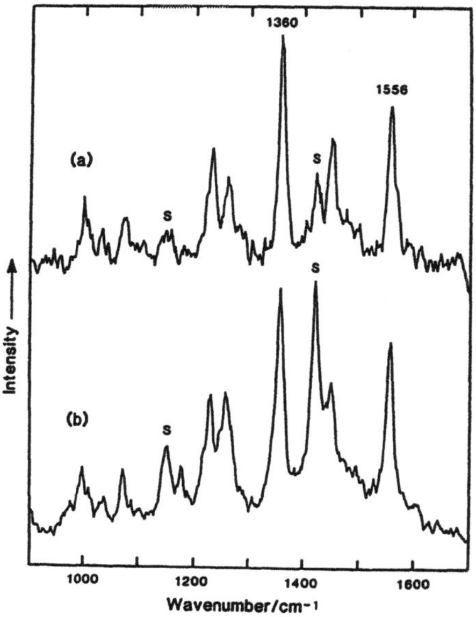

Fig. 1. Single-laser (408 nm) RR spectrum of Fe(TPP)Cl (10^{-4} mol dm^{-3} in CH_2Cl_2) (a) at low laser irradiance and (b) at high laser irradiance.

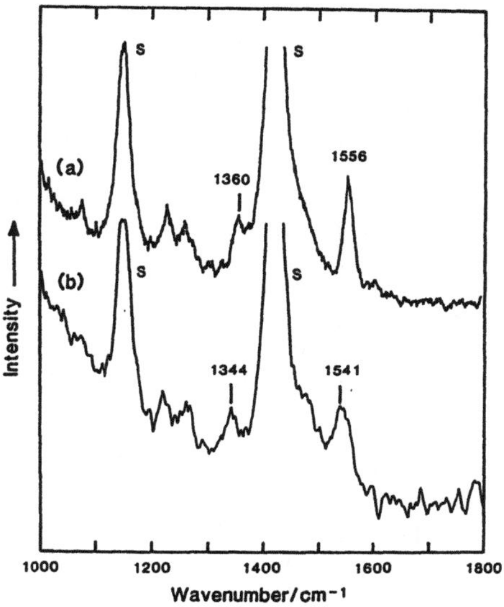

Fig. 2. Two-colour (406 nm pump, 448 nm probe) RR spectra of Fe(TPP)Cl (10^{-4} mol dm^{-3} in CH_2Cl_2) (a) probe only, (b) pump and probe, $\Delta t = 0$.

mJ/pulse) gave a spectrum containing bands due to the complex which were very much diminished with respect to the solvent bands (marked S) but no discernable new bands appeared (Fig. 1b). The ground-state complex bands were not appreciably broadened under high laser fluence. Similar results were obtained with 424 nm irradiation. When 448 nm irradiation was used, no significant loss of ground-state complex bands was observed even at high laser fluence, presumably due to the low absorbance of the complex at this wavelength (see Fig. 2(a)).

The use of two-laser pump (406 nm)-probe (448 nm) experiments again showed large ground state depletion, but only when the two pulses were overlapped in time, thus confirming the short-lived (< 10 ns) nature of the excited state. As seen in Fig. 2, in these two-colour pump-probe experiments the strong ground-state marker bands at 1360 (ν_4) and 1556 cm^{-1} (ν_2) gave way to new bands at 1344 and 1541 cm^{-1}, respectively, arising from the phototransient state.

The use of 448 nm radiation to probe the RRS of this transient species was indicated by the flash photolysis data [7]. The *ca.* 15 cm^{-1} down-shifts in the RR marker band wavenumbers are consistent with a porphyrin core expansion in this excited state species [11]. This contrasts with the results for Zn(TPP) [3,4] and indicates an absence of significant Jahn-Teller distortion but probably the presence of coupling between the unfilled copper(II) d-orbital and the porphyrin $\pi\pi^*$ orbitals in the case of Fe(TPP)Cl [12].

3. Copper porphyrin-DNA exciplex formation

It is fairly well established that planar (four-coordinated) porphyrins (e.g. M = H_2, Cu(II) or Ni(II)) form strong complexes with DNA by intercalation at G-C sites, while porphyrins with axial ligands (5- or 6-coordinate, e.g. M = Zn(II), Co(III), Fe(III), or Mn(III)) form groove-bound complexes at A-T sites [13]. A recent RR study revealed the novel fact that the planar metalloporphyrin Cu(T4MPyP) binds selectively to A-T sites when the complex is in its electronically excited state [14]. We have extended this earlier study by determining the RR excitation profiles for the ground and excited state complexes when bound to poly(dA-dT) in aqueous solution [10].

Our two-colour pump (545 nm; Q-band)-probe (445, 450, and 457 nm) TR3S data were obtained with the pump and probe pulses overlapped in time. Mixtures of Cu(T4MPyP) and poly (dA-dT) in water gave ν_4 and ν_2 porphyrin marker bands at 1370 and 1574 cm^{-1}, respectively, for the ground state species and at 1353 and 1558 cm^{-1}, respectively, for the excited state. The excitation profiles for the ground state bands both peak at *ca.* 426 nm, in agreement with the B(Soret) absorption maximum. Unexpectedly, the excitation profiles for the excited state bands both were found to peak at a wavelength shifted only some 4-7 nm to the red. The excited state transient absorption has not been measured directly but it may confidently be inferred that it too will peak at *ca.* 430 nm. Only ground state RR bands were observed if the pump and probe beams were not exactly overlapped in both space and time, clearly indicating a lifetime for the excited state shorter than the width of the probe laser pulse (i.e. < 10 ns). Previous estimates have put this lifetime at *ca* 10 ps.

For porphyrins and metalloporphyrins the "normal" behaviour of electronically excited triplet states is to show UV/Vis absorptions substantially red-shifted from the ground state B-band wavelength. For example, the Cu(TPP) triplet absorbs at 470 nm [15]. Its anomolous absorption wavelength (*ca.* 430 nm) and exceptionally short lifetime lead us to support an earlier suggestion [14] that the Cu(T4MPyP) excited state is modified by complexation with the poly(dA-dT) (exciplex formation). The downshifts of the exciplex RR band wavenumbers are of comparable magnitude to those observed on electro-oxidation of Cu(TPP) to form its radical cation [16]. The overall evidence points to the observed "exciplex" being a charge-transfer (CT) complex between metalloporphyrin (P) and the nucleic acid (N), of the form P^+N^-. This, in turn, indicates that the binding is not to the DNA phosphate groups but rather to the A and/or T base residues, which quench the electronically excited metalloporphyrin by acting as electron acceptors in a CT reaction mechanism.

4. Acknowledgements

This paper is based on work done in collaboration with others named in references 9 and 10. Much of any scientific credit is due to them, but I take full responsibility for any distortions introduced in this brief summary.

References

1. J. A. Shelnutt, K. Alston, E. Q. Findsen, M. R. Ondrias and J. M. Rifkind, in *Porphyrins, excited states and dynamics*, eds. M. Gouterman, P. M. Rentzepis and K. D. Straub (Am. Chem. Soc., Washington, 1986) p. 232.
2. D. L. Rousseau and J. M. Friedman, in *Biological applications of Raman spectroscopy*, Vol. 3, ed. T. G. Spiro (Wiley-Interscience, New York, 1987) p. 133.
3. V. A. Walters, J. C. de Paula, C. T. Babcock and G. E. Leroi, *J. Am. Chem. Soc.* 11 (1989) 8300.

4. R. A. Reid, R. Purello, K. Kushmeider and T. G. Spiro, in: *Proceedings of the 4th International Conference on Time-Resolved Vibrational Spectroscopy*, eds. T. G. Spiro and R. S. Czernuszewicz (1989).
5. J. Darwent, P. Douglas, A. Harriman, G. Porter and M. C. Richoux, *Coord Chem. Rev.* 44 (1982) 83.
6. A. Harriman *J. Chem. Soc. Faraday Trans. I* 77 (1981) 369.
7. P. A. Cornelius, A. W. Steele, D. A. Chernoff and R. M. Hochstrasser, *Chem. Phys. Letters* 82 (1981) 9; Y Liang, D. K. Negus, R. M Hochstrasser, M. Gunner and P. L. Dutton, *Chem. Phys. Letters* 84 (1981) 236.
8. B. M. Dzhaganov, V. S. Chirvonyi and G. P. Curinovich, in: *Laser picosecond spectroscopy and photochemistry of biomolecules*, ed. V. S. Letokhov (Hilger, Bristol, 1987) p. 137.
9. J. C. Austin, S. E. J. Bell and R. E. Hester, *Chem. Phys. Letters* 169 (1990) 342.
10. L. Chinsky, P.-Y. Turpin, A. Al-Obaidi, S. E. J. Bell and R. E. Hester, *J. Phys. Chem.*, in press (1991).
11. N. Parthasarathi, C. Hanson, S. Yamaguchi and T. G. Spiro, *J. Am. Chem. Soc.* (109) 3865.
12. R. L. Ake and M. Gouterman *Theoret. Chim. Acta* 15 (1969) 20.
13. R. F. Pasternak, E. J. Gibbs and J. J. Villafranca, *Biochemistry* 22 (1983) 2406.
14. P.-Y. Turpin, L. Chinsky, A. Laigle, M. Tsuboi, J. R. Kincaid and K. Nakamoto, *Photochem. Photobiol.* 51 (1990) 519.
15. J. Rodriguez, C. Kirmaier and D. Holton, *J. Am. Chem. Soc.* 111 (1989) 6500.
16. R. S. Czernuszewicz, K. A. Macor, X. Y. Li, J. R. Kincaid and T. G. Spiro, *J. Am. Chem. Soc.* 111 (1989) 3860.

Rate of Translation of a Ligand Molecule Along a DNA-Duplex Chain

G.D. Strahan, D. Lu, M. Tsuboi, and K. Nakamoto

Todd Wehr Chemistry Building, Marquette University,
535 N. 14th Street, Milwaukee, WI 53233, USA

1. Introduction

In any decoding process of the genetic information in a biological system, there must be a specific ligand-DNA binding. Lac-repressor (a protein) of Escherichia coli, for example, must find a particular sequence called lac-operator in its long DNA chain to achieve a specific binding there. Because one cannot suppose any long-range force between these repressor and operator, we are compelled to consider that this is a diffusion-driven reaction. In 1970, the in-vitro kinetics of the association reaction of lac-repressor and lac-operator was examined by Riggs et al.[1], and the second-order rate constant was determined to be $7\times10^9 M^{-1}sec^{-1}$, in aqueous solution at 24°C. This value was found to be more than one order of magnitude greater than that expected from the known diffusion constants of macromolecules. At that time, Riggs et al.[1] already presented an idea that the lac-repressor searches for the operator not by performing a three-dimensional random walk, but rather by binding to DNA and rolling or hopping along it, thus reducing the search for the operator to lower dimensions. Although they reached a rather negative conclusion on this idea at that time, it gradually became more significant. In 1981, for example, Winter et al.[2] have re-examined the repressor-operator kinetics in detail, and found that the rate constant depends not only on the ionic strength of the solution but also on the length of the DNA chain containing the operator. They showed that their experimental results are well explained by the above two-step mechanism (binding and then sliding), if the one-dimensional diffusion constant (D_1) in the sliding is assumed to be 8×10^{-10} cm^2/sec[2]. This D_1 value corresponds to a random-walk rate of 10^8 base-pairs/sec, or one base-pair/μsec.[2] Thus, it is desirable to design an experiment in which one can directly observe such a translocation.

Quite recently, Turpin et al.[3] found that a transient complex is formed between a water-soluble Cu porphyrin, namely copper-tetrakis[4-N-methylpyridyl]-porphyrin (abbreviated as Cu(TMPy-P4)) and the AT site of a DNA duplex, on pulsed laser excitation at 428 nm. Because this porphyrin, in its ground electronic state, is known to be intercalated selectively at the GC site[4], it has been tacitly assumed that a translocation of this molecule from the GC to AT site must occur on its electronic excitation[3]. Here, we present spectroscopic evidence to support the concept of translocation of Cu(TMPy-P4) along the DNA chain (Fig. 1).

Springer Proceedings in Physics, Vol. 68 Time Resolved Vibrational Spectroscopy V
Ed.: H. Takahashi

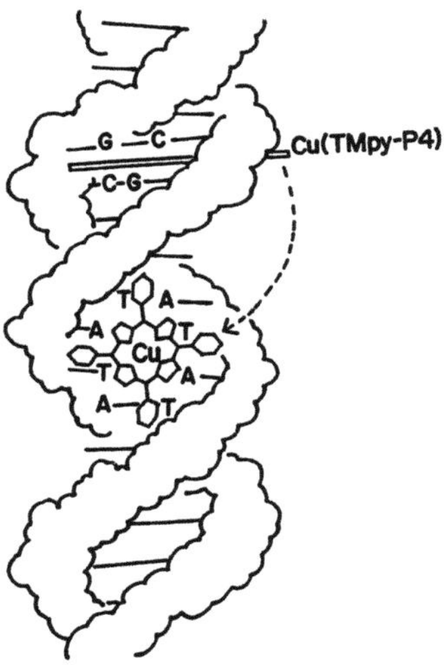

Fig. 1 Translocation of Cu(TMPy-P4) from GC to ATAT site upon pulsed laser excitation.

2. Experimental Results

Experimental results obtained from our extensive resonance Raman study[5] on the Cu-porphyrin-DNA system may be summarized as follows:

(1) A Raman criterion of the intercalation of the ground-state Cu(TMPy-P4) in DNA was established by a frequency shift of +5 cm^{-1} of the C_β-H in-plane deformation band at 1100 cm^{-1}.[4] This frequency shift is observed when Cu(TMPy-P4) is mixed with DNA having a GC or CG sequence, i.e., natural DNA, poly[d(G-C)].poly[d(G-C)], $d(CGTACG)_2$, $d(GCATATATGC)_2$, $d(GCGCGCATATGCGCGC)_2$, or $d[(GC)_7ATAT(GC)_7]_2$(32-mer), but not when mixed with poly[d(A-T)].poly[d(A-T)].

(2) The "transient complex" is represented by two prominent Raman bands at 1550 and 1346 cm^{-1}, which are absent in the Raman spectrum of ground-state Cu(TMPy-P4). These extra bands occur only when a pulsed laser excitation (peak power 10^5W) is made, and only when Cu(TMPy-P4) is mixed with oligonucleotides having AT sites. They do not occur, for example, when it is mixed with poly[d(G-C)].poly[d(G-C)].

(3) The efficiency in exciting these extra Raman bands is a function of the wavelength of the pulsed laser, and this coincides nearly with the absorbance profile of Cu(TMPy-P4); both have a maximum at about 426 nm.

(4) The transient complex is formed preferentially at the ATAT site rather than the isolated single-AT site, as is seen in Figure 2 (a)--(d).

(5) A counter-argument, that the transient complex may come from the ground-state Cu(TMPy-P4) already bound to the ATAT site (although its amount must be very small), rather than from that bound to the GC site, has been invalidated. As seen in Figure 2 (d), even when the relative amount of the

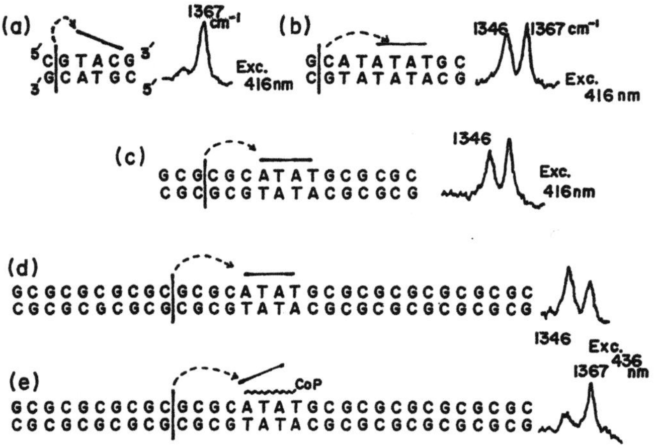

Fig. 2 Schematic diagram showing the relationship between translocation and appearance of the transient complex band at 1346 cm^{-1}.

ATAT site (with respect to that of the GC site) is lowered, the occurrence of the transient complex is not lowered at all, as far as the absolute amount of the ATAT site is kept sufficiently high.

(6) When the ATAT site is blocked prior to pulsed laser irradiation, the occurrence of the transient complex is markedly reduced (Figure 2)(e). The blocking was made by adding Co(TMPy-P4) which is known to be bound selectively at the AT site in its ground state.

(7) When the Cu(TMPy-P4)-DNA solution is frozen to -85°C, the occurrence of the transient complex is prevented. This observation indicates that the transient complex is not formed through a totally electronic processes. Instead its formation must involve an atomic and/or molecular movement.

3. Kinetics of Photochemical Process

In the photochemical process caused by pulsed laser irradiation of the Cu(TMPy-P4)-DNA mixture, at least the following three chemical species are involved: (G) ground-state Cu(TMPy-P4) intercalated at GC site of DNA, (B) electronically excited (at 436 nm, Soret) Cu(TMPy-P4)*, whose lifetime is supposed to be very short (1 ps or so), and (E) the transient complex formed between Cu(TMPy-P4) and ATAT site of DNA. E is not necessarily an exciplex; this might be a ground-state metastable charge-transfer complex which is formed through an excited-state Cu(TMPy-P4)[6]. In any case, E must revert to G, within a short period of time (ns), (see Figure 3). For a kinetic analysis, the rate constants, λ, k_b, k', and k_e, are defined as follows:

$$d[G]/dt = -\lambda[G] + k_b[B] + k_e[E] \quad (1)$$
$$d[B]/dt = \lambda[G] - k_b[B] - k'[B] \quad (2)$$
$$d[E]/dt = k'[B] - k_e[E] \quad (3)$$

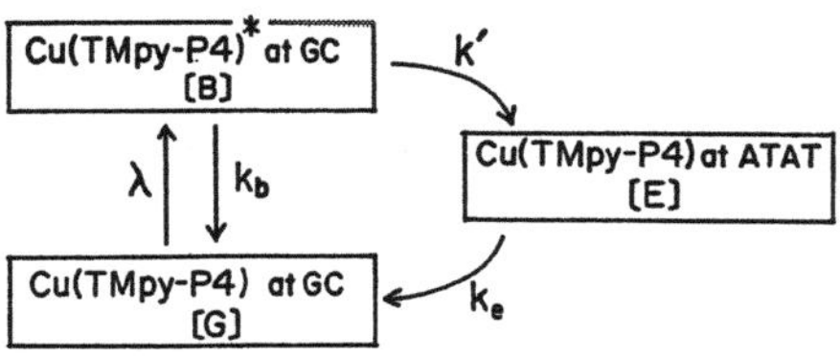

Fig. 3 Kinetics of photochemical process.

where [G], [B], and [E] are the concentrations of G,B, and E, respectively. Let us assume that within the pulsed laser duration (8 ns) a steady-state is reached, so that d[G]/dt=d[B]/dt=d[E]/dt=0. Then,

$$[G] = [(k_b/k')+1]k_e[E] \tag{4}$$

Under our experimental conditions, the relative intensities of the extra Raman bands became nearly equal to the corresponding ground-state Raman bands, when the average power of the pusled laser reached 15 mW. This power corresponds to the pumping rate, $\lambda=6\times10^{11}s^{-1}$, which was estimated from the following values of parameters: pulse duration=8ns, repetition=20Hz, wavelength=436nm, total concentration of Cu(TMPy-P4)=1×10^{-5}M, its molar extinction coefficient at 436 nm = $2.3\times10^{5}M^{-1}cm^{-1}$, and the cross section of the laser beam = $2\times2\times\pi/4$ mm^2. By substituting [G]=[E] and $\lambda=6\times10^{11}s^{-1}$ in Eq.(4), we find that

$$k_e = (k'/k_b) \times 6\times10^{11} \tag{5}$$

By our one-color pump-probe experiment, it was shown, at 20°C, that

Lifetime of E < 35 ns, i.e., $k_e > 3\times10^7s^{-1}$.

The lifetime of the Soret-excited state would be 10^{-12}~10^{-13} in the order of magnitude. If so, from Eq.(5),

$$k'/k_e = (1/6) \times 10^{1\sim2}.$$

Thus, the rate of translocation, k', must be in the same order of magnitude with that of k_e, which is now known to be greater than $3 \times 10^7s^{-1}$.

Acknowledgments - This work was supported by the National Institute of Health (GM-37608). M. T. was partly supported by a grant (No. 02670985) from Ministry of Education, Science, and Culture of Japan.

References

1. A.D. Riggs, S. Bourgeois, and M. Cohn, J. Mol. Biol, 53, 401 (1970).
2. R.B. Winter, O.G. Berg, and P.H. von Hippel, Biochemistry, 20, 6961 (1981).
3. P.Y. Turpin, L. Chinsky, A. Laigle, M. Tsuboi, J.R. Kincaid, and K. Nakamoto, Photochem. Photobiol., 51, 519 (1990).
4. J.H. Schneider, J. Odo, and K. Nakamoto, Nucl. Acids Res., 16, 10323 (1988).
5. G.D. Strahan, D. Lu, M. Tsuboi, and K. Nakamoto, to be published.
6. L. Chinsky, P.Y. Turpin, A.H.R. Al-Obaidi, S. Bell and R.E. Hester, J. Phys. Chem. in press.

Resonance Raman and CARS Investigations of Short-Living Transients in Metalloporphyrin Solutions

P.A. Apanasevich, S.G. Kruglik, V.V. Kvach, and V.A. Orlovich

Institute of Physics, Byelorussian Academy of Sciences, Leninsky prospekt 70, 220602 Minsk, Byelorussia

1. Introduction

There has been renewed interest in spectroscopic investigations of metalloporphyrins (Me-P) because of versatile physico-chemical phenomena they are participating in. We have investigated different porphyrins complexes with transition metals having unfilled inner $3d^m$ electron shell. Below the results are presented on nickel (m=8) and copper (m=9) porphyrins excited state (ES) obtained with time-resolved Resonance Raman (RR) and Resonance Coherent anti-Stokes Raman scattering (RCARS) techniques.

2. Experimental

The porphyrins solutions were investigated using 10...20 ns duration laser pulses. The Me-P transients were produced under tight focusing of the laser beams into the sample.

The Raman spectrometer was designed on the basis of repetitively pulsed and Q-switched YAG:Nd laser. The narrow band tunable dye lasers pumped by the second and the third harmonics of the master laser output were used in RR/RCARS experiments for excitation of the samples or/and together with YAG:Nd outputs were mixed in nonlinear optical media (crystals, compressed molecular hydrogen) to cover necessary excitation frequency region.

The RR spectra were recorded using diffraction gratings double monochromator and rotating cell technique. The spectrometer control, data acquisition and processing were performed with personnel computer.

3. Results

Previously transient Raman spectra from the $^3B_{1g}$ or d-d ES of Ni-P have been investigated by authors [1] and E.W.Findsen et. al. [2]. The results of the frequencies and symmetry species measurements for the modes of Ni- octaethylporphyrin (OEP) are tabulated. The analysis of the data in Table 1 proves that Ni-P molecule in the excited $^3B_{1g}$ state retains its original D_{4h} symmetry. Although it is noteworthy that the negative shifts for the number of modes in the ES respectively to their ground state (GS) values undoubtedly indicate the structural changes (core expansion) in the $^3B_{1g}$ state.

In the present study the particular consideration have been given to the intermediate π-electronic $^3T_1$ state and $^1B_{1g}$ state being another one d-d state lying between the lowest excited singlet $^1S_1$ and $^3B_{1g}$ state. Earlier that states were detected

Table 1. Vibrational frequencies and symmetry species of the Ni-OEP modes in the ground and low-lying $^3B_{1g}$ excited states. Mode designations according to T.Kitagawa et.al. [3].

Mode	ν_{10}	ν_{2}	ν_{19}	ν_{11}	ν_{3}	ν_{29}	ν_{20}	ν_{4}	ν_{21}	ν_{13}
GS	1655	1604	1603	1583	1519	1410	1397	1383	1310	1222
$^3B_{1g}$	1627	1581	1562	1565	1494	1383	1364	1375	1308	1210
Symmetry	b_{1g}	a_{1g}	a_{2g}	b_{1g}	a_{1g}	b_{2g}	a_{2g}	a_{1g}	a_{2g}	b_{1g}

with picosecond transient absorption spectroscopy [4]. Previously we have detected the drop of the scattering efficiency from the $^3B_{1g}$ state in RCARS spectra [1], related with stepwise population of the $^3T_1$ and $^1B_{1g}$ states due to laser radiation absorption by the molecules excited to $^3B_{1g}$ state. Therefore more comprehensive investigation of the RCARS spectra were performed at different photon densities of resonance laser fluxes. The spectra obtained under carefull control of the excitation intensities are depicted in Fig.1. Besides the GS spectrum (a) the rest ones contain ES originated Raman bands. The peaks at 1364, 1383 cm^{-1} are related with the population of the $^3B_{1g}$ state. The pumping intensity rise causes appearance of another peaks at 1390, 1404 cm^{-1} which are tentatively assigned to more short-living upper states. The most probable candidate is the $^1B_{1g}$ state due to the similarity of electronic enhancement conditions for RCARS process as it follows from transient absorption measurements [4]. Small extinction coefficients of transient triplet-triplet absorption in the green-yellow spectral region prevent transient RCARS spectra detection from the $^3T_1$ state even under its high population levels.

The copper atom having unpaired electron located at the d-orbital strongly perturb the electronic states manifold of the porphyrins. The phenomena taking place in the ES of Cu-P have been widely investigated with transient absorption and luminescence techniques. The outstanding feature of the

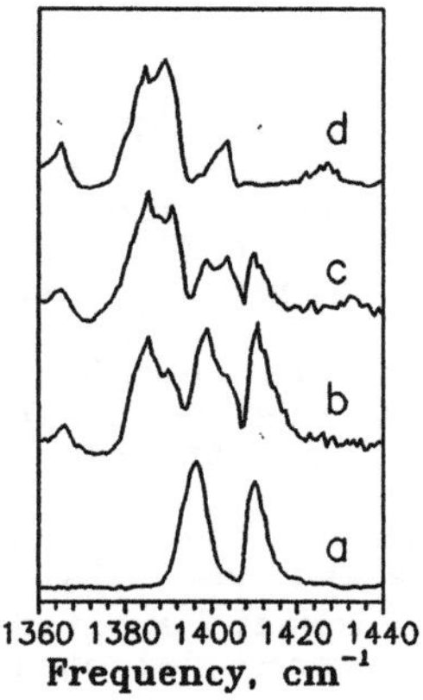

Figure 1.

Resonance CARS spectra of Ni-OEP in toluene obtained at different laser intensities starting from 1 MW/cm^2 (spectrum a) to 1 GW/cm^2 (spectrum d). The spectra (b,c) were recorded under intermediate excitation powers.

Table 2. Frequencies and symmetry species of the Raman-active modes of Cu-OEP in THF solution in the ground and d-d excited electronic states.

Mode	ν_{10}	ν_{2}	ν_{19}	ν_{11}	ν_{3}	ν_{29}	ν_{20}	ν_{4}	ν_{21}	ν_{13}
GS	1639	1594	1586	1571	1504	1406	1392	1380	1316	1214
d-d	1612	1592	1541	1557	1491	1397	1375	1363	1300	1198
Symmetry	b_{1g}	a_{1g}	a_{2g}	b_{1g}	a_{1g}	b_{2g}	a_{2g}	a_{1g}	a_{2g}	b_{1g}

transient RCARS spectra we have observed for various Cu-P in oxygen-containing solvents like tetrahydrofuran (THF) is the presence of strong isolated Raman bands which can not be assigned solely to the GS vibrations. The above mentioned analysis of the frequencies and symmetry species of the Raman-active modes leads to the conclusion that the regularities observed in transient spectra of Cu-P in THF (Table 2) are close to that of Ni-P solutions (Table 1). Thus it is implied that the transient peaks in Raman spectra borrow their intensity from the π-electronic transitions under excited d-electronic configuration of the central atom. The iterative extended Huckel's calculations [5] predict that the intermolecular interactions with oxygen-containing solvent molecule can strongly perturb the Cu-P ES manifold resulting in significant lowering of the charge transfer (CT) and d-d states. The calculations place these levels close to the lowest triplet state T_1. As it was previously observed with picosecond transient absorption spectroscopy [6] some oxygen-containing solvents cause significant (up to 3 order of the magnitude) shortening of the T_1 state lifetime. That is due to the lowering down of the d-d state being the lowest ES and having lifetime constant in subnanosecond time domain.

The analysis of the modes frequencies in Table 2 reveals that the most dramatic shifts (13...45 cm^{-1}) were observed for the modes ν_3, ν_{10}, ν_{19} known as a highly sensitive markers of the porphyrin core size [7]. The large negative shifts correlate with the core expansion and probably reflect the doming of the porphyrin planar structure under the effect of the solvent molecule interaction with the electronically excited porphyrin molecule via oxygen heteroatom of the solvent and the metal atom of the porphyrin molecules respectively.

Another "challenge" of the Cu-P spectroscopy is the observation of the transient RR spectra from the lowest π-electronic triplet state. The Cu-P solutions in benzene were studied by "pump-probe" technique. Typical neutral solvent has been chosen to avoid the complication of the ES manifold due to the strong shifting of the CT or/and d-d state. The YAG:Nd laser second harmonic was used as a "pump" pulse to excite porphyrin molecules to the relatively long-living ($\approx$ 100 ns) triplet state T_1. The transient RR spectra were recorded using 10 ns delayed "probe" pulses at 416 nm. Figure 2 depicts experimental spectra patterns. The spectrum (a) is the GS originated Raman spectrum which was obtained without "pumping" pulse. It does not display any indication of the lines distortions caused by partial depopulation of the GS or the bands manifestation from the ES because the probe pulse

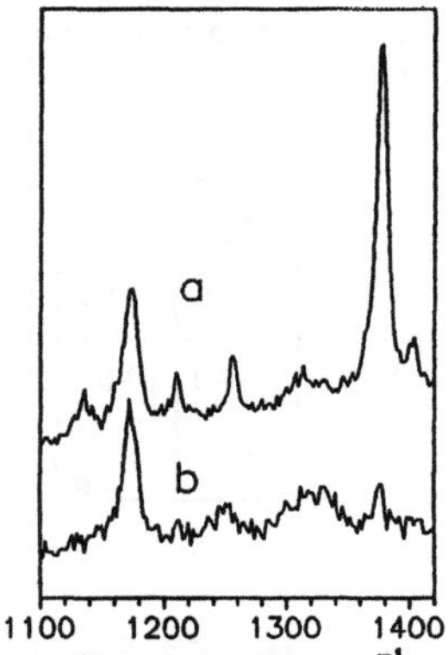

Figure 2.

RR spectra of Cu-OEP in benzene obtained with 416 nm excitation.
a - spectrum from the GS,
b - transient spectrum recorded with 10 ns delay after pumping up with intense laser pulse at 532 nm. Peak at 1178 cm^{-1} belongs to the solvent.

intensities were below 10^5 W/cm^2. That is significantly lower than the saturation intensity values. Spectrum (b) is the transient one in which strong band at 1380 cm^{-1} belonging to the totally symmetric vibration practically disappeared. Otherwise the broad bands centered at 1320 and 1250 cm^{-1} are displayed. Similar behaviour was observed in vibrational frequency range 500...900 cm^{-1}. The broad bands in transient RR spectra can be assigned to the lowest triplet state. The study of the reasons underlying the strong broadening of the Raman bands in the T_1 state are in progress now.

References

1. P.A.Apanasevich, V.V.Kvach and V.A.Orlovich, J. Ram. Spectr., 20, 125 (1989)
2. E.W.Findsen, J.A.Shelnutt and M.R.Ondrias, J. Chem. Phys., 92, 307 (1988)
3. T.Kitagawa, M.Abe and H.Ogoshi, J.Chem.Phys., 69, 4516 (1978)
4. D.Kim, C.Kirmaier and D.Holten, Chem. Phys., 75, 305 (1983)
5. J.A.Shelnutt, K.D.Straub, P.M.Rentzepis, M.Gouterman and E.R.Davidson, Biochemistry, 23, 3946 (1984)
6. P.A.Apanasevich, R.Gadonas, V.V.Kvach, V.Krasauskas, V.A.Orlovich and V.Chyrvony, Sov. Chem. Phys., 7, 21 (1988)
7. N.Parthasarathi, C.Hanson, S.Yamaguchi and T.G.Spiro, J. Am. Chem. Soc., 109, 3865 (1987)

Time-Resolved Resonance Raman Spectra of Photoreduction Intermediates of Iron(III)-Octaethylporphyrin Complex

S. Sato[1], *K. Kamogawa*[1], *K. Aoyagi*[2], *and T. Kitagawa*[1]

[1]Institute for Molecular Science, Okazaki National Research Institutes, Okazaki 444, Japan

[2]Fukushima National College of Technology, Taira, Iwaki 970, Japan

1.Introduction

Photoreduction of ferric porphyrin is a very common phenomenon both for heme-proteins in biological system and protein free complexes in organic solvents or in H_2O[1], although elucidation of its mechanism is left unexplained. Octaethylporphyrinato-iron(III)-(2-methyl-imidazole)-(methanol) [OEPFe(III)(2-MeIm)(MeOH)] was found to be photoreduced to the ferrous high spin complex OEPFe(II)(2-MeIm) upon illumination of UV or near UV light in the absence of O_2[2]. The previous flashphotolysis experiment[3] indicated that it takes microseconds to complete the photoreduction. However, in the presence of O_2 reoxidation takes place in milliseconds and as a result, apparently no photoreduction occurs. We have investigated the photoreduction process by using time resolved resonance Raman spectroscopy and pointed out from the transient spectra in high frequency region that a four coordinated ferrous intermediate-spin porphyrin [OEPFe(II)] is generated within laser pulse width(7ns) after photoexcitation [4]. In this study we extended the measurement to a lower frequency region where Fe-ligand stretching bands [Fe-MeOH, Fe-(2-MeIm)] were expected.

2.Experiment

The TR^3 experiment was carried out with two ns lasers ($\Delta\tau$ =7-17ns) operated at 10Hz, and an intensified photodiode array detector attached to a triple-polychromator(Spex 1877). The probe beam was shifted by a home made H_2 or D_2 Raman shifter. Sample was stirred magnetically in a cuvette. In order to reveal small differences between successive transient spectra, we developed a new computer controlled system; a single computer controls firing of two pulse lasers so that they could have a given delay time(Δt_d) and also timing of reading data from the detector. This device enabled us to measure the probe only spectrum and plural pump/probe spectra with different Δt_d for an identical setting of the sample at a few second periods in turn; each spectrum was obtained as accumulation of 100-200 measurements of individual spectra with 3.2sec exposure. Thus the probe only spectrum can be always monitored during the TR^3 experiments and served as a standard for successive TR^3 data. We emphasize that this device allows us to pursue reliably an intensity change among the transient spectra.

3.Results and Discussion

The transient spectrum in high frequency region probed at 416nm, which was observed for 20ns after photoexcitation at 308nm, gave RR bands at 1636, 1592, 1504, and $1366cm^{-1}$. The former three are typical for the OEPFe(II)[4], while the $1366cm^{-1}$ band are lower by $11cm^{-1}$ than that of chemically generated OEPFe(II). Since $1366cm^{-1}$ band has a broad width, this band would be a superposition of a strong band of OEPFe(II)(2-MeIm) at $1360cm^{-1}$ and that of OEPFe(II) around $1377cm^{-1}$. At 1μs after photoexcitation prominent peaks appear at 1581, 1466,and $1360cm^{-1}$ which are typical for OEPFe(II)(2-MeIm). The same spectral features were obtained when the probe wavelength was shifted to 449nm.

Previously we reported that intensity of the Fe-MeOH stretching band of OEPFe(III)(2-MeIm)(MeOH) was enhanced upon excitation around 585nm[5]. However, it was difficult to obtain the TR^3 spectra by the excitation at this wavelength due to relatively lower resonance enhancement. In this study we noticed that the corresponding band was also observable at $542cm^{-1}$ upon excitation at 416nm. This assignment was confirmed through observation of an isotopic shift(*ca.* $-16cm^{-1}$) for perdeuterated MeOD. The Fe-MeOH stretching band observed for parent molecules decreased in intensity in the TR^3 spectra observed at 20ns-1μs after photoexcitation. On the other hand, the Fe-(2-MeIm) stretching band of the photoproduct was identified at $203cm^{-1}$ in this time interval; this band was weak at Δt_d=20ns but sufficiently intense at Δt_d=1μs. These observations are compatible with the proposal[4] that photoreduction intermediate is the four coordinated ferrous complex.

References

[1] Kitagawa, Y. Orii, J. Biochem.(*Tokyo*) 84, 1245(1978). (b)D. N. Hendrickson, M. G. Kinnaird, and K. S. Suslick, J. Am. Chem. Soc. 109, 1243(1987). (c) C. Bartocci, F. Scandola, A. Ferri, V. Carassiti, J. Am. Chem. Soc. 102, 7067(1980)

[2] Y. Ozaki, K. Iriyama, H. Ogoshi, and T. Kitagawa, J. Am. Chem. Soc. 109, 5583 (1987).

[3] V. Fidler, T. Ogura, S. Sato, K. Aoyagi, and T. Kitagawa, Bull. Chem. Soc. Jpn., in press.

[4] S. Sato, T. Ogura, V. Fidler, K. Kamogawa, and T. Kitagawa, Proceedings of ICORS XII, p.626(1990).

[5] T. Ogura, V. Fidler, Y. Ozaki, and T. Kitagawa, Chem. Phys. Lett. 169, 457(1990)

Resonance Raman Spectra of Copper Porphyrins in the Excited Triplet State

M. Asano, S. Sato, and T. Kitagawa

Institute for Molecular Science, Myodaiji, Okazaki 444, Japan

1. Introduction

Copper porphyrins emit from the lowest excited porphyrin (π, π^*) triplet state even at room temperature. An unpaired electron in the metal $d_{x^2-y^2}$ orbital interacts with porphyrin π electron system. Due to this interaction, emission from the lowest excited (π, π^*) singlet state is quenched by prompt intersystem crossing and "phosphorescence" is observed with a 10 ~100 ns lifetime in toluene at 300 K.

In typical metalloporphyrins, the lowest excited triplet states are approximately described as a single configuration of $^3(a_{2u}e_g)$ or $^3(a_{1u}e_g)$ depending on porphyrin peripheral substituents. The emission spectra of copper porphyrins with the lowest excited $^3(a_{2u}e_g)$ congifuration are red-shifted in fluid solution compared with those in rigid media, while such shift is not recognized for $^3(a_{1u}e_g)$-type complexes[1]. An extent of the red shift in solution depends on meso-substituents among $^3(a_{2u}e_g)$-type complexes. The excited state structure of these complexes in solution must be different from that in rigid media. In order to reveal an orgin of the medium dependent photophysical behavior of $^3(a_{2u}e_g)$-type complexes, resonance Raman spectra of the excited triplet state were measured by using the pump-probe technique.

2. Experimental

The Raman experiments were carried out with two 10Hz Nd:YAG lasers with 8ns duration. Porphyrins were brought to the S_1 excited state by the pump pulse at 532nm. Resonance Raman spectra of excited spieces were observed by the probe pulse at 450nm (2nd Stokes of D_2) or 416nm (1st Stokes of H_2) from 3rd harmonic, which is close to the maximum of T-T absorption band. Scattered light was detected by an intensified photodiode array attached to a triplepoly-chromator(Spex 1877). Sample solution was sealed in a cuvette under Ar atmosphere. Both benzene and benzene-d_6 solution were used to avoid an error from subtraction of the solvent spectrum.

3. Results and Discussion

Difference spectra between pump-probe and probe-only spectra of copper porphyrins decreased in intensities without any change of the spectral pattern

Springer Proceedings in Physics, Vol. 68 Time Resolved Vibrational Spectroscopy V
Ed.: H. Takahashi

with a time constant corresponding to their triplet lifetimes determined previously[1]. For the delay time of 15ns, the triplet state resonance Raman spectra of TPPCu (TPP: tetraphenylporphin), which exhibits much red-shifted emission spectrum in solution, showed prominant bands at 1596, 1566, 1516, 1299, 1181 cm^{-1}, a broad band around 800 cm^{-1}, and other several weak bands in 750∼1650 cm^{-1} region. By perdeuteration of meso phenyl groups, a strong band at 1596 cm^{-1} showed a downshift to 1561 cm^{-1}. Therefore, this band is assigned as a phenyl mode(ν_{8a}). The 1299 cm^{-1} band was also shifted to 1256 cm^{-1} upon phenyl deuteration and accordingly assigned to Cm-Ph stretching mode (ν_1). The phenyl mode at 1596 cm^{-1} in the excited states was downshifted only by 3 cm^{-1} from that in the ground states but was considerably intensified. On the contrary, the ν_1 band was upshifted by 65 cm^{-1} from the ground state frequency.

TMPCu(TMP: tetramesitylporphin) shows slightly red-shifted emission in solution. The ν_{8a} band at 1609 cm^{-1} was not intense in the excited state with no shift from the ground state. A band at 1264 cm^{-1}, which is assignable to ν_1 band, was upshifted by 33 cm^{-1} from the ground state. This shift is smaller than that for TPPCu. A band around 800 cm^{-1} was much weaker for TMPCu than for TPPCu. With TFPPCu (TFPP: tetra(pentafluorophenyl)porphin) whose lowest triplet configuration is $^3(a_{1u}e_g)$, the ν_{8a} band was observed at 1649 cm^{-1} in the triplet state and was neither intensified nor shifted from the ground state.

For the porphyrin which gives red-shifted emission in solution, the ν_{8a} mode gain more Raman intensity and ν_1 band was large up-shifted in T_1 than in S_0 state. The Cm-Ph bond seems to increase its bond order. Torsional distortion of phenyl rings toward coplanar structure in the lowest excited triplet state of TPPCu, however, can not fully explain little frequency shifts of phenyl mode in the T_1 state. Planar structure of porphyrin macrocycle might be deformed in the excited triplet state in TPPCu-type complexes.

References

1) M.Asano, Y.Kaizu, H.Kobayashi, J.Chem.Phys.,89, 6567 (1988).; M.Asano, Y.Kaizu, to be published.

Time-Resolved Resonance Raman Spectroscopy of the Excited States of Ru(II) Complexes with Novel Polynucleating Ligands

S.E.J. Bell[1], *K.C. Gordon*[1], *J.J. McGarvey*[1], *D.I.C. Martin*[1], *A. Kirsch-de Mesmaeker*[2], *and J.-P. Lecomte*[2]

[1]School of Chemistry, The Queen's University of Belfast, Belfast BT9 5AG, Northern Ireland
[2]Université Libre de Bruxelles, Faculté des Sciences, CP 160, B-1050 Bruxelles, Belgium

1.Introduction

For more than a decade transient resonance Raman spectroscopy using pulsed laser excitation has been an established technique[1] for probing the excited states of metal complexes, though there are relatively few reports of true time-resolved resonance Raman (TR^3) studies involving two lasers in a pump-delayed probe arrangement. Some recent examples are listed [2,3]. We have used such two laser methods on the nanosecond timescale to investigate the decay processes following metal-ligand charge-transfer (MLCT) excitation of some heteroleptic complexes of Ru(II) with the novel ligand 1,4,5,8,9,12-hexaazatriphenylene (HAT) and either 2,2'-bipyridyl (bpy), 1,4,5,8-tetraazaphenanthrene (TAP) or 1,10-phenanthroline (phen). This report is largely concerned with the complex $[Ru(bpy)_2HAT]^{2+}$ (A)

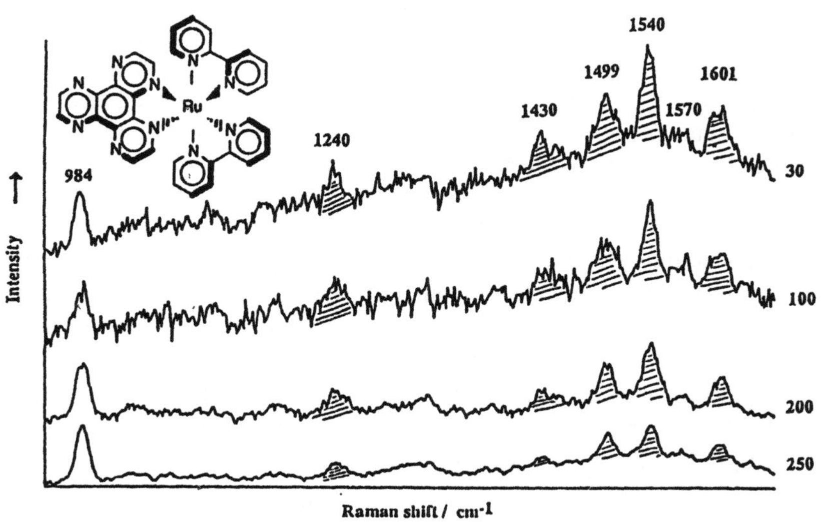

Figure 1. Time-resolved RR spectra of $[Ru(bpy)_2HAT]^{2+}$ (inset) probed at 447 nm following excitation at 354.7 nm. Pump-probe delays in ns are shown on the right. 984 cm^{-1} peak : SO_4^{2-} internal standard.

Springer Proceedings in Physics, Vol. 68 Time Resolved Vibrational Spectroscopy V
Ed.: H. Takahashi

(see figure 1, inset). Time-resolved luminescence studies using the same multichannel detector complement the Raman data.

2.Experimental

Two Q-switched Nd/YAG lasers (Quanta-Ray DCR2 and GCR3) with pulse durations *ca* 8ns were used as the pump and probe sources. The gated diode array multichannel detector and triple spectrometer used have been described previously[4]. For the luminescence measurements, the pump laser source alone was used, the time resolution in this case being achieved by varying the delay between the laser and the detector gate pulse. The complexes $Ru(bpy)_2(HAT)Cl_2$, $Ru(HAT)_3Cl_2$ and $Ru(phen)_2(HAT)Cl_2$ were prepared as described elsewhere [5]. Solutions (10^{-3} M for TR^3 studies and (10^{-3}-10-5) M for luminescence measurements) were made up in distilled water with 0.5 M Na_2SO_4 as internal standard for the Raman intensity measurements.

3. Results and Discussion

Previous flash photolysis experiments on (A) have shown that excitation within the visible absorption bands produces a relatively long-lived charge-transfer excited state with a lifetime around 100 ns. Transient absorption difference measurements show that this charge-transfer excited state has a strong absorption band near 350 nm but is less absorbing than the ground state at *ca* 450 nm. Time-resolved Raman experiments would therefore most logically be carried out using a photolysis wavelength near 450 nm (lambda max for the ground state) and probing the transient at *ca* 350 nm. Unfortunately with probe laser pulses of this wavelength secondary photolysis meant that time-resolved Raman experiments could not be carried out. For this reason experiments were instead carried out with a probe wavelength of 447 nm, where secondary photolysis was less of a problem. Figure 1 shows Raman spectra, probed at 447 nm, at a series of delays following excitation of (A) at 355 nm. The spectra are very similar to the ground state except that several bands are more intense, when measured with respect to the internal standard and decay on a timescale of 100's of ns as shown. This data is different from that obtained in single colour experiments at 447 nm despite the fact that we would expect to populate the same excited state at this wavelength. Single-colour spectra at high laser fluence (not shown) have a much stronger 1430 cm^{-1} band than either the ground-state or intermediate(100-200 ns) time-delay spectra. This apparent anomaly is resolved by the two-colour spectra recorded at short pump-probe delays. The most prominent feature in these spectra is again a band at 1430 cm^{-1} (possibly due to an in-plane CH mode of the HAT ligand [6]) ; in addition, the band at 1540 cm^{-1} (possibly a ring stretching mode) is significantly reduced with respect to other bands common to both spectra. The 1430 cm^{-1} band decays rapidly, within 20 ns or less, and there is a concomitant build-up of the 1540 cm^{-1} band. These results imply that the decay of the initially-excited species is biphasic with a fast component, which is resolved with single-laser (in effect ,zero time-delay) or short time-delay two-colour experiments and a more slowly decaying component which is observed at longer time-delays. It is this second component which has previously been characterized by flash-photolysis.

Time-resolved fluorescence measurements provide independent confirmation of the biphasic nature of the decay, a two-component decay being observed over the concentration range 10^{-3} - 10^{-5} M. The time constant of the slower decay was found to be concentration- dependent but the values found at the extremes of the concentration range studied were consistent with the observed decay kinetics of the Raman experiments (carried out at 10^{-3} M) and the flash photolysis studies (carried out at 10^{-5} M). Both fluorescence and (preliminary) TR^3 experiments on the related heteroleptic complex $[Ru(phen)_2HAT]^{2+}$ (B) showed that this complex also has a two-component decay. In contrast, gated luminescence studies on the homoleptic complex $[Ru(HAT)_3]^{2+}$ (C) showed that the relaxation following MLCT excitation was characterised by a single exponential trace, with a concentration independent lifetime of 165 ns, in agreement with a recent literature value[5]. It appears therefore that the two stage decay observed with (A) and (B) but not with (C) is due to some sort of internal rearrangement following excitation. It seems likely that the more slowly decaying transient in the case of (A) and (B) is due to the ^{3}MLCT state. If this follows the pattern which has now been established [7] for a number of both homo- and heteroleptic complexes the optical electron is localized on a single ligand and in the case of heteroleptic complexes it is generally the most readily reduced ligand which carries the optical electron. Although it initially seems puzzling that the resonance Raman spectra of the ground state and ^{3}MLCT state of (A) are very similar (apart from absolute intensity differences) especially when the ground state spectrum is dominated by HAT ligand modes and the ^{3}MLCT state would be expected to contain not HAT but the radical anion $HAT^{\cdot -}$. However it has been shown that ligands with extended pi systems such as phenanthroline show only small band shifts on reduction[2] and it is likely therefore that the even more delocalised HAT ligand would also be expected to show only small shifts which we have been unable to observe at our presently attainable signal-to-noise levels.

The short-lived transient species is more difficult to assign but it appears to be one in which the optical electron is not in the most thermodynamically stable position. This could either be because it lies on the the bpy ligand (or is delocalised over the bpy_2-HAT ligand set) . Alternatively, it could lie not in the LUMO of the ligand but in a higher energy orbital ,i.e. the reduced ligand is electronically excited. The fact that there are no $bpy^{\cdot -}$ features in the spectrum of the short-lived species suggests that the first possibility above can be eliminated but at this time we cannot distinguish between the remainder. In any case this series of experiments demonstrates that the use of single-colour transient Raman techniques does not necessarily give information on the longest-lived transient species formed. Although the strategy of using single-colour pulsed laser excitation to obtain the resonance Raman spectra of transient species with lifetimes considerably longer than the pulse duration has proved extremely successful in the case of $Ru(bpy)_3^{2+}$

[1,7] the resolution of problems associated with short-lived, initially formed species, requires the use of two laser pump and probe methods.

Acknowledgement

We thank the Science & Engineering Research Council for support (Grant GR/F 42980) and The Department of Education (N. Ireland) for a research studentship and a Distinction Award (to D.I. C. M.)

References

[1] P. G. Bradley , N. Kress, B. A. Hornberger, R. F. Dallinger, and W. H. Woodruff, J. Am. Chem. Soc. **103**, 7441 (1981).
[2] Y. J. Chang, T. Xiaobing Xu , Yu Soo-Chang, D. R. Anderson , L. K. Orman, and J. B. Hopkins, J.Phys. Chem. **94** , 729 (1990).
[3] V. A. Walters, J. C. de Paula, G. T. Babcock, and G. E. Leroi, J. Am. Chem. Soc. **111**, 8300 (1989)
[4] J. J. McGarvey, S. E. J. Bell, and K. C. Gordon, Inorg. Chem. **27**, 4003 (1988)
[5] A. Kirsch-de Mesmaeker, G. Orellana, J. K. Barton, and N. J. Turro, Photochem. Photobiol. **52**, 461 (1990).
[6]F. Vanhecke, K. Heremans, A. Kirsch-de Mesmaeker, L. Jacquet, and A. Masschelein, J. Raman Spectrosc. **20**, 617 (1989).
[7] G. D. Danzer, and J. R. Kincaid, J.Phys. Chem. **94**, 3976 (1990).

Resonance Raman Spectra of the ^{3}MLCT Excited States of Ruthenium Complexes with Asymmetric Ligands

G.D. Danzer and J.R. Kincaid

Chemistry Department, Marquette University, Milwaukee, WI 53233, USA

1. Introduction

Complexes of ruthenium(II) with 2,2'-bipyridine and related ligands are of interest not only as structurally diverse systems for the study of fundamental photophysics and photochemical processes [1,2] but also as potentially useful components of solar energy conversion devices [3,4]. During the past decade, a vast amount of effort has been focused on understanding the structure and reactivity of the metal-to-ligand charge-transfer (^{3}MLCT) excited states of tris-ligated ruthenium(II) polypyridine complexes. The pioneering work of Dallinger, Woodruff and co-workers [5] first demonstrated the utility of time-resolved resonance Raman (TR^3) spectroscopy as an especially effective probe of the relatively long-lived ^{3}MLCT states of such complexes. Subsequent studies by our group [6,7] and others [8] have illustrated the utility of TR^3 methods in documenting selective population of individual ligand localized excited states in heteroleptic tris-ligated complexes.

In our previous study [7] of tris-ligated complexes comprised of 2,2'-bipyridine (bpy) and 2,2'-bipyrazine (bpz), the TR^3 spectra provided unambiguous evidence for selective localization of the optical electron on the bipyrazine fragment in the ^{3}MLCT excited state. Thus the TR^3 spectrum of the complex $Ru(bpy)_2(bpz)^{2+}$ exhibits one set of modes, the components of which are ascribable to the bpy^0 "neutral ligand" fragment of the ^{3}MLCT state. A second set of modes is also observed which is ascribable to the bipyrazine anion radical (bpz^-) fragment of the ^{3}MLCT state. Therefore, the ^{3}MLCT state is properly formulated as $[Ru(bpy^0)_2(bpz^-)^{2+}]^*$.

Our most recent efforts have focused on similar studies involving inherently asymmetric ligands with respect to the C_2 axis of the bidentate ligand. Two such ligands have been utilized. They are 4-methyl-2,2'-bipyridine (4-me-bpy) and 2,2-pyridyl-pyrazine (pypz).

2. Ground State RR Studies of Tris-Ligated PYPZ Complexes

The RR spectra of the tris-ligated homoleptic complex of pypz (i.e.; $Ru(pypz)_3^{2+}$) taken in resonance with the ^{1}MLCT absorption band is given in Figure 1. Here the observed bands can be grouped into two sets. One set appears to correlate with the observed modes of $Ru(bpy)_3^{2+}$ (Figure 1). The other set appears to correlate with the observed modes of $Ru(bpz)_3^{2+}$ (Figure 1). Therefore, the ground state RR spectrum is readily interpreted as a composite of "pyridyl-based" and "pyrazine-based" modes. It is interesting to note that the mode at 1320 in the $Ru(bpy_3^{2+}$ RR spectrum has been shown to contain a large contribution

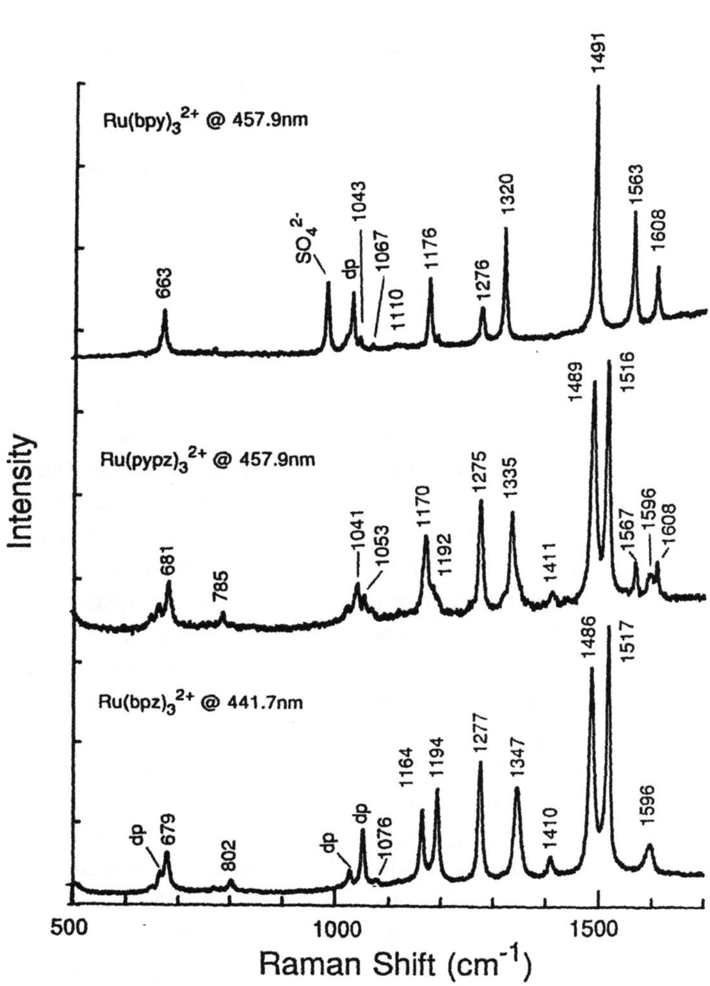

Figure 1 Resonance Raman spectra of tris-ligated ruthenium(II) complexes using ^{1}MLCT excitation.

from the inter-ring stretching coordinate (i.e.; $\nu(C_2\text{-}C_{2'})$)[9]. The mode at 1347 in the $Ru(bpz)_3^{2+}$ RR spectrum is also believed to contain a large contribution from the inter-ring stretching coordinate. In the case of $Ru(pypz)_3^{2+}$ a band observed at 1335, midway between 1320 and 1347, may reasonably assigned to the inter-ring stretch. Although a complete normal-coordinate analysis would be required to verify these assignments this observation is consistent with the concept of pyridine and pyrazine based modes.

3. TR^3 Studies of Tris-Ligated PYPZ Complexes

The TR^3 Spectrum of $Ru(pypz)_3^{2+}$ is given in Figure 2. No modes ascribable to the pyridine anion radical fragment are observed (i.e.; modes which are similar to those observed for bpy$^-$

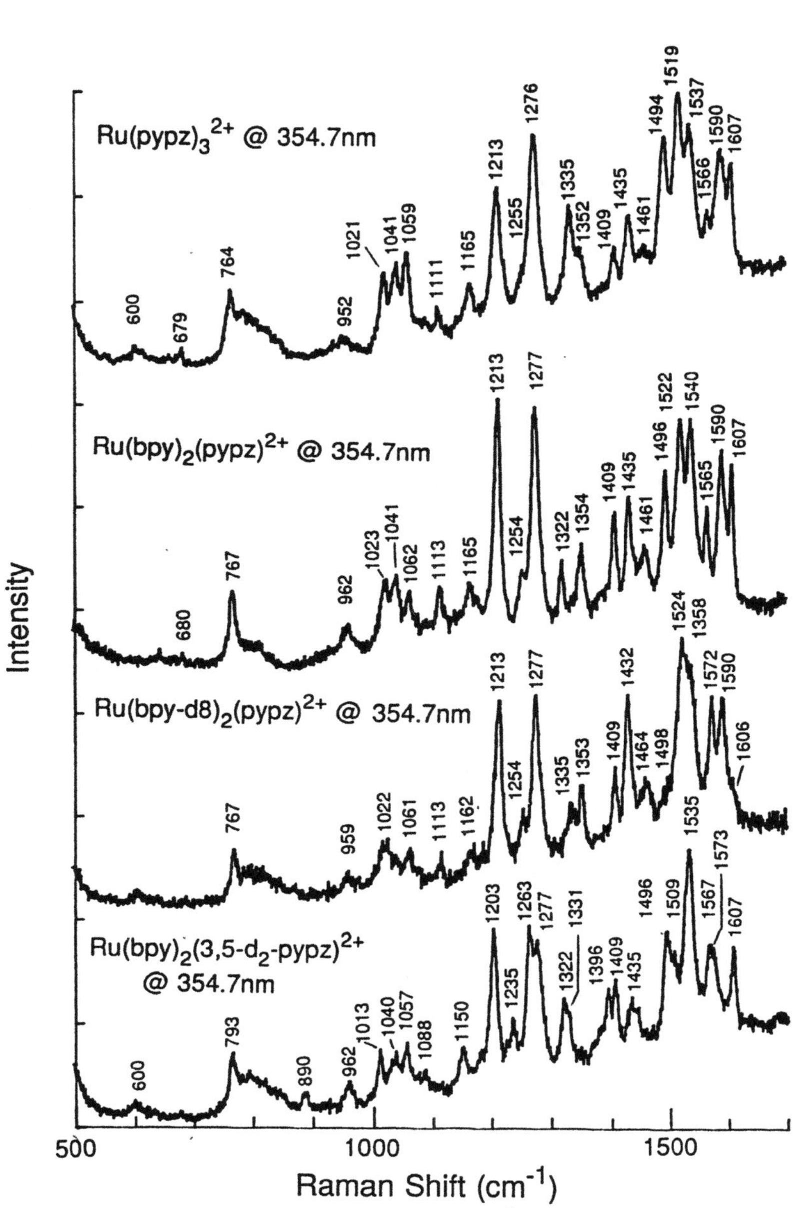

Figure 2 Time-resolved resonance-Raman (TR^3) spectra of pyridyl-pyrazine complexes.

in the $Ru(bpy)_3^{2+}$ TR^3 spectrum [9]) whereas bands which appear similar to the bipyrazine radical fragment [7] of $Ru(bpz)_3^{2+}$ are evident. The TR^3 spectra of $Ru(bpy)_2(pypz)^{2+}$ (Figure 2) confirm the selective localization of the optical electron on the pypz ligand. Likewise the TR^3 spectrum of $Ru(bpy\text{-}d_8)_2(pypz)^{2+}$ (Figure 2), where bpy-d_8 is perdeuteriated bipyridine, exhibits modes ascribable to the bpy-d_8 neutral ligand fragments (bpy-$d_8^{\,0}$)[9], modes similar to those of the bipyrazine radical fragment (bpz^-) [7] and modes similar to the bpy ground state spectrum (bpy^o)[9]. The TR^3 spectrum of $Ru(bpy)_2(3,5\text{-}d_2\text{-}pypz)^{2+}$ (Figure 2), where 3,5-d_2-pypz is pypz which has been deuteriated in the 3 and 5 positions of the pyridine portion of the ligand, exhibits deuterium shifts in the bpy^o like modes which are consistent with those observed for $Ru(3,5\text{-}d_4\text{-}bpy)_3^{2+}$ [9]. Therefore, the observed TR^3 spectra are consistent with the interpretation of the optical electron being predominately localized on the pyrazine half of the pypz ligand.

4. Summary

The RR spectrum of tris-ligated ruthenium(II) complexes containing the pypz ligand can be readily interpreted as being a composite of "pyridyl"-based and "pyrazine"-based modes. Likewise the TR^3 spectra of the ^3MLCT state can be interpreted as "pyridyl" based modes (i.e.; similar to bpy^o) and "pyrazine anion radical" based modes (i.e.; similar to bpz^-). Similar results have also been observed for complexes containing the 4-me-bpy asymmetric ligand.

References

[1] K. Kalyanasundaram, Coord. Chem. Rev. **46**, 159 (1982).

[2] M.K. DeArmond, V. Balzani, F. Barigelletti, S. Campagna, P. Belser, A. VonZelewsky, Coord. Chem. Rev. **84**, 85 (1988).

[3] T.J. Meyer, Accts. Chem. Res. **22**,163 (1989).

[4] M. Gratzel, Accts. Chem. Res. **14**, 376 (1981).

[5] P.G. Bradley, N. Kress, B.A. Hornberger, R.F. Dallinger, W.H. Woodruff, J. Am. Chem. Soc. **103**, 7441 (1981).

[6] S.F. McClanahan, R.F. Dallinger, F.J. Holler, J.R. Kincaid, J. Am. Chem. Soc. **107**, 4853 (1985).

[7] G.D. Danzer, J.R. Kincaid, J. Phys. Chem. **94**, 3976 (1990).

[8] P.A. Mabrouk, M.S. Wrighton, Inorg. Chem. **25**, 526 (1986).

[9] D.P. Strommen, P.K. Mallick, G.D. Danzer, R.S. Lumpkin, J.R. Kincaid, J. Phys. Chem. **94**, 1357 (1990).

Time-Resolved Resonance Raman Spectral Analysis of Ruthenium Tris(bipyridine) Excited State: Revisited

S. Umapathy and R.E. Hester*

Department of Chemistry, University of York, York, YO1 5DD, UK
*Present address: Department of Inorganic and Physical Chemistry, Indian Institute of Science, Bangalore 560012, India

Time-resolved resonance Raman spectroscopy (TR3S) has widely been used to establish the vibrational structure and the nature of the metal-ligand charge transfer (MLCT) triplet excited state of d^6 transition metal complexes [1-6]. In particular, bipyridyl complexes of ruthenium(II) have been studied extensively by TR3S [1, 3-6]. Recently however, certain aspects of the procedures used to interpret the TR3S data have been questioned [3-5]. In this paper, we reconsider those aspects with particular reference to ruthenium(II)<u>tris</u>(bipyridine) (RBY).

The structure of the MLCT triplet excited state of ruthenium (II) <u>tris</u>(bipyridine) has been shown by TR3S studies to have an asymmetric structure with the odd electron being localised on one of the three ligands. This electron localisation is indicated in the Raman spectra by (a) the presence of two sets of Raman bands,one for the neutral ligands (akin to ligands in the unexcited complex) and one for the reduced ligand (where the odd electron is resident), (b) the observation of Raman bands due to the neutral ligands in the excited state spectrum of the complex having 2/3 of the intensity of that observed in the ground state spectrum, (c) the band positions in the spectra due to the reduced ligand being consistent with those of the uncomplexed radical anion of the ligand, and (d) the shift in band positions corresponding to the change in electron density distribution due to the additional electron within the ligand [1]. Of these four main criteria (a) and (b) have been contradicted by Hopkins <u>et al</u>.[3-5] but (c) and (d) have been well supported by various authors[1-6].

The criteria (a) and (b) depend solely on whether the photoconversion of the ground state to the excited state is complete, ie. whether the population saturation has been achieved. This can be tested by plotting the band intensities of the excited state against the laser power (or pulse energy). If the band intensities are normalised against an internal standard then the intensity values will have been corrected for sample absorption and laser power fluctuations. Thus a plot of this normalised intensity against laser power will be linear, as the concentration of the excited state population increase with laser power. On the other hand, if the population saturation has been achieved at low laser power then the plot should be linear at low laser powers and reach a pleatu at high laser power. Thus the RBY excited state band intensities (normalised against an internal standard) at various laser pulse energies were plotted and it was observed that the excited state population saturation has been reached with high energy pulses, ie. the photochemical conversion to the MLCT excited state is essentially 100%. The wavelength of the laser excitation was ca. 363.8 nm, which is in resonance with the reduced ligand absorption leading to specific enhancement of the reduced ligand band.

From the data and the spectra obtained at other wavelengths, we can show that the resonance Raman spectra of the RBY MLCT excited triplet state do contain bands due to the neutral ligands in addition to the bands due to the reduced ligand. This is in contrast to conclusions reached by Hopkins <u>et al</u>. in a series of recent publications [3-5]. A close study of their results shows that the spectra observed are consistent with results reported here and in earlier publications [1,6]. However, their conclusions are contradictory due to the inappropriate use of spectral subtraction routines, without correction for absorption of the laser by the sample. Thus, in spite of various earlier publications [7,8,9] discussing the various aspects of laser absorption by the sample, we would like to reiterate the fact that when absorbing samples are studied under varying laser powers then the use of an internal standard is a necessity. If the sample has a very narrow spectral absorption bandwidth then it also becomes necessary to consider more than one internal standard band in order for the resonance Raman spectrum to have accurate intensity values. Otherwise one may employ a 180° sampling geometry to minimise laser absorption by the sample.

The presence of 2/3 intensities (the criterion (b)) of the Raman bands due to the neutral ligands present in the excited state of the complex in comparison with that of the intensities in the ground state of the complex is observable, with the use of an internal standard, as demonstrated earlier [1]. It is important to note that this criterion depends on the Raman cross-section of the neutral ligands before (in the ground state of the complex) and after (in the excited state of the complex) the photoexcitation of the ruthenium(II) complex.

In the case of RBY, the absorptions (in the ultraviolet region) responsible for the resonance Raman bands due to the neutral ligands present in the ground and the excited state of RBY is MLCT in character with comparable extinction coefficients. Thus it is apparant that the criteria (b) is valid in this case. If we consider Hopkins <u>et al</u> [3] results it is evident that their spectra is unlikely to have such a fixed ratio of intensities as the extinction coefficient is vastly different for the ground and excited state of the complex at the wavelength of the probe laser (532 nm). Therefore it is not surprising that Hopkins <u>et al</u> did not observe the 2/3 intensity. In the case of their spectra recorded with ultraviolet excitation the use of spectral subtraction routines without appropriate internal standard leads to misleading interpretation of the intensities. In the light of the above discussion it is important to recognise that the relative intensities are dependent upon (i) the exciting laser wavelength and (ii) the corresponding absorption at that wavelength and not on the chemical system employed (as in RBY) as implied by Hopkins et al. [3-5].

References:

[1] P. G. Bradley <u>et al</u>. JACS, 103, 7441 (1981). [2] S Umapathy <u>et al</u>. JPC, 94, 8800 (1990).[3] L K Orman and J B Hopkins, CPL, 49, 375 (1988).[4] L K Orman <u>et al</u>, JCP, 90, 1469 (1989).[5] Y J Chang et al, JPC, 94,729 (1990). [6] M Forster and R E Hester, CPL, 81, 42 (1981).[7] W Rauch and H Bettermann, Appl.Spectrosc., 42, 520 (1988). [8] I Tsukamnto <u>et al</u>, J. Raman Spectrosc. 17 313 (1986). [9] M Ludwig and S A Asher, Appl. Spectrosc., 42, 1458 (1988).

Part III

Photochemical Reactions

Photochemical Ring Openings are Complete in Picoseconds: A Time-Resolved UV Resonance Raman Study

P.J. Reid, S.J. Doig, and R.A. Mathies

Department of Chemistry, University of California, Berkeley, CA 94720, USA

1. Introduction

A classic electrocyclic ring-opening reaction is the photoconversion of 1,3-cyclohexadiene (CHD) to *cis*-hexatriene (*cis*-HT)[1]. Our time-resolved resonance Raman studies have shown that the timescale for *cis*-HT formation is 8 ± 1 ps[2]. To determine if the kinetics of CHD are representative of other photochemical ring-opening reactions, we have studied the photoconversions of 1,3,5-cyclooctatriene (COT) to octatetraene and of α-phellandrene (α-PHE) to 3,7-dimethyl-1,3,5-octatriene.[3] COT is known to photochemically ring open with a disrotatory motion of the methylene portion of the ring as opposed to the conrotatory ring opening of CHD[4]. α-PHE is a substituted 1,3-cyclohexadiene in which one of the carbons undergoing conrotatory rotation has a hydrogen substituted for an isopropyl group.[5]

2. Experimental Methods and Results

Time-resolved UV resonance Raman difference spectra were obtained using a two-color, pump-probe apparatus with 2.6 ps time resolution.[4] Difference spectra of the photoconversion of COT to octatetraene are presented in the left half of Figure 1. Appearance of the octatetraene photoproduct on the ground state surface was observed by monitoring the intensity of the photoproduct ethylenic at 1590 cm^{-1}. The peak height of the photoproduct ethylenic was plotted as a function of time as depicted in the left half of Figure 2. The best fit to a single exponential resulted in a time constant of 12 ± 2 ps for the formation of ground state octatetraene. Difference spectra of the photoconversion of α-PHE to 3,7-dimethyl-1,3,5-octatriene are presented in the right half of Figure 1. At 3 ps, a positive peak at 1613 cm^{-1} is clearly observed which is assigned as the ethylenic stretch of the photoproduct. The intensity of this line was plotted as a function of time as depicted in the right half of Figure 2. The best fit to a single exponential resulted in a time constant of 11 ± 2 ps for the appearance of 3,7-dimethyl-1,3,5-octatriene.

3. Discussion

The similar ring-opening times for CHD and COT demonstrates that the rate of ground state photoproduct formation is insensitive to differences in the stereoselectivity of the reaction. Similarly, the correspondence in photoproduct formation rates between CHD and α-PHE shows that the rate of photoproduct formation is insensitive to the presence of alkyl substituants. Also, these studies demonstrate that photochemical electrocyclic ring-opening reactions are characterized by extremely fast, ~10 ps repopulation of the ground state surface. The ~10 ps photoproduct formation time is apparently a general feature of electrocyclic ring-opening reactions.

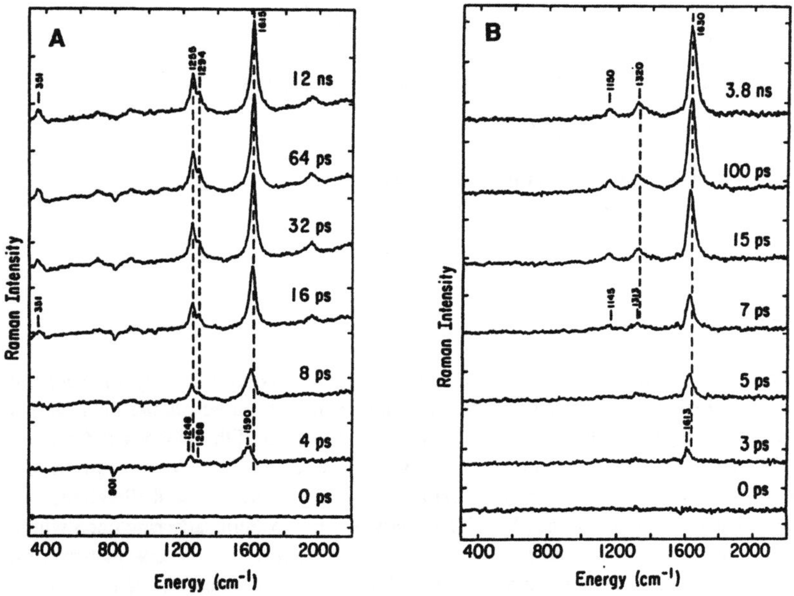

Figure 1. (A) Resonance Raman difference spectra of the photoconversion of cyclooctatriene to *cis,cis*-octatetraene. Positive peaks are due to the appearance of photoproduct. The line at 801 cm^{-1} is due to the cyclohexane solvent. (B) Resonance Raman difference spectra of the photoconversion of α-phellandrene to 3,7-dimethyl-1,3,5-octatriene.

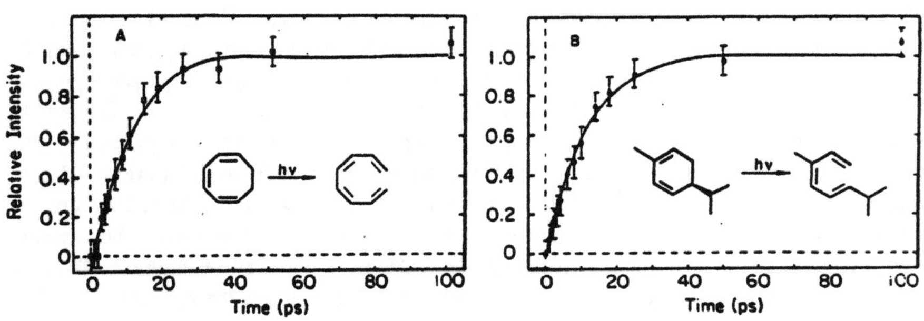

Figure 2. Intensity of the ethylenic stretch of (A) *cis,cis*-octatetraene and (B) 3,7-dimethyl-1,3,5-octatriene as a function of delay time.

References

1. R.J. de Kock, N.G. Minnard, E. Havinga: Rec. Trav. Chim. Pays-Bas **79**, 922 (1960)
2. P.J. Reid, S.J. Doig, R.A. Mathies: Chem. Phys. Lett. **136**, 163 (1989)
3. P.J. Reid, S.J. Doig, R.A. Mathies: J. Phys. Chem. **94**, 8396 (1990)
4. T. Goldfarb, L. Lindquist: J. Am. Chem. Soc. **89**, 4588 (1967)
5. J.E. Baldwin, S.M. Krueger: J. Am. Chem. Soc. **91**, 6444 (1969)

Combined Transient Absorption and Resonance Raman Studies of the Photoreduction of 2,2'-Bipyridine by 1,4-Diazabicyclo(2.2.2)octane (DABCO) in Acetonitrile

B. Bayri[1], *G. Buntinx*[2], *O. Poizat*[1], *P. Valat*[3], *and V. Wintgens*[3]

[1]Laboratoire de Spectrochimie Infrarouge et Raman, CNRS, 2, rue Henri Dunant, F-94320 Thiais, France

[2]Laboratoire de Spectrochimie Infrarouge et Raman, USTLFA, Bat. C5, F-59655 Villeneuve d'Ascq, France

[3]Laboratoire des Matériaux Moléculaires, CNRS, 2, rue Henri Dunant, F-94320 Thiais, France

In this paper we show that 2,2'-bipyridine (22BPY) is photoreduced by DABCO via electron transfer to the lowest triplet state ^{3}22BPY•. The flash photolysis system for transient absorption (crossed-beam arrangement) included a Nd-Yag laser (266nm, 7ns, 0.2mJ), a pulsed xenon lamp and a spectrometer equipped with a digitalized oscilloscope (resolution better than 4.6nm). Raman measurements were performed using an Excimer laser (248nm, 15ns, 1.5mJ) as pump source, a Nd-Yag laser coupled to a Dye laser (370-390nm, 8ns, 1.5mJ) as probe source, and a multichannel spectrometer comprising a gated intensified photodiode array (20ns) as detector (resolution ~7cm^{-1}). A pulse and delay generator (10ns steps from 0 to 10ms) triggers the two lasers and the detector gate (total jitter<4ns).

Figure 1A shows the transient absorption spectra observed for a deaerated solution of 22BPY (10^{-4}M) in acetonitrile at different times after 266nm photolysis. These are characteristic of the ^{3}22BPY• triplet species [1]. Figure 1B shows the spectra obtained in the presence of 4×10^{-4}M DABCO. The triplet band at 355nm is rapidly replaced by a transient spectrum with two groups of bands in the ranges 350-400nm (maximum at 390nm, shoulder around 370nm) and 500-600nm (maxima at 540 and 580nm). This spectrum can be assigned to the 22BPY$^{-\cdot}$ anion by analogy with spectra

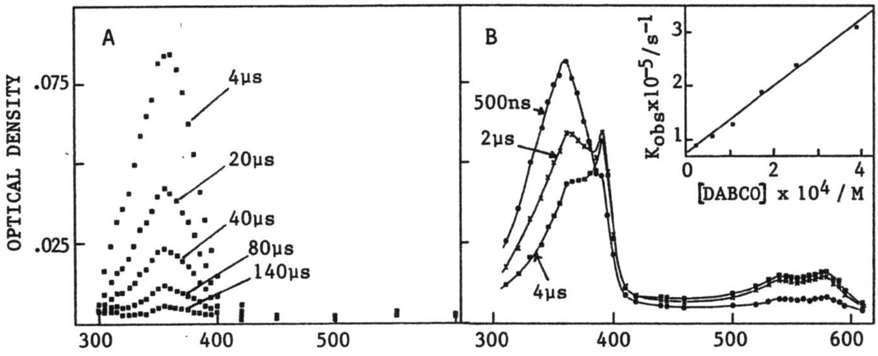

Figure 1 - Transient absorption spectra of 22BPY (10^{-4}M) in deaerated solutions of (A) acetonitrile and (B) acetonitrile containing 4×10^{-4}M DABCO. Pump excitation is at 266nm (0.2mJ). The insert shows the plot of the rate of the triplet decay as a function of the concentration of DABCO.

Springer Proceedings in Physics, Vol. 68 Time Resolved Vibrational Spectroscopy V
Ed.: H. Takahashi

reported for this species engaged in alkali metal salts (cis conformation) [2-4] or produced by gamma irradiation in a polymer film [5] or in a glassy matrix [4] (trans conformation). The triplet absorption decay and the anion absorption rise follow the same pseudo-first-order kinetics. Accordingly we can conclude that ^{3}22BPY* is quenched by DABCO via electron transfer. The rate constant for quenching of ^{3}22BPY* is obtained from a plot of the pseudo-first-order decay rate constant of the triplet absorption (monitored at 330nm) as a function of DABCO concentration (insert in Fig. 1B). The slope of this plot gives a quenching rate constant $k_Q=6\times10^8 M^{-1}s^{-1}$. This value is significantly weaker than that found [6] for 4,4'-bipyridine ($k_Q=8.8\times10^9 M^{-1}s^{-1}$). As predicted by the Rehm-Weller theory on electron transfer [7], k_Q is lowered in less polar solvents ($k_Q\sim2\times10^6 M^{-1}s^{-1}$ in cyclohexane) and varies in the opposite way as the oxydation potential of the quencher : on going from DABCO ($E_{1/2}$ vs. SCE= 0.87V) to triethylamine ($E_{1/2}$=0.96V) and tetramethyl-para-phenylenediamine ($E_{1/2}$=0.32V), the k_Q value is found to move to 2×10^8 and $\sim10^{10} M^{-1}s^{-1}$, respectively.

Figure 2, trace (a), presents the time-resolved resonance Raman spectrum of ^{3}22BPY* in deaerated acetonitrile for pump excitation at 266nm (0.5mJ). A comparable spectrum is observed in cyclohexane. As seen in Figure 2, trace (b), new Raman lines appear on pump excitation at 248nm (1.5mJ) which are not observed in basicified solutions and are thus ascribed to the photoionized species $22BPY^{+\cdot}$. Trace (c) shows the cation spectrum resulting from subtraction of the triplet bands. The dependance of the intensity ratio of the triplet state and radical cation spectra on the laser power indicates that photoionization arises via a two-photon process.

Figure 3A presents the time evolution of the Raman spectra (probed at 390nm) following 248nm photolysis of a deaerated acetonitrile solution of 22BPY (5×10^{-4}M) and DABCO (10^{-1}M). The same results are obtained with CD_3CN solutions. The spectrum at 10ns corresponds essentially to the triplet state ^{3}22BPY*. This spectrum decays rapidly whereas new transient lines rise. At 250ns the triplet features (e.g. the typical line at $1322cm^{-1}$) have completely disappeared and the remaining spectrum is obviously characterizing the anion radical $22BPY^{-\cdot}$ (trace (c) in Figure 3A). In air-saturated solutions both the triplet state and anion radical bands are rapidly quenched by oxygen. Note that the intensity of the $22BPY^{-\cdot}$ spectrum is significantly reduced upon changing the probe excitation from 390 to 370nm, as expected from the lower resonance accordance of this laser line with the anion absorption.

The Raman frequencies of the photoproduced $22BPY^{-\cdot}$ anion are somewhat different from those reported [3,8] for cis-$22BPY^{-\cdot}$ coordinated to Li^+ (main lines at 1554, 1485, 1412, 1350, 1275, 1205, 991, 745 and $645cm^{-1}$). The relative band intensities cannot either be compared in the two types of spectra although the probe excitations (390 and 351-406nm, respectively) are in resonance with the UV absorption in both cases. A better correspondance is found with the spectrum obtained with 586nm probe excitation for the trans-conformer in a glassy matrix [4] for which seven of the observed bands have nearly common frequencies (1572, 1495, 1470, 1438, 1080, 1015 and $972cm^{-1}$) with the $22BPY^{-\cdot}$ spectrum (c) in Figure 3A. These results suggest that the photoreduction of 22BPY in acetonitrile yields mainly the trans-$22BPY^{-\cdot}$ anion. Drastic variations in the relative band intensities, due to resonance effects, arise in going from 390 to 586nm probe excitations which can be accounted for by changes in the vibrational activity in the related resonant UV and visible absorptions of the anion (these have been ascribed to $\phi_6 \rightarrow \phi_7$ and $\phi_7 \rightarrow \phi_{10}$ $\pi\pi^*$ transitions, respectively, from molecular orbital calculations [2-4]).

Finally, in acetonitrile solutions containing water or methanol (10^{-2} to 10^{-1}M), the rapid decay of the triplet state Raman spectrum is still observed in the presence of DABCO but we do not detect the corresponding rise of the anion spectrum. Instead, a different transient spectrum is observed on probing at 370nm (Fig. 3B) but not at 390nm. This spectrum resembles notably the anion radical spectrum (c) in Figure 3A (frequencies,

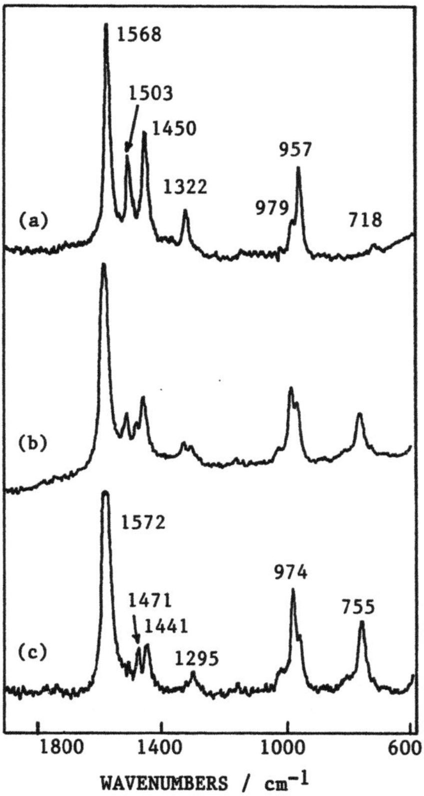

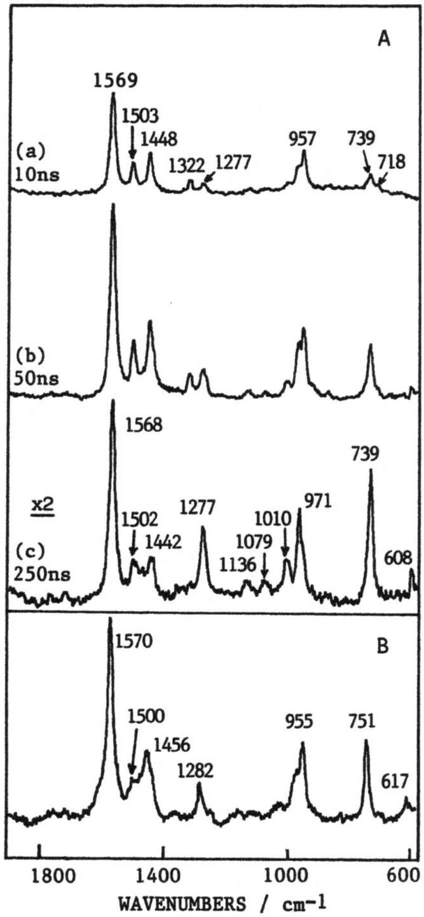

Figure 2 (left) - Time-resolved resonance Raman spectra of a deaerated solution of 22BPY (5×10^{-4}M) in acetonitrile probed at 390nm after a 50ns delay following pulse photolysis. (a) pump excitation at 266nm (0.5mJ) ; (b) pump excitation at 248nm (1.5mJ) ; (c) = (b) - 0.5 (a). Solvent bands have been subtracted. Identical spectra are obtained with deuterated acetonitrile solvent.

Figure 3 (right) - Time-resolved resonance Raman spectra following 248nm (1.5mJ) pulse photolysis of deaerated solutions of 22BPY (5×10^{-4}M) in (A) acetonitrile containing 10^{-1}M DABCO (probe excitation at 390nm) ; (B) acetonitrile containing 10^{-1}M DABCO and 10^{-1}M H_2O (probe excitation at 370nm). Identical spectra are obtained with deuterated acetonitrile solvent. The pump/probe time delay is indicated for each spectrum. Spectra (a), (b) and (c) are normalized with respect to the solvent lines which have then been subtracted.

resonance band activity and relative intensities). It is ascribed to the 22BPYH' radical resulting from protonation of the anion (pka>14). The shift in frequency of the resonance Raman effect on going from $22BPY^{-\cdot}$ and 22BPYH' is consistent with the shift of the absorption maximum from 390 to 365nm as found from pulsed radiolysis studies [9]. The assignment of the transient spectrum in Figure 3B to 22BPYH' is also in accord with the fact that comparable spectra are observed on photolysis of 22BPY in the presence of hydrogen-atom donor species [10]. As noted above ^{3}22BPY* is also quenched by TEA in acetonitrile. However, in this case, the anion radical is not observed even when the pump and probe excitations are coincident. The transient absorption of the product of reduction is not clearly distinguishable from the triplet absorption but its Raman spectrum is comparable to that shown in Figure 3B and displays the same specific resonance enhancement at 370nm [10]. Therefore electron transfer from TEA to ^{3}22BPY* is followed by rapid protonation of the anion radical by TEA^+ to form the 22BPYH' radical.

These results can be compared to those obtained with 4,4'-bipyridine [6] and reveal a notable analogy of these aza-aromatic compounds and of carbonyl compounds regarding the photoreduction by aliphatic amines.

References

1 R.D. Saini, S. Dhanya and P.K. Bhattacharyya, J. Photochem. Photobiol. 43A, 91 (1988)
2 E. König and S. Kremer, Chem. Phys. Lett. 5, 87 (1970)
3 B.C. Noble and R.D. Peacock, Spectrochim. Acta 46A, 407 (1990)
4 H. Kawashima, T. Kato and T. Shida, J. Raman Spectrosc. 22, 187 (1991)
5 H. Hiratsuka, K. Sekiguchi, Y. Hatano, Y. Tanizaki and Y. Mori, Can. J. Chem. 65, 1185 (1987)
6 G. Buntinx, P. Valat, V. Wintgens and O. Poizat, to be published
7 D. Rehm and A. Weller, Isr. J. Chem. 8, 259 (1970)
8 P.G. Bradley, N. Kress, B.A. Hornberger, R.F. Dallinger and W.H. Woodruff, J. Am. Chem. Soc. 103, 7441 (1981) ; G.D. Danzer, J.A. Golus, D.P. Strommen and J.R. Kincaid, J. Raman Spectrosc. 21, 3 (1990)
9 Q.G. Mulazzani, S. Emmi, P.G. Fuochi, M. Venturi, M.Z. Hoffman and M.G. Simic, J. Phys. Chem. 83, 1582 (1979)
10 B. Brahim, O. Poizat, G. Buntinx, P. Valat and V. Wintgens, in preparation

Solvent Effects on the Structures and Dynamics of the T_1 State and Radical Anion of Benzil Investigated by Time-Resolved Resonance Raman Spectroscopy

K. Ebihara, H. Hiura, and H. Takahashi

Department of Chemistry, School of Science and Engineering, Waseda University, Tokyo 169, Japan

1. Introduction

Benzil is one of the most typical aromatic α-dicarbonyls. A number of studies have been carried out on its photochemistry as well as the structures of its ground and excited states. In this paper the structures of the T_1 state and the radical anion were investigated using time-resolved resonance Raman spectroscopy.

2. Results and Discussion

Time-resolved resonance Raman spectra of the T_1 state of benzil were obtained by pumping at 308 nm and probing at 480 nm in deoxygenated methanol, and those of the radical anion by pumping at 308 nm and probing at 620 nm in dimethylsulfoxide (DMSO). In Fig. 1 are compared the Raman spectra of the ground state S_0, the radical anion and the T_1 state. It is seen that the band at 1685 cm^{-1} which is assignable to the C=O stretch shifts to 1518 cm^{-1} in the radical anion and to 1428 cm^{-1} in the T_1 state. The assignment of these bands to the C=O stretch is based on ^{18}O and ^{13}C isotopic substitutions of the two carbonyl groups. In contrast, the down-shifts of the phenyl vibrations are not very large; the mode 8a shifts from 1598 cm^{-1} to 1587 cm^{-1} and to 1579 cm^{-1}, and the mode 12 (trigonal vibration) from 999 cm^{-1} to 995 cm^{-1} and to 997 cm^{-1} (the assignments are based on deuteration of the phenyl groups). The very large down-shifts of the C=O stretch and the small shifts of the phenyl vibrations suggest that the spin density is almost localized on the two carbonyl groups in both the radical cation and the T_1 state, leading to a very large weakening of the carbonyl groups, particularly in the T_1 state.

The most striking feature of the spectra is that the central C-C stretch appearing at 1050 cm^{-1} in the ground state spectrum shifts to 1315 cm^{-1} in the radical anion and to 1162 cm^{-1} in the T_1 state. The very high C-C stretching frequency of the radical anion indicates that the central C-C bond of the radical anion has a considerably large double-bond nature. It is known that in the ground state benzil takes a twisted structure with the two benzoyl planes twisted 111°36' about the central C-C bond [1]. However, considering the large double-bond nature of the central C-C bond one may reasonably conclude that the radical cation takes an almost planer structure and most probably the T_1 state also. A support for this conclusion is provided by the fact that while in the ground state spectrum both symmetric and anti-symmetric C-Ph stretch are observed at 1284 and 1208 cm^{-1}, respectively, the band due to the anti-symmetric C-Ph stretch is not detectable in the spectrum of the radical cation nor in the spectrum of the T_1 state.

Springer Proceedings in Physics, Vol. 68 Time Resolved Vibrational Spectroscopy V
Ed.: H. Takahashi

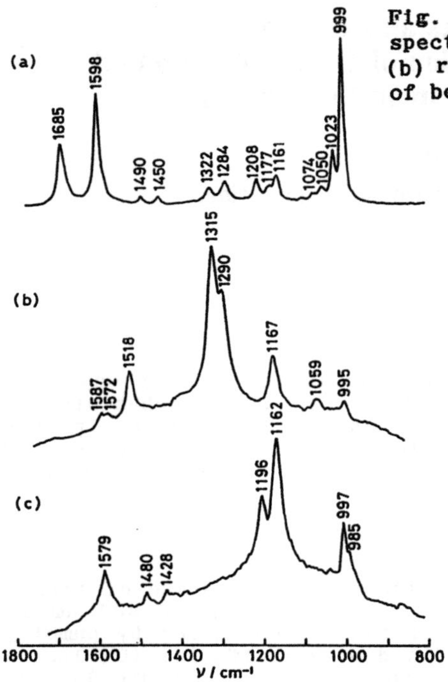

Fig. 1. Comparison of the Raman spectra of the (a) S_0 state, (b) radical anion, and (c) T_1 state of benzil.

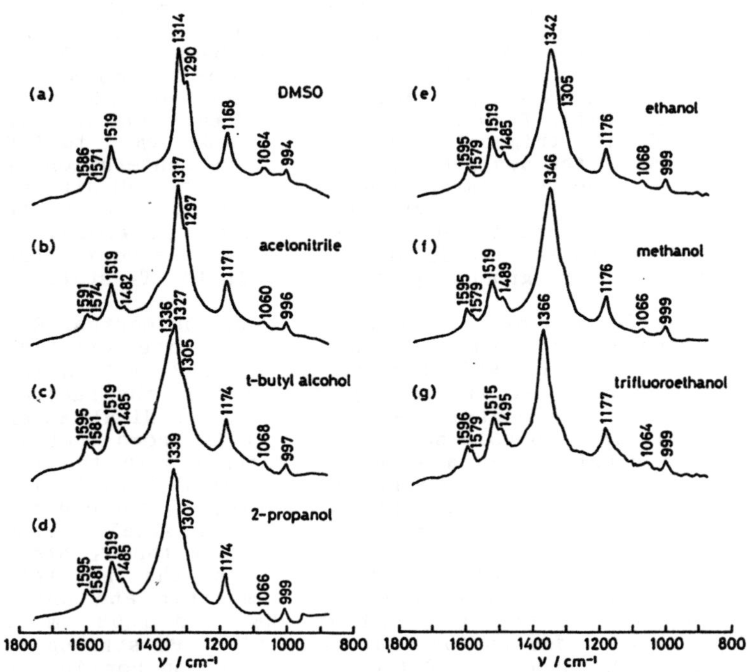

Fig. 2. Solvent dependence of the resonance Raman spectra of benzil radical anion.

Another feature of interest is that the C-C stretch of the radical anion appearing at 1315 cm^{-1} in dimethylsulfoxide (DMSO) solution shifts to higher frequencies as the hydrogen-bond-donor ability of the solvent increases and in 1,1,1-trifluoroethanol it goes up to 1366 cm^{-1} as shown in Fig. 2. The high frequency shifts indicate that the bond order of the central C-C bond increases in hydrogen-bond-donor solvents.

Reference

1. C.J. Brown and R.Sadanaga, Acta Cryst. **18**, 158 (1965)

Structures and Dynamics of Transient Species of Diphenylacetylene from Time-Resolved Resonance Raman Spectroscopy

H. Hiura, H. Yoneda, and H. Takahashi

Department of Chemistry, School of Science and Engineering, Waseda University, Tokyo 169, Japan

1. Introduction

In order to investigate the structural changes occurring when a molecule is electronically excited or when an electron is removed from or attached to a molecule, we have measured time-resolved resonance Raman spectra of diphenylacetylene and its deuterated and ^{13}C-substituted analogues. Dynamics of the transient species have also been investigated.

2. Results and Discussion

The Raman spectra of the S_0 state, the radical cation $R^{\cdot+}$, the radical anion $R^{\cdot-}$, and the T_1 state are compared in Fig. 1. The spectra reveal that the triple bond C≡C stretch (2217 cm^{-1}) exhibits large down-shifts in the order S_0, $R^{\cdot+}$, $R^{\cdot-}$ and T_1. The phenyl ring vibrations, namely, the modes 8a (1596 cm^{-1}), 19a (1482 cm^{-1}), 12 (999 cm^{-1}) and 1 (755 cm^{-1}) also exhibit the same trend, though the shifts are much smaller than those of the C≡C stretch. The low-frequency shifts, which indicate that the skeleton of diphenylacetylene weakens in the order S_0, $R^{\cdot+}$, $R^{\cdot-}$, and T_1, may qualitatively be explained on the basis of the nature of HOMO and LUMO; π-bonding nature of HOMO and π-anti-bonding nature of LUMO with respect to the central C≡C and phenyl C=C bonds. The electron configuration of $R^{\cdot+}$ is such that an electron is lost from HOMO having π-bonding nature leading to a decrease in the bond order of the C≡C and C=C bonds. In $R^{\cdot-}$ an electron is added to LUMO having π-anti-bonding nature leading to a decrease in the bond order of the C≡C and C=C bonds which is expected to be somewhat larger than that of $R^{\cdot+}$. The electronic state of T_1 can best be approximated by the electron configuration in which an electron is elevated from HOMO to LUMO; at least the contribution of this electronic configuration is the largest. Therefore, a combined effect of $R^{\cdot+}$ and $R^{\cdot-}$ must be operative in T_1 and very large decreases in the bond order of the multiple bonds are expected.

In order to obtain information on the dynamics of the transient species, time-resolved resonance Raman spectra were measured by varying the wavelength of the pumping beam to produce the transient species. With the pump wavelength at 308 nm both T_1 and $R^{\cdot+}$ are produced. With the pump wavelength longer than 360 nm, neither T_1 nor $R^{\cdot+}$ is produced. Addition of benzophenone as a triplet sensitizer, however, the both species are observed even under the 360 nm pumping (Fig. 2). These facts indicate that the photo-ejection of an electron from diphenylacetylene takes place via the T_1 state. This does not, of course, exclude the possibility that the electron ejection takes place also via the S_1 state. Dependence of the intensity of the Raman band at 2142

Springer Proceedings in Physics, Vol. 68 Time Resolved Vibrational Spectroscopy V
Ed.: H. Takahashi

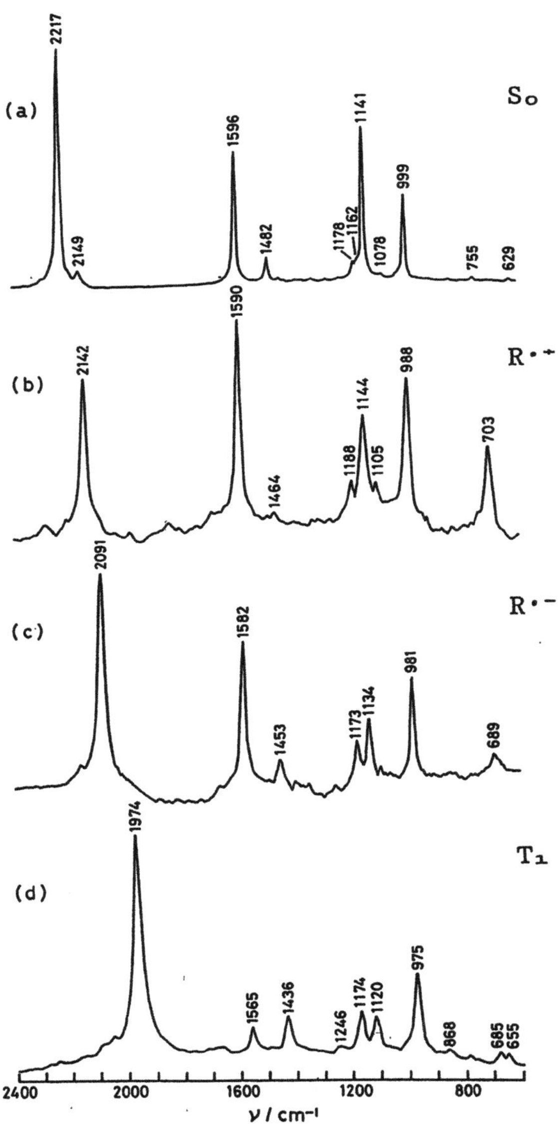

Fig. 1. Comparison of the Raman spectra of the (a) S_0 state, (b) radical cation, (c) radical anion, and (d) T_1 state.

cm^{-1} of $R^{\cdot +}$ on the pump laser power shows that two photons are involved in the photoionization.

The relative concentrations of $R^{\cdot +}$ and $R^{\cdot -}$ are strongly dependent on the nature of solvent. In acetonitrile the concentration of $R^{\cdot +}$ seems to be larger than that of $R^{\cdot -}$. With the probe wavelength at 455 nm which is the absorption maximum of $R^{\cdot -}$, all three species $R^{\cdot +}$ (λ_{max} = 421 nm), $R^{\cdot -}$ and T_1 (λ_{max} = 412nm) are

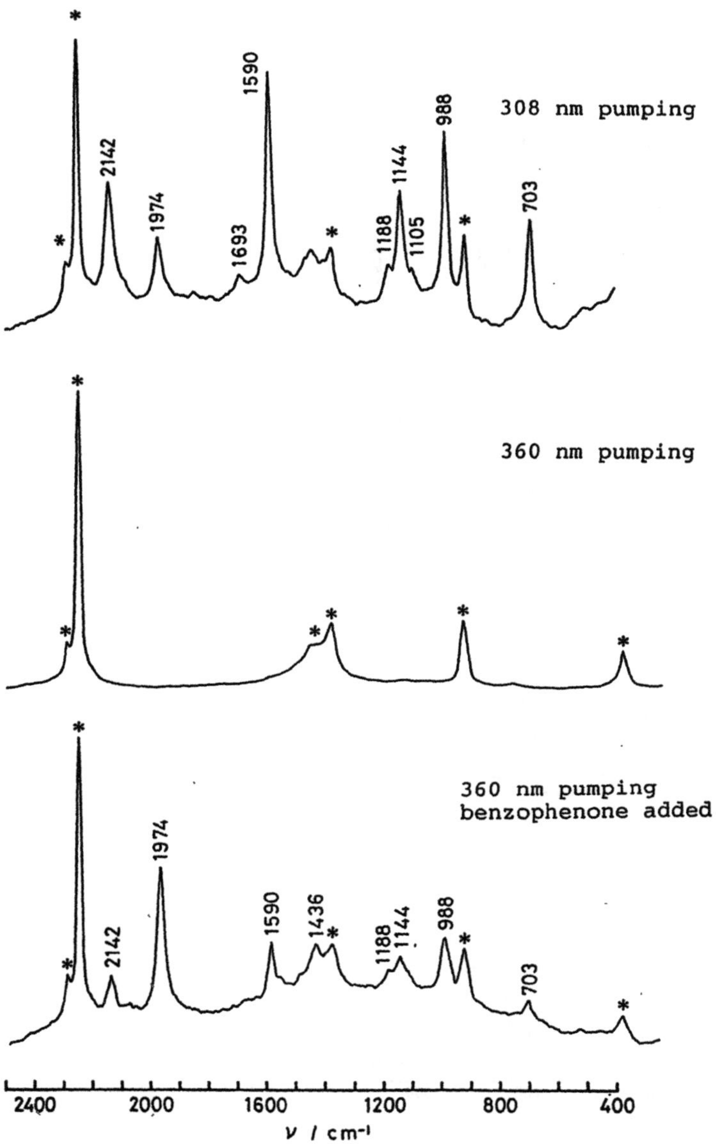

Fig. 2. Pump-wavelength dependence of the resonance Raman spectra of DPA in acetonitrile measured 100 ns after UV irradiation.

simultaneously observable at 100 ns after UV (308 nm) irradiation. In dimethylformamide, however, $R^{\cdot +}$ is not detectable at 100 ns after UV irradiation even under 420 nm probing.

Transient Raman Spectra and Structure of Chloranil in T_1 Chloranil-Alkylbenzene Complexes

T. Tahara and H. Hamaguchi

Kanagawa Academy of Science and Technology (KAST),
KSP East 301, 100-1 Sakato, Kawasaki 213, Japan

It is well known that chloranil in the lowest excited triplet state forms complexes with electron donating alkylbenzenes. This T_1 complex (TC) attracts much interest as a prototypical charge transfer complex. It has been made clear by now that complete charge transfer takes place in the TCs with strong electron donors while only partial charge transfer occurs in those with weak donors. However, the molecular structure of the component molecule in these complexes is not well known. Time-resolved vibrational spectra afford detailed information about the molecular structure of short-lived excited species and are expected to shed some new light on the character of the TCs from the view point of molecular structure.

Transient resonance Raman spectra of the TCs were measured with a nanosecond time-resolved Raman system constructed at KAST. In figure 1, the transient Raman spectra of TCs with several alkylbenzenes are compared with that of the free T_1 state in chloroform. These spectra were measured at delay time of 20 ns by use of pumping at 355 nm and probing at 532 nm. The electron donors were used also as solvents. The probe laser at 532 nm is in resonance with a T-T electric transition due to chloranil and hence only Raman bands due to vibrations of chloranil part are enhanced. Therefore spectral features of TCs are very similar to that of the free T_1 state, though several vibrational frequencies of chloranil in TCs are significantly lower than those of the free T_1 state. The ionization potential is a good indicator of the electron donating ability of donors. Two typical vibrational frequencies of chloranil in TCs are plotted as a function of the ionization potential of donors in

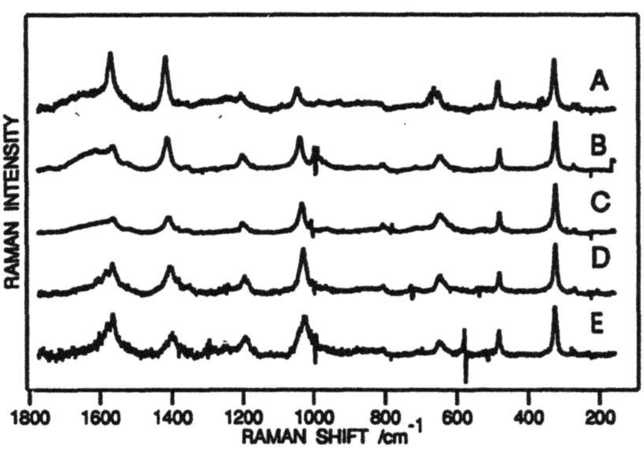

Figure 1. Transient Raman spectra of chloranil in chloroform (A), benzene (B), toluene (C), m-xylene (D), mesitylene (E).

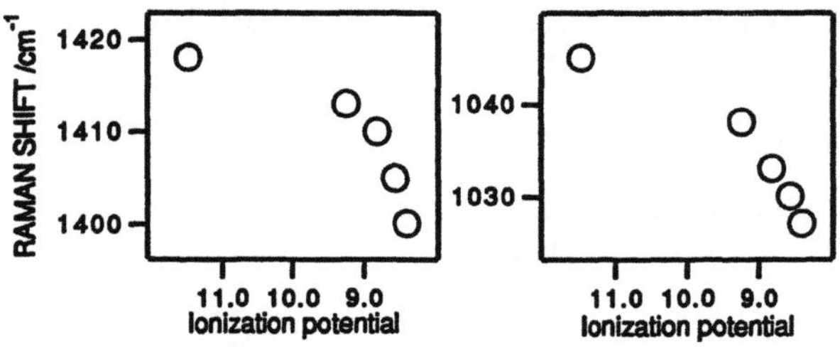

Figure 2. Vibrational frequensies of chloranil in T_1 complexes and the ionization potential of donors.

figure 2. They show gradual downshift in accordance with the decrease of the ionization potential of donors (i.e. the increase of the electron donating ability of donors). It is concluded that observed frequency shifts are due to structural change of chloranil caused by charge transfer interaction.

It has been proposed that the partial charge transfer takes place from donors to chloranil in these TCs [1]. The partial charge transfer results in the mixing between the locally excited T_1 state and the fully charge transfer state [2]. This mixing causes structural change of the component molecule. Our data indicate that the molecular structure of chloranil in a complex is intermediate between those of the pure T_1 state and anion radical and that it varies with the electron donating ability of the donors.

References

[1] K. Kawai, Y. Shirota, H. Tsubomura and H. Mikawa, Bull. Chem. Soc. Jpn, 45, 77(1972).

[2] R. S. Mulliken, J. Am. Chem. Soc., 74, 811(1952).

Time-Resolved Infrared Study of the Photoinduced Energy Transfer Process in the Phthalocyanine-Aromatic Amine System

C. Kato[1;2] *and H. Hamaguchi*[1]

[1]Kanagawa Academy of Science and Technology (KAST), KSP East 301, 100-1 Sakato, Takatsu-ku, Kawasaki 213, Japan
[2]Yamazaki Works, Hitachi Chemical Co., Ltd., 4-13-1, Higashi-cho, Hitachi-shi 317, Japan

Phthalocyanines are well known to provide organic photoconductors of high photoelectric conversion efficiencies. In order to elucidate the mechanism which gives rise to this high efficiency, we need to know the characters of these materials in the excited states. Previously, we reported time-resolved IR (TRIR) absorption spectra of τ-form metal-free phthalocyanine (τ-H_2P_C) fine particles using a newly developed IR spectrometer with μs time-resolution [1]. We assigned the observed TRIR bands to the triplet exciton of τ-H_2P_C. In the present study, we further improved our TRIR spectrometer (detection limit of absorbance change down to $3x10^{-7}$) and reexamined the TRIR absorption of H_2P_C.

An evaporated thin film of α-H_2P_C formed on an NaCl substrate was used for the measurement. Thickness of the film was 1.1 μm. A frequency doubled output of a cw Q-switched Nd:YAG laser (Spectron Laser Systems SL-902TQ, 532 nm, 2 kHz, laser energy at the sample point : 19 μJ / pulse, pulse duration : 100 ns) was used for the photoexcitation.

Figure 1 shows the TRIR absorption spectra of the α-H_2P_C thin film with different gate timings (upper : 50 - 122 μs, middle : 0 - 30 μs after the photoexcitation), and IR absorption spectrum in the ground state (lower). Figure 2 shows the decay curves of the two TRIR bands at 1240 cm^{-1} and 1335 cm^{-1} which

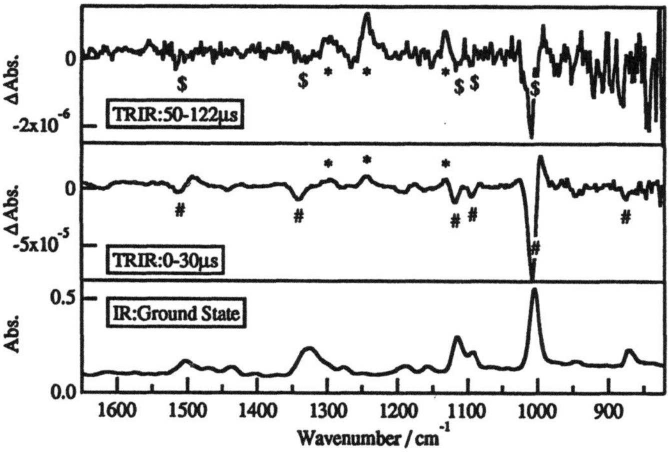

Fig.1. TRIR spectra of α-H_2P_C with different gate timings, and IR spectrum in the ground state.

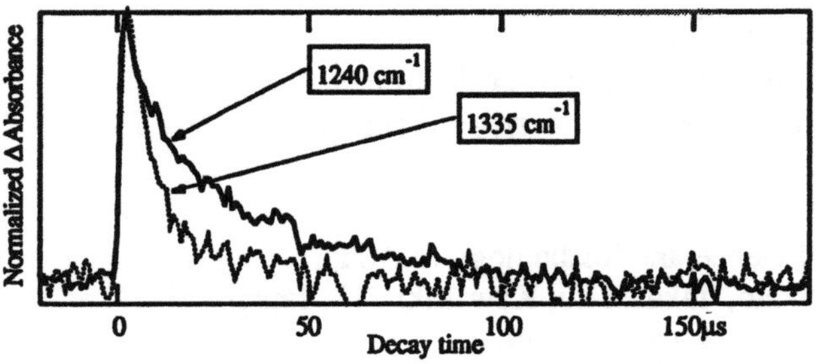

Fig.2. Decay curves of two TRIR bands of α-H_2P_C.

appear in the middle trace of fig. 1. The upper and middle traces in fig. 1 show different spectral patterns. The TRIR bands at 1240 cm^{-1} and 1335 cm^{-1} decay with different time constants. These findings indicate that more than one transient phenomena are involved in these observations.

TRIR bands marked by "#" in the middle trace of fig. 1 appear just at the same position of the IR bands in the ground state, and have differential shape spectral patterns which are also observed by changing the temperature. They are attributable to the reflectance changes at the interface of the sample which are caused by thermally induced refractive index jump (thermal bands). On the other hand, temporal and spectral behaviors of TRIR bands marked by "*" and "$" are different from those of the TRIR bands marked by "#", indicating that these bands are assignable to the generation of transient species and the depletion of the ground state, respectively. Thus, the transient signals we reported previously proved to be the superpositions of the signals due to the "thermal" bands and the signals due to the "real" excited species, which is most probably the triplet exciton of H_2P_C.

We thank Mr. A. Yokota, Dr. Y. Furukawa and Prof. M. Tasumi of the University of Tokyo for showing us their results of temperature change of the FTIR spectra of H_2P_C.

[Reference] K. Iwata, C. Kato and H. Hamaguchi, Appl. Spectrosc., 43, 16(1989).

Transient Resonance CARS Study on Photoionization of Shorter Diphenylpolyenes in Various Polar Solvents

T. Kamisuki[1], *T. Dudev*[2], *and C. Hirose*[1]

[1]Tokyo Institute of Technology, Research Laboratory of Resources Utilization, Nagatsuta, Midori-ku, Yokohama 227, Japan
[2]Department of Chemistry, University of Sofia, BG-1126 Sofia, Bulgaria

1. Introduction

Coherent anti-Stokes Raman spectroscopy (CARS) is a useful means of vibrational investigation of electronically excited molecules or photochemical intermediates. This is based on the high conversion efficiency and the excellent capability of fluorescence elimination. The transient resonance CARS technique by using ns pulsed lasers was applied to the study of solvent-induced photoionization of shorter diphenylpolyenes [DPP] ($(\phi-CH=CH)_n-\phi$; n=2: DPB, n=3: DPH, and n=4: DPO) in various polar solvents.

Very recently, the photoinduced ion (both cation and anion) radicals of DPB [1] and DPH [2] have been detected in polar solvents and in exciplex-forming systems by using the technique. Simultaneous formation of both ion radicals in solution by photolysis has been reported in a few cases. [3] Detailed investigation of other relevant systems is needed to generalize or to promote a better understanding. This paper presents the results of the observed spectra of $DPP^{\cdot +}$ and temporal behavior of the ionic species ($DPH^{\cdot +}$) population in the solvents with widely distributed solvent properties of polarity and electron-donating (accepting) ability.

2. Experimental

The method consists of generating excited state population through the pulsed irradiation by an N_2 laser (337 nm, 10 ns) and observing the CARS signal by scanning an ω_2 dye laser with a fixed ω_1 dye laser tuned to a transient absorption. For 'time-resolved' CARS measurement, 355 nm pulsed light (15 ns) of a Nd:YAG laser was used for sample excitation with synchronously definite delay-time being taken between the excitation pulses and the ω_1 and ω_2 pulses. CARS spectra of DPB, DPH and DPO solutions in various polar solvents having different donor number (acetonitrile(AN), acetone(AC), 2-methyltetrahydrofuran(MTHF), dimethylformamide(DMF), dimethylsulfoxide(DMSO), and N,N-dimethylaniline(DMA)) were measured under resonance conditions of the corresponding absorption bands of the ion radicals.

Springer Proceedings in Physics, Vol. 68 Time Resolved Vibrational Spectroscopy V
Ed.: H. Takahashi

3. Results and discussion

3.1 Solvent effect of the ion radical formation of DPB, DPH and DPO

The photoproduced ion radicals of DPB and DPH in polar solvents have been identified by comparing with the transient resonance CARS spectra and the fluorescence spectra in exciplex-forming systems such as in a p-dioxane solution containing DCNB or DMA as an electron accepting or donating molecule[1,2]. The comparison with the resonance CARS spectra of the DPH and DPO anion radicals produced chemically by metallic Na reduction confirms the identification to the ion radicals. The resonance conditions of the present CARS measurements are that ω_1 is set coincident to the observed absorption peaks of the ion radicals.[4] We focussed our attention to the solvent effect on the most intense CARS band at 1598 $cm^{-1}(DPB^{\cdot +})$, 1588 $cm^{-1}(DPH^{\cdot +})$, 1574 $cm^{-1}(DPB^{\cdot -})$, and 1570 $cm^{-1}(DPH^{\cdot -})$, because of distinct separation between the bands of the corresponding anion and cation radicals. For these diphenylpolyenes clear solvent effects on the formation of the ion radicals were observed. Donor number (DN) of the solvents, which are proposed in the literature[5] as a measure of electron donating ability, are found to have a correlation to the observed solvent effects on the formation of the ion radicals by the UV photolysis: the solvent with lower DN's (high electron accepting power) such as AN and AC produces dominantly $DPP^{\cdot +}$, the solvent with higher electron-donating power (DMF) gives both $DPP^{\cdot +}$ and $DPP^{\cdot -}$, and the strongest one in the series, DMA, yields only $DPP^{\cdot -}$.

3.2 Time-resolved measurements of the CARS signals

Figure 1 shows the time-resolved resonance CARS signals of the DPH ion radicals in various deaerated solvents when delay-time of 0 ns - a few μs was imposed between the UV pulse excitation and dye laser beams for CARS generation.

From the distinct dependence of the spectra on solvents, following features are derived. There is no appreciable signal of $DPH^{\cdot -}$ at the instant of UV irradiation in the AN and AC solutions but the signal becomes detectable at 50 ns after the excitation. In the DMSO solution the initially minor signal of $DPH^{\cdot +}$ grows by time becoming of comparable intensity after 50 ns. In the DMF solution both the positive and negative ions are produced within the duration-time of the UV pulse and no significant change in the spectral pattern was observed between 0 and 50 ns.

The qualitative features of the solvent effects for DPB and DPO are similar to those for DPH. The observed spectral features of DPB near 1550 cm^{-1} on the delay-time of 0 ns in AN, AC, DMF, and

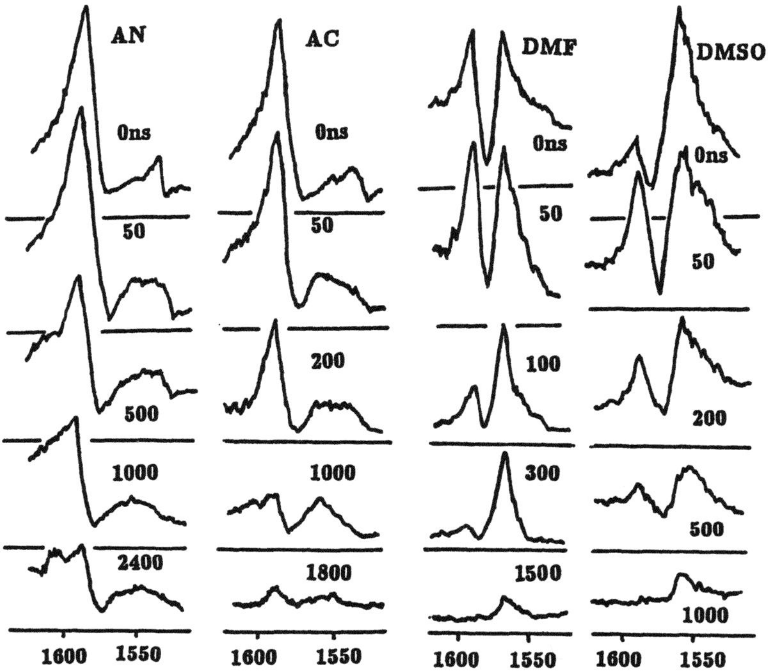

Figure 1. Time-resolved transient resonance CARS spectra of DPH in acetonitrile(AN), acetone (AC), dimethylformamide(DMF), and dimethylsulfoxide(DMSO) in the region of 1520 -1620 cm^{-1} by using 355 nm (337 nm for 0 ns) excitation; ω_1= 16050 cm^{-1}.

DMSO are similar to those of DPH, and in DMF both $DPB^{\cdot+}$ and $DPB^{\cdot-}$ were observed with comparable intensity.

A significant change of the lifetime was observed in air-saturated from that in deaerated DPH solutions. The presence of oxygen shortens remarkably the lifetime of the anion radical and prolongs that of the cation radical in the solvents studied.

In view of the above results and taking into account that the photoionization of DPH in acetone can be achieved by a monophotonic process of 337 nm light absorption under lower UV power, the photoionization mechanism may be proposed as follows. The first process is the rapid formation (within pulse width of 10 ns) of either cation or anion radical via CT type interaction of the excited solute molecule with solvent molecules, and the second process has slow rise-time (more than pulse width, 50 -10 ns) which is interpreted by the subsequent capture of solvated electron (or hole) by a S_0 solute molecule.

References

1. T. Dudev, T. Kamisuki, N. Akamatsu, and C. Hirose, J. Phys. Chem. in press .
2. T. Kamisuki and C. Hirose, J. Phys. Chem. in press, M. Watanabe, T. Kamisuki, N. Akamatsu, and C. Hirose, Chem. Phys. Lett. 170, 451(1990).
3. H. Leonhardt, and A. Weller, Ber. Bunsenges. Phys. Chem. 67, 791(1963), R. Konuk, J. Cornelisse, and S. P. McGlynn, J. Chem. Phys. 82, 3929(1985).
4. M. Almgren and J. K. Tomas, Photochem. Photobiol. 31, 329(1980), G. J. Hoijtink and P. H. van der Meij, Z. Phys. Chem. (Munich), 20, 1(1959), Y. Yamamoto, T. Aoyama, K. Hayashi, J. Chem. Soc. Faraday Trans. 1, 84, 2209(1988), T. Shida and W. Hamill, J. Chem. Phys. 44, 4372(1966).
5. V. Gutmann, Chem. Brit. 7, 102(1972), T. Kagiya, Y. Sumida, and T. Inoue, Bull. Chem. Soc. Jpn, 41, 767(1968).

Time-Resolved Resonance Raman and Surface-Enhanced Resonance Raman Scattering Study on Monocation Radical Formation Processes of Heptyl Viologen at Silver Electrode Surfaces

Y. Mineo, K. Shibasaki, N. Yamasawa, and K. Itoh

Department of Chemistry, School of Science and Engineering, Waseda University, Shinjuku-ku, Tokyo 169, Japan

1. Introduction

Heptyl viologen ($HV^{2+}(X^-)_2$, X^-= a counter anion; see Figure 1) has been known to perform a one-electron redox reaction with half-wave potential near -0.4 V (vs. Ag/AgCl). Monocation radical ($HV^{+\bullet}$ X^-), which is formed on stepping electrode potential to a value negative of -0.4 V, is insoluble in water and forms a red-purple film giving a broad band centered around 550 nm on an electrode. On reversing the potential to a value positive of -0.4 V, the colored film disappears. This property can be applied to electrochromic display devices. Therefore, a number of studies have been made to elucidate the mechanism of the film formation process.[1,2] These studies have proved that the initial film formation consists of nucleation, nuclear growth and subsequent film formation processes. In the present paper we applied time-resolved resonance Raman scattering (RRS) and surface-enhanced resonance Raman scattering (SERRS) spectroscopies to analyze these processes. Time-resolved RRS spectra have already been reported by Osawa and Suetaka[3] for $HV^{+\bullet}$ films on a Pt electrode; the spectra, however, were observed with time resolution of 5 msec and recorded after the initial processes had already terminated. So, we confined the measurement mainly within several milliseconds since the application of reduction potential.

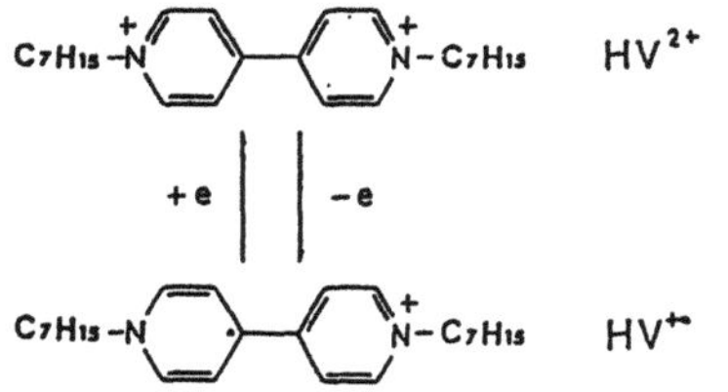

Fig. 1 Structures of HV^{2+} and $HV^{+\bullet}$

2. Experimental

A Raman spectroelectro-chemical apparatus used in this paper consists of a usual electrochemical cell containing a silver working electrode, a spectrometer (SPEX Industries, Inc. Triplemate 1877) equipped with a multichannel detector (Princeton Instruments Inc. model ICCD-576) and a Q-switched Nd-YAG laser (Spectron Laser Systems, SL404) as an

Springer Proceedings in Physics, Vol. 68 Time Resolved Vibrational Spectroscopy V
Ed.: H. Takahashi

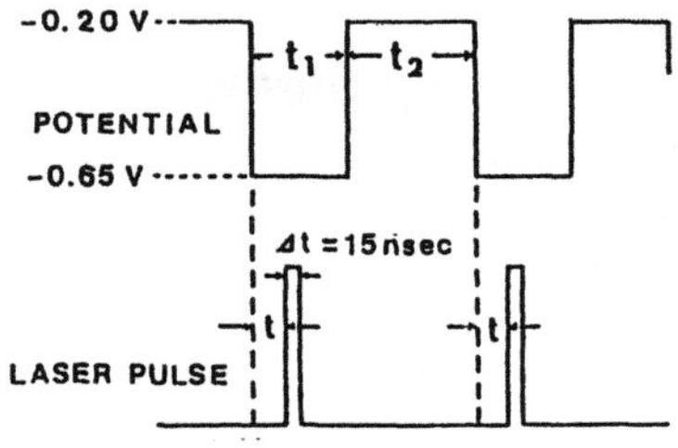

Fig. 2 Step potentials used for time-resolved RRS and SERRS measurements

excitation source; the laser has a pulse width of 15 nsec and a maximum repetition rate of 15 Hz. Applying step potentials to the electrode, firing an excitation laser pulse and opening the detector gate were triggered by a sequence of TTL pulses from a pulse generator (Stanford Research System Inc., model DG535). In the time-resolved measurements, we applied step potentials as shown in Figure 2; t_1 and t_2 were adjusted so that a $HV^{+\bullet}$ film formed during the t_1 step (-0.65 V) totally disappears during the t_2 step (-0.2 V). The spectra were taken by accumulating signals excited by thousands of laser pulses (532 nm). All the time-resolved spectra were recorded as a function of time (t), after the application of the reduction potential.

3. Results and discussion

3.1 Time-resolved RRS spectra observed for the $HV^{+\bullet}$ formation on a smooth Ag electrode surface

Figure 3 shows the time-resolved RRS spectra of $HV^{+\bullet}$ formed on a smooth Ag electrode from 10^{-3} mol/L HV^{2+} containing 0.3 mol/L NaH_2PO_2; the step potentials were applied under a condition of t_1=30 msec and t_2=170 msec. The RRS bands near 1530 and 1030 cm^{-1} are assigned to $HV^{+\bullet}$.[3] The ratio of the 1530-cm^{-1} band intensity to the maximum intensity at t = 29 msec is plotted against $t^{1/2}$ in Figure 4. From the figure it is clear that the curve can be divided into three regions; in the first region the intensity increases with $t^{3/2}$ (The data can be fitted to a function, $3.84t^{3/2}$, as shown by a solid line.), in the second region the relative intensity increase is a linear function of $t^{1/2}$ and in the third region the rate of the intensity increase is reduced compared to that observed in the former two regions. We call these regions as A, B and C, respectively.

3.2 Time-resolved SERRS spectra observed for a few monolayers of $HV^{+\bullet}$ on a rough Ag electrode

Time-resolved SERRS spectra of $HV^{+\bullet}$ formed on a rough Ag electrode were also measured under a condition of t_1=50 msec and t_2=50 msec. After ordinary oxidation-reduction cycle treatment being performed to activate the electrode immersed into 10^{-3} mol/L HV^{2+} containing 0.3 M KBr, the bulk solution was completely replaced by 0.3 mol/L KBr solution. So, the spectra monitored the the monocation radical formation of a few monolayers of HV^{2+} adsorbed on the Ag electrode. The spectra gave features similar to those in Figure 3, showing SERRS bands near 1525 and 1025 cm^{-1} due to $HV^{+\bullet}$. The relative

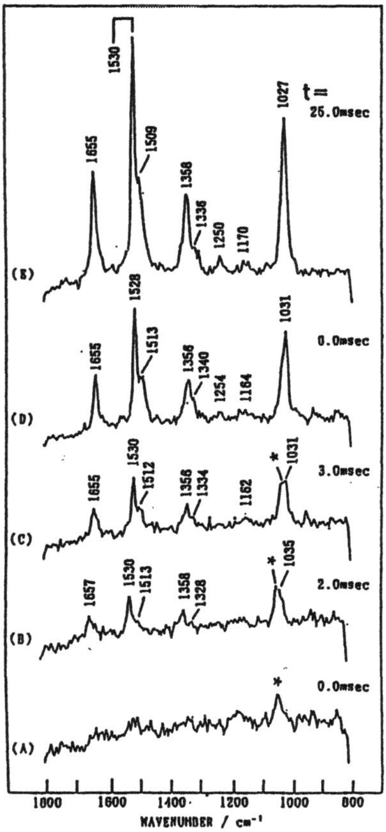

Fig. 3
Time-resolved RRS spectra observed for 1mM HV^{2+}/ a smooth Ag-E/0.3M NaH_2PO_2 λ_{ex}=532 nm(*;due to $H_2PO_2^-$)

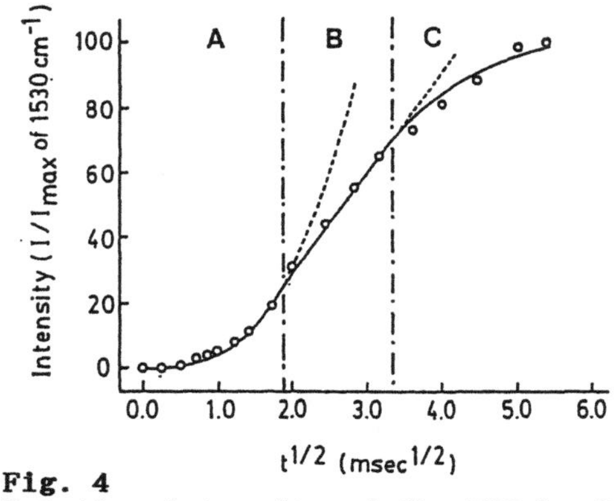

Fig. 4
Relative intensity of the RRS band at 1530 cm^{-1} due to $HV^{+\cdot}$ on a smooth Ag electrode vs. $t^{1/2}$

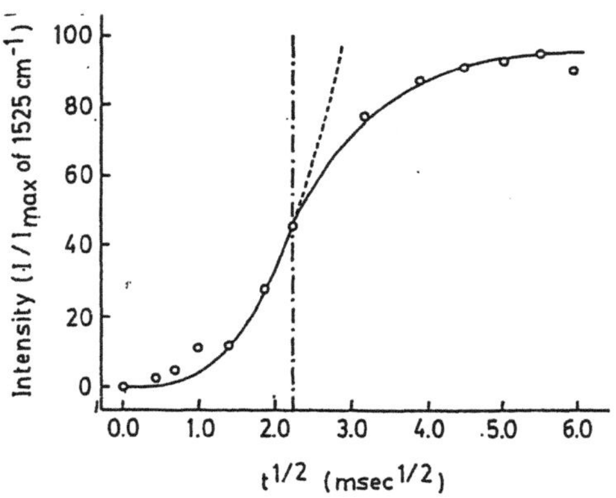

Fig. 5
Relative intensity of the SERRS band at 1525 cm^{-1} due to $HV^{+\cdot}$ on a rough Ag electrode vs. $t^{1/2}$

intensity of the 1525-cm^{-1} band (normalized by the maximum intensity at t = 49 msec) is plotted as a function of $t^{1/2}$ in Figure 5. In contrast to the former case, the radical formation process can be divided into two regions. In the first region the intensity increases with $t^{3/2}$ (A solid curve in this region represents a function, $4.13t^{3/2}$, fitted to the observed points by a least square method.), as observed in the region A for the $HV^{+\cdot}$ formation on the smooth Ag surface. The result suggests that at the beginning of the film formation a similar process takes place both for the smooth and rough Ag surfaces. Another characteristic feature of Figure 5 is the absence of the region corresponding to the region B in Figure 4, where the intensity increase is a linear function of $t^{1/2}$

3.3 Kinetic features of the film formation revealed by the time-resolved RRS and SERRS spectra

Osawa and Suetaka[3] performed a chronoamperometric measurement for the formation of a $HV^{+\bullet}$ film on a Pt electrode surface, in which transient current was measured as a function of time after the potential step from -0.2 to -0.55 V (vs. Ag/AgCl). The result indicated that at the beginning of the formation (t = 0 - 1.3 msec) the transient current increases with $t^{1/2}$, and after reaching to a plateau it decreases with $t^{-1/2}$ (t > 2 msec). Electrochemical studies[1,2] have indicated that the $t^{1/2}$ dependence of the transient can be interpreted by an instantaneous nucleation mechanism, which assumes an instantaneous formation of nuclei and a hemispherical mass-transfer controlled growth of the nuclei. After a certain period the nuclei overlap with each other and forms an effectively planar monocation radical film. As a result, the process is switched over to a linear diffusion controlled film growth, which conforms to the $t^{-1/2}$ dependence of the transient current. The transient current is proportional to the number of $HV^{+\bullet}$ formed per unit time at time, t, while the RRS intensity is proportional to the number of total $HV^{+\bullet}$ molecules formed at t; then, the RRS intensity should be proportional to the integration of the transient current in terms of t. In consequence, the region, where the transient current shows the $t^{1/2}$ dependence, corresponds to the region A and the region, where the transient shows the $t^{-1/2}$ dependence, corresponds to the region B. Thus, the kinetic feature of the $HV^{+\bullet}$ film formation on the smooth Ag electrode surface conforms well to the results of the chronoamperometric measurement. The absence of the region B in the intensity vs. time curve for the time-resolved SERRS spectra (Figure 5) is reasonable, because in the absence of HV^{2+} in the bulk solution there should be no diffusion-controlled film growth process.

3.4 Structures of $HV^{+\bullet}$ molecules in the initial film formation processes

The time-resolved RRS and SERRS spectra are expected to reflect the structures of $HV^{+\bullet}$ in the initial film formation processes. The spectra (B) and (C) in Figure 3 corresponds to the region A and the spectrum (D) to the region B in Figure 4. All the spectra are similar to each other, giving a prominent band near 1530 cm^{-1} and a shoulder band near 1510 cm^{-1}. According to Osawa and Suetaka,[3] the 1510-cm^{-1} band is ascribed to a monocation radical dimer and the 1530-cm^{-1} band to either a monocation radical monomer or a monocation radical salt. It is not reasonable, however, to assign these bands to different species, because the intensities of the doublet bands increase with t, keeping a constant intensity ratio. The doublet feature near 1530 and 1510 cm^{-1} is observed also for all the time-resolved SERRS spectra. A thick $HV^{+\bullet}$ film, which is formed by keeping the potential at -0.65 V for several minutes, gives rise to a strong band at 1526 cm^{-1}, in which the 1510-cm^{-1} shoulder is buried; the result suggests that a structural transformation takes place during the film formation process. Time-resolved RRS spectra were also measured for the monocation radical of a benzyl viologen analogue. The spectra also gave the doublet in an early stage of the monocation radical formation. On the basis of these results we concluded the following; (i) the doublet is assigned to a species, which is present during the nucleation and nuclear growth processes and also in an early stage of the film formation process, and (ii) the bipyridiniumyl moiety in a monocation radical state assumes a specific structure on interaction with the electrode surfaces and keeps its structure in the early stage of the film formation.

An excitation profile of the time-resolved RRS spectra of $HV^{+\bullet}$ in the 550 - 650 nm region, where the radical film gives a broad band, will give further insight into the structure of the species. This is now being carried out in our laboratory.

References

[1] J.Bruinink and C.G.A.Kregting, J. Electrochem. Soc. 125, 1397 (1978).

[2] R.Cieslinski and N.R.Armstrong, J. Electroanal. Chem. 161, 59 (1984).

[3] M.Osawa and W.Suetaka, J. Electroanal. Chem. 270, 261(1989).

Development of Infrared Diode Laser Kinetic Spectroscopy and Its Applications to Chemical Reactions

E. Hirota

The Graduate University for Advanced Studies, 4259 Nagatsuta, Midori, Yokohama 227, Japan and
The Institute for Molecular Science, Okazaki 444, Japan

1. Infrared Diode Laser Kinetic Spectroscopy

Infrared diode laser spectroscopy is versatile; it allows us to observe vibration-rotation transitions of transient molecules, free radicals, and molecular ions in nearly entire infrared region. We need 10^{10} to 10^{12} molecules to observe their spectra in most favorable cases, provided that the pathlength can be extended as long as 20 m or so. This method has been providing us with very detailed spectroscopic data on such short-lived molecules as reaction intermediates; the number is approaching one hundred [1]. It is a natural extension to try to make this versatile method time-resolved, in order to get more dynamical information in real time on systems where chemical reactions are taking place. We have achieved this by simply introducing a transient digitizer in our signal detection circuitry and by controlling it with a personal computer [2]. Using this system we collect the signal for a pair of time gates, one after and the other before a trigger to induce a chemical process, and take the difference between the two signals, while we scan the wavelength of the source diode slowly. We normally set three pairs of such gates to record time-resolved spectra. We may employ the system also as a signal averager to follow the time evolution of a spectral line. The time resolution of our system is 300 ns, which is determined mainly by the infrared detector and its peripherals. We have applied this system to a few photochemical reactions and also photochemically induced reactions.

2. Results

Three recent examples are given below to illustrate the method.

2.1 The Photolysis of Cl_2SO at 248 nm [3]

The system proceeds in two main channels: ClSO + Cl [a] and 2Cl + SO [b] with the available energy of 241 and 28 kJ mol^{-1}, respectively [4]. We monitored the fine structure transition of Cl and found a broad emission superimposed on a sharp absorption feature. The observed lineshape did not change when the polarization of the excimer laser light was rotated. We thus ascribed the lineshape to the two channels; the broad emission was mainly from [a] and a sharp absorption primarily due to [b].

Springer Proceedings in Physics, Vol. 68 Time Resolved Vibrational Spectroscopy V
Ed.: H. Takahashi

We then analyzed the observed spectrum by simulating the lineshape with a square function convoluted with a Gaussian function. The width of the emission was found to agree well with the value from the TOF result [4]. The narrow absorption peak had the width that corresponded to a velocity less than twice the thermal velocity at room temperature. The ratio of the integrated intensities of the two components, emission to absorption, was 4.6.

We then recorded the time evolution of the signal by relaxing the system with a large amount of helium. The signal started as an emission just after the laser shot came in and then turned to absorption. We fitted the time profile to a rate equation in which we took into account quenching between the two fine structure levels and also the reaction of chlorine atoms in these two states. The least-squares analysis yielded the quenching constant of 2.82×10^{-11} cm^3 mol^{-1} s^{-1} and 1.0×10^{-13} cm^3 mol^{-1} s^{-1} for Cl_2SO and He, respectively, while the reaction was fond to be too slow to determine the rate constant. The analysis yielded the ratio of the nascent populations in the upper and lower fine structure levels to be 0.79, which may be compared with the statistical value of 0.5. If we assume that the upper-state population generated by channel [b] is negligible in view of its very small availbale energy, we may calculate that the nascent population ratio of the two levels is 0.94 for channel [a] and the branching ratio of the two channels [a]/[b] is about 10.

2.2 The Photolysis of CH_3I at 248 nm [5]

Previous TOF and infrared emission studies [6-9] have shown that the 248 nm photolysis of CH_3I proceeds in two channels that correspond to iodine fragments in the two fine structure levels, with the relative contribution of the upper level being 70 to 80 %, and that methyl fragments generated by the photolysis are distributed over the ν_2 states peaking at v =2.

We have observed the v =1 ← 0 up to 4 ← 3 of CH_3 radicals as functions of time for the sample pressure ranging from 9 to 100 mTorr. The observed Doppler lineshape showed a clear correlation between the transition dipole moment and the recoil velocity in the photodissociation. The observed linewidth indicated that CH_3 fragments were predominantly prepared through the channel with iodine atoms in the upper fine structure level, i.e. I* adiabatic channel, as reported by previous TOF studies [6-8]. The relative population in each vibrational level was determined to be 0.66, 0.26, 0.08, and 0.004 for v_2 = 0, 1, 2, 3, in sharp contrast to the inverted vibrational distribution reported by the TOF studies [6-8].

The present result suggests that the geometry of the CH_3 moiety in the precursor CH_3I is relaxed gradually to its equilibrium planar structure before the cleavage of C-I bond takes place.

2.3 The $O(^1D)$ + CH_4 Reaction [10]

This reaction is one of the most fundamental reactions. However, spectroscopic investigations of this reaction have been directed only to OH fragments [11]. We have applied infrared diode laser kinetic spectroscopy to methyl fragments in order to get more detailed information on the reaction.

We generated O(^{1}D) by the excimer laser photolysis of N_2O, NO_2, and O_3 with the total pressure ranging from 100 to 200 mTorr, and monitored the ν_2 bands v= 1 ← 0 up to 4 ← 3 of CH_3 as functions of time. We found that the vibrational distribution was non-inverted and was much less excited than a prior distribution. The fraction of the available energy released to the ν_2 vibration was about 0.019. On the other hand, we observed about ten vibration-rotation lines of OH of which the lower-state energy ranges from 7400 to 14500 cm^{-1}. This observation is in line with previous laser-induced fluorescence studies on OH [11]. A similar non-inverted vibrational distribution of CH_3 was obtained for the reaction of O(^{3}P) with CH_4, which has been known as a direct hydrogen abstraction reaction; here we generated O(^{3}P) by the 193 nm photolysis of sulfur dioxide.

The present results suggest that the CH_3 moiety deforms fast enough to follow the hydrogen atom abstracted in these reactions.

3. Discussion

The unique feature of the present method is that, while keeping its inherent high spectral resolution, it provides us with information on the dynamical behavior of the reaction system. In particular, it directly measures the difference between the populations in two vibrational states as a function of time. Although it suffers from the drawbacks of infrared spectroscopy such as the low sensitivity and the slow response time, it is quite versatile, and yet yields very detailed information which is very difficult, or almost impossible to obtain with other methods presently available.

[1] E. Hirota, Int. Rev. Phys. Chem. 8, 171 (1989); High-Resolution Spectroscopy of Tansient Molecules, Springer, Heidelberg, 1985.
[2] H. Kanamori, J. E. Butler, K. Kawaguchi, C. Yamada, and E. Hirota, J. Mol. Spectrosc. 113, 262 (1985); H. Kanamori and E. Hirota, J. Chem. Phys. 86, 3901 (1987).
[3] E. Tiemann, H. Kanamori, and E. Hirota, to be published.
[4] M. Kawasaki, K. Kasatani, H. Sato, H. Shinohara, N. Nishi, H. Ohtoshi, and I. Tanaka, Chem. Phys. 91, 285 (1984).
[5] T. Suzuki, H. Kanamori, and E. Hirota, J. Chem. Phys. 94, xxx (1991).
[6] R. K. Sparks, K. Shobatake, L. R. Carlson, and Y. T. Lee, J. Chem. Phys. 75, 3838 (1981).
[7] G. N. A. van Veen, T. Baller, A. E. de Vries, and N. J. A. van Veen, Chem. Phys. 87, 405 (1984).
[8] M. D. Barry and P. A. Gorry, Mol. Phys. 52, 461 (1984).
[9] H. W. Hermann and S. R. Leone, J. Chem. Phys. 76, 4759 (1982); 76, 4766 (1982).
[10] T. Suzuki and E. Hirota, to be published.
[11] A. C. Luntz, J. Chem. Phys. 73, 1143 (1980).

Part IV

Transient Species

The Potential Energy Surface of the T_1 State of 1,3,5-Hexatrienes and Methylated Derivatives: A Survey of Present Knowledge

R. Wilbrandt[1], *S. Møller*[1], *F.W. Langkilde*[2], *A.M. Brouwer*[3], *F. Negri*[4], *and G. Orlandi*[4]

[1]Department of Environmental Science and Technology, Risø National Laboratory, DK-4000 Roskilde, Denmark
[2]Department of Physical Chemistry A, Royal Danish School of Pharmacy, 2 Universitetsparken, DK-2100 Copenhagen, Denmark
[3]Laboratory of Organic Chemistry, University of Amsterdam, NL-1018 WS Amsterdam, The Netherlands
[4]Dipartimento di Chimica G. Ciamician, University of Bologna, I-40126 Bologna, Italy

1. Introduction

In previous studies[1,2] we considered the time-resolved resonance Raman (TRRR) spectra of 1,3,5-hexatriene (HT) in its lowest excited triplet state in order to obtain information about the geometry of the T_1 intermediate and the mechanism of the cis-trans photoisomerization in T_1.

In the present paper some results of the most recent experimental and theoretical studies are presented, and the present picture of the potential energy surface in the lowest excited T_1 state of 1,3,5-hexatrienes evolving from these and our previous investigations is summarized. A more detailed account of this work is to be published[3-5].

In order to obtain information on the T_1 state of HT, several experimental and theoretical methods are used:

Time-resolved resonance Raman spectroscopy: The experimental method giving the most detailed structural information on equilibrium geometries in the excited T_1 state is TRRR spectroscopy. Vibrational *frequencies* in the T_1 state provide characteristic fingerprints of the equilibrium geometry or geometries in the T_1 state. Thereby, detailed structural information on the relative minima of the potential energy surface is obtained. *Intensities* of resonance Raman bands are dependent on the relative populations of scattering molecules and characteristic of the changes in molecular geometry occurring upon excitation from the T_1 to the T_n state of the resonance transition. Hence, information not only on the T_1, but also the T_n state is available from the spectra.

By variation of the *temperature* in TRRR experiments information on the relative populations among different minima and hence on the *relative energy of different minima* is obtained.

The *kinetics of equilibration* among different minima on the T_1 potential energy surface at various temperatures, as revealed by TRRR spectra, allows to draw conclusions about the *height of barriers* separating different minima from each other.

If barriers are sufficiently small, equilibration occurs rapidly, i.e. within the lifetime of the T_1 state; if they are large, on the other hand, no equilibration will be observed. Equilibration along torsional coordinates around formal C=C double and C-C single bonds can be studied by *varying the ground state configurations and conformations* of the HT derivatives by studying different isomers and rotamers.

The study of *deuteriated molecules* provides a means to further support the assignment of vibrational spectra.

Time-resolved absorption spectroscopy: Time-resolved absorption spectroscopy allows to determine triplet-triplet absorption spectra and kinetics of triplet state formation and decay. In particular, $T_1 \rightarrow T_n$ *absorption spectra* are important for an assignment of the T_n state active in the resonance transition, which is necessary to calculate TRRR intensities. *Kinetics* of triplet decay as a function of temperature yield activation energies and frequency factors of Arrhenius plots and are interpreted in terms of *torsional barriers* on the T_1 potential energy surface. The measurement of rate constants of *triplet energy transfer* using HT as either donor or acceptor yields information on energies of vertical and relaxed triplet states of HT.

Photostationary triplet state photochemistry: The measurement of *photostationary states* and *quantum yields of photoisomerization* by sensitized stationary photochemistry for a variety of donors and acceptors with different triplet energies is an additional, more indirect, method to determine triplet state energies and minima on the T_1 potential energy surface.

Theoretical quantum chemical calculations: The understanding of the results of the above mentioned experimental investigations are crucially dependent on a comparison with results from theoretical calculations. The experimental TRRR spectra were interpreted by quantum mechanical calculations (QCFF/PI and CNDO/S) of equilibrium geometries and vibrational frequencies at the trans (E), cis (Z) and centrally twisted (P) geometries in T_1 and in the T_n state responsible for the TRRR activity. On the basis of these calculations the T_1 normal modes showing the strongest Franck-Condon activity in the TRRR spectra can be identified for each geometric isomer of HT.

2. Results and Discussion

The observed TRRR spectra for unsubstituted HT are shown[1-3] to result from a superposition of the spectra of the Z and E forms with possible weak contributions from the P form. It was concluded that the transient T_1 intermediate in the photoisomerization of HT, monitored by TRRR spectra, exists as an equilibrium between several geometric structures, in which planar molecules represent a substantial fraction, and that a possible minimum of the potential energy curve at the centrally twisted geometry must be rather shallow. An example of experimental and calculated spectra is shown in Figure 1 for 3-d-HT.

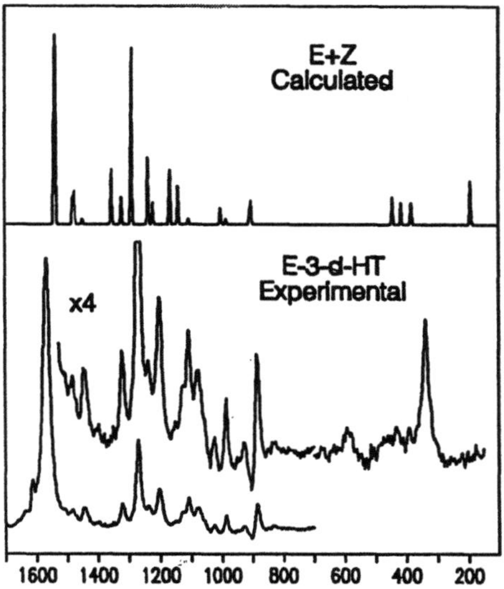

Figure 1: Calculated and experimental T_1 resonance Raman spectrum of 3-d-HT.

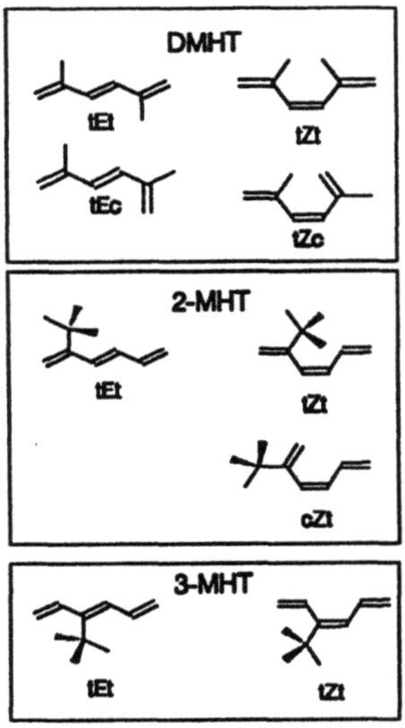

Figure 2: Planar geometries of DMHT, 2-MHT and 3-MHT.

In this paper we consider the combined results of TRRR spectra and decay kinetics of the T_1 intermediate of HT, some of its deuteriated derivatives, and of its 2,5-dimethyl- (DMHT), 2-methyl- (2-MHT), and 3-methyl- (3-MHT) derivatives. The T_1 TRRR specra of the two isomers of non-deuteriated DMHT have been reported previously[6]. We now report the spectra of E- and Z-3,4-d_2-DMHT and of E-3-d-DMHT together with remeasured spectra of DMHT-d_0. Furthermore, resonance Raman spectra of E-2-MHT, E-4-d-2-methyl-1,3,5-hexatriene (E-4-d-2-MHT), Z-2-methyl-1,3,5-hexatriene (Z-2-MHT), E-3-methyl-1,3,5-hexatriene (E-3-MHT), and Z-3-methyl-1,3,5-hexatriene (Z-3-MHT) in the lowest triplet state T_1 are reported and discussed. QCFF/PI calculations are performed to determine energies and optimized geometries in the S_0, T_1, and T_n electronic states. For each T_1 species, $T_1 \rightarrow T_n$ transitions, vibrational frequencies, and RR intensities are calculated and compared with the observed frequencies and intensities (as exemplified in Fig.1). An example of determination of the activation energy of the decay of the T_1 state is shown in Fig. 3 for 2-MHT and 3-MHT.

Finally, in Fig. 4 the qualitative potential energy surfaces in the T_1 state of the methyl substituted HT's are shown, resulting from different pieces of information as discussed above. It can be seen that characteristic differences are found for the different molecules. The T_1 potential energy surface of 3-MHT is very similar to that of unsubstituted HT[1]. Steric interaction is responsible for the high energy of the Z isomers of DMHT. The barrier along the torsional coordinate of the central CC bond at the P geometry is seen to vary, being 0.7-0.8 Kcal/mol for HT, cPt-DMHT and 3-MHT, while 2-MHT shows a lower barrier of 0.4 kcal/mol and no barrier is found for the tPt form of DMHT.

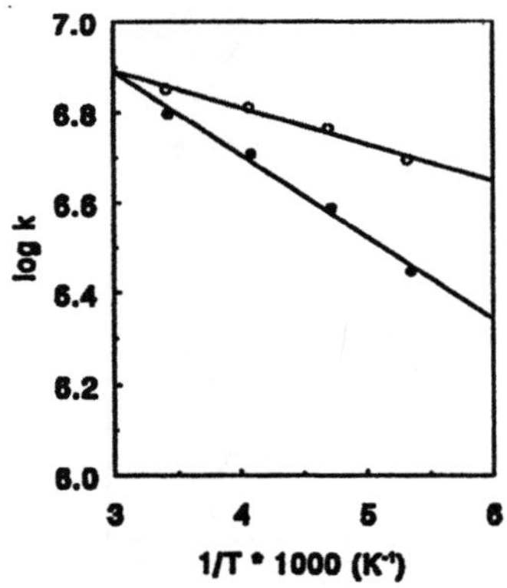

Figure 3: First order rate constant for T_1 decay for 2-MHT (open circles) and 3-MHT (filled circles).

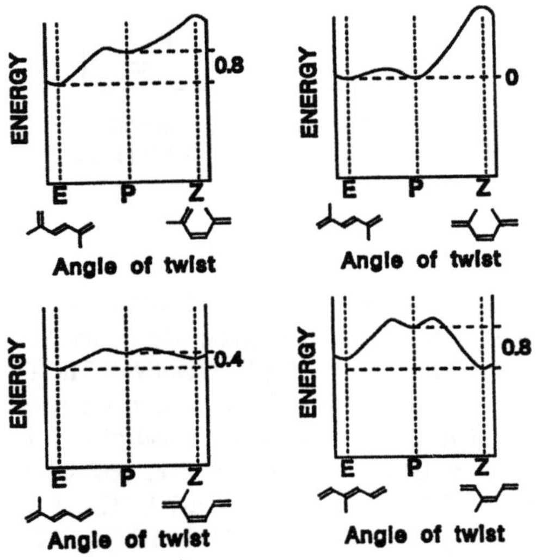

Figure 4: Potential energy curves in T_1 for DMHT, 2-MHT and 3-MHT.

References

1. F.Negri, G.Orlandi, A.M.Brouwer, F.W.Langkilde and R.Wilbrandt, J.Chem.Phys. 90, 5944-5963 (1989)
2. R.Wilbrandt, F.W.Langkilde, A.M.Brouwer, F.Negri and G.Orlandi, J.Mol.Struct. 217, 151-160 (1990)
3. F.W.Langkilde, R.Wilbrandt, S.Møller, A.M.Brouwer, F.Negri and G.Orlandi, J. Phys. Chem. 95 (1991) (in press)
4. F.Negri, G.Orlandi, A.M.Brouwer, F.W.Langkilde, S.Møller and R.Wilbrandt, J. Phys. Chem. 95 (1991) (in press)
5. F.W.Langkilde, R.Wilbrandt, A.M.Brouwer, H.J.C.Jacobs, F.Negri and G.Orlandi, "Steric control of excited state molecular structure. Resonance Raman and absorption spectroscopy of 2-methyl- and 3-methyl-1,3,5-hexatriene in the T_1 state" (submitted)
6. F.W.Langkilde, N.-H.Jensen, R.Wilbrandt, A.M.Brouwer and H.J.C.Jacobs, J.Phys.Chem. 91, 1029 (1987)

Transient Raman Study of the Low-Lying Singlet Manifold (S_1 and S_2) of Biphenyl

Y. Sasaki[1,2], *Y. Koike*[2], *and H. Hamaguchi*[1]

[1]Molecular Spectroscopy Laboratory,
Kanagawa Academy of Science and Technology (KAST),
KSP East 301, 100-1 Sakato, Takatsu-ku, Kawasaki 213, Japan
[2]Mitsubishi Kasei Corporation, Research Center,
1000 Kamoshida-cho, Midori-ku, Yokohama 227, Japan

1. Introduction

Time-resolved Raman spectrum of the S_1 state of biphenyl was first reported by Kato *et al.* [1]. Several vibrational frequencies were determined and assigned, and a twisted S_1 tructure was suggested on the basis of a vibrational analysis [2]. However, the broad feature in the 1300 ~ 1600 cm^{-1} region in the S_1 Raman spectrum was left unassigned.

It is known that there exist two closely-lying states in the lowest exited singlet manifold of bipheyl in crystal at 1.8 K [3] and in supersonic jet beam [4]. The energy separation between these two states in solution is not known at present, but it is highly likely that strong interaction between these two states manifests itself in the S_1 state Raman spectrum. The purpose of this study is to clarify the origin of the broad band.

2. Results and discussion

Time-resolved Raman spectra of S_1 biphenyl under various conditions were measured and, as a result, the following information has been obtained. (1) The band exhibits a frequency shift upon deuteration. The transition which causes the band is related to a specific vibrational mode. (2) Excitation-wavelength-dependence of the band differs from those of the sharp S_1 bands (Fig.1). The intensity of the broad band has an

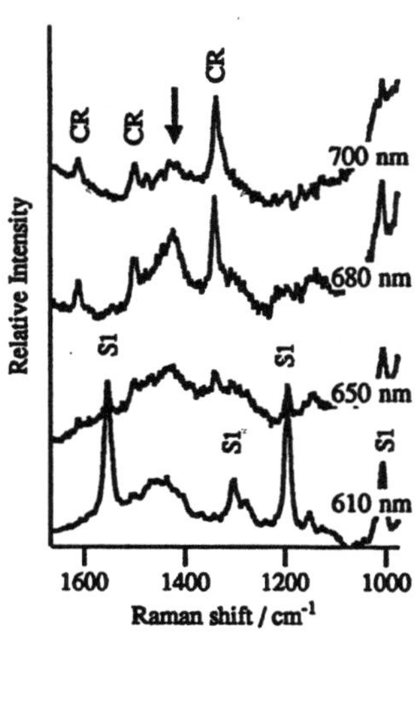

Fig. 1. Transient Raman spectra of photoexcited biphenyl in n-heptane obtained at the indicated excitation wavelength (delay time 0 nsec). CR designates bands due to the cation radical.

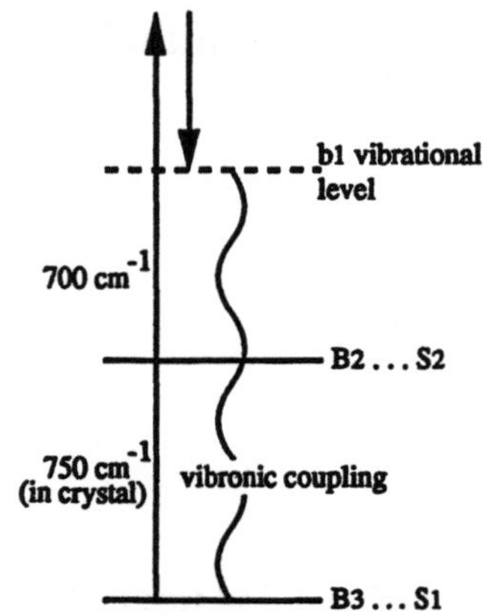

Fig. 2. A scheme of vibronic coupling between S1 and S2 through a b1 vibration.

origin which is different from that of the sharp S_1 bands. (3) Decay behavior of the band is similar to those of the S_1 bands. The initial state of the transition appears to be the 0 vibrational state of S_1 state. (4) Depolarization ratio of the band is much less than 0.75. This transition holds a totally symmetric character.

Putting these results together, we propose a scheme which assumes a vibronic coupling between the S_1 and S_2 states (Fig. 2). The symmetry species of the S_1 and S_2 states are B_3 and B_2, respectively, in D_{2h} point group and the energy separation of these states is 750 cm^{-1} in crystal [3]. Therefore the 0 vibrational state of S_1 and the b_1 vibrational state of S_2 are likely to be strongly vibronic coupled. The vibronic transition $B_2 \times b_1 \leftarrow B_3 \times a$ holds a totally symmetric character and is expected to have a considerable intensity borrowed from the $B_3 \leftarrow B_3$ transition (Rayleigh scattering). The large band width is ascribable to the fast electronic dephasing time associated with the $S_2 \leftarrow S_1$ transition. Further investigations to confirm this scheme are in progress.

References

[1] C. Kato, H. Hamaguchi and M. Tasumi, Chem. Phys. Lett., 120, 183 (1985).
[2] H. Hamaguchi in "Vibrational spectra and structure", Vol. 16, (J. R. Durig ed.) pp. 227, Elsevier (1987).
[3] R. M. Hochstrasser, R. D. McAlpine and J. D. Whiteman, J. Chem. Phys., 58, 5078 (1973).
[4] J. Murakami, M. Ito and K. Kaya, J. Chem. Phys., 74, 650 (1981).

Resonance Raman Spectra and Excited States of Psoralen and Coumarin

K. Tamagake[1], *M. Tsuboi*[2], *and K. Nakamoto*[3]

[1]Faculty of Pharm. Sciences, Okayama University, Tsushima, Okayama 700, Japan
[2]Department of Chemistry, Iwaki-Meisei University, Chuodai-Iiono, Iwaki 970, Japan
[3]Department of Chemistry, Marquette University, 535N 14th St., Milwaukee, WI 53233, USA

Crosslinking of double helix DNA by photocycloaddition of psoralen(PS) to pyrimidine bases is considered to be responsible for photochemical therapy of certain skin disorders. The action spectra for the therapy show that the excitation to the S_1 is more effective than that to the S_2 state. However, differences between these two excited states have not been well understood. Here we report on the RRS and PPP-CI of PS and coumarin relating to the structural changes of the photochemically active double bonds, 3,4 and 4',5', in these excited states.

Figure 1 shows the Raman spectra of coumarin and psoralen measured with 6 different excitation lines at 488, 355, 320, 309, 282 and 274nm. Most Raman lines are enhanced strongly when excited at 320 and 309nm in the 1st absorption band(S_1). On the other hand, only a few lines in the C=C and C-C stretcing regions are enhanced by excitation at 282 and 274nm in the 2nd absorption band(S_2). As a result, the spectra of PS and coumarin become very simple and similar to each other. This implies that the excitation energy in the S_1 state is well delocalized all over the molecule whereas it is quite localized around the pyron moiety in the S_2 state.

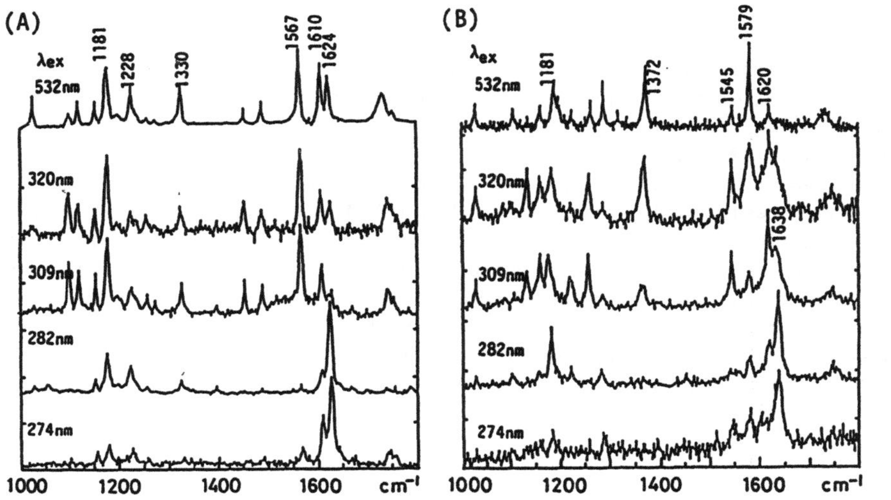

Fig.1 Raman spectra of coumarin(A) and psoralen(B) excited at various wavelength.

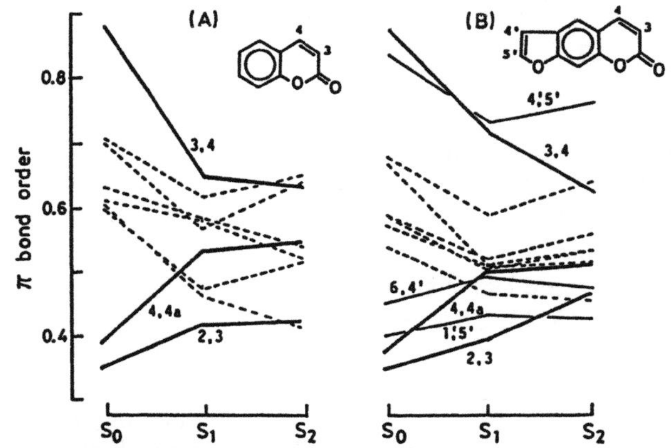

Fig.2 Dependence of bond orders on the electronic states of coumarin(A) and psoralen(B). Bonds belonging to pyron, benzene and furan groups are presented by thick solid, broken and thin solid lines, respectively.

Our MO study revealed this trend more clearly. Figure 2 shows the bond orders in the S_0, S_1 and S_2 states calculated by PPP-CI. The 3,4 double bond in the pyron moiety which is the strongest in the S_0 state become monotonically weaker as the molecule is excited to the higher states. The adjacent 2,3 and 4,4a single bonds also show monotonic changes but become stronger in this case. In contrast with the pyron ring, the most bonds belonging to the benzene and furan groups show maxima or minima in bond order at the S_1 state.

Based on such properties and assuming that Abrecht's A-term is most responsible for the Raman intensities at rigorous resonance, the band at 1624cm^{-1} of coumarin and 1638cm^{-1} of PS were assigned to the 3,4 C=C str. since these bands become the strongest when excited to the S_2 states by 282 and 274nm lines. The band at 1176cm^{-1} of coumarin and 1181cm^{-1} of PS are assigned to the 2,3 and/or 4,4a C-C str. by the same reason. Unlike the 3,4 C=C str., the 4'5' C=C str. is expected to be the strongest in the S_1 but it was difficult to distinguish this vibration from the benzene modes in 1500-1600cm^{-1} which are also expected to have similar excitation profiles. We tentatively assign the 1545cm^{-1} band to the 4',5' C=C str. by taking account of a normal mode analysis performed with an MNDO force field.

We have shown that the 4',5' and 3,4 bonds become the weakest in the S_1 and S_2 state, respectively, based on both RRS and PPP-CI studies. In general, the reduction in bond orders is attributed to an increase in the antibonding π^* character which is responsible for the photocycloadditive capability. Hence, we anticipate that the most reactive bonds in the S_1 and S_2 states are the 4',5' and 3,4 double bonds, respectively. This result may explain why the S_1 state is more effective in photochemical therapy than the S_2 state because the 4',5' bond must react first to form crosslinking with DNA as pointed out by Song et al.(Photochem.Photobiol., 29,1177(1979)).

Raman Spectra and Normal Coordinate Analysis of Radical Ions of trans-Stilbene and Some Ring-Fluorinated Derivatives

S. Schneider

Institut für Physikalische und Theoretische Chemie,
Universität Erlangen-Nürnberg, Egerlandstr. 3,
W-8520 Erlangen, Fed. Rep. of Germany

INTRODUCTION

The resonance Raman spectra of anion and cation radicals of trans-Stilbene (TS) generated by photoinduced electron transfer exhibit unexpectedly large differences in band location and intensity /1, 2/. Extensive studies performed later on with several deuterium substituted isotopomers /3/ allowed an empirical assignment of most of the "key vibrations". Nevertheless, a great number of open questions remained, e.g. whether the radical ions exhibit a non-planar geometry. Therefore, an attempt was made to establish a force field for the radical ions of TS and, later on, for the radical ions of ring-fluorinated trans-stilbenes (F-TS) /4/ in order to clarify band assignments.

FORCE FIELD CALCULATION

The force field calculations rely on one hand on the observed molecular frequencies of 13 d-substituted isotopomers of TS and of 4 ring-fluorinated derivatives, respectively (fig.1). On the other hand, the reliable force field established for ground state TS° /5/ is used as starting point in the calculations by an adapted Schachtschneider procedure /6/.
In accordance with suggestions in the literature we assume that both anions and cations of TS exhibit a planar geometry, and that all vibrations observed in the resonance Raman spectra are in plane vibrations. The set of internal (local) coordinates is chosen as in /5/.
It proved useful and sufficient to restrict the set of force constants subject to optimization to those describing the motions of the atoms in the ethylenic part of the molecules, because this is the most sensitive region affected by oxidation and reduction of TS and F-TS, respectively. The resulting force fields (cf. table 1) reproduce the experimental frequencies in general within $\pm$ 10 cm^{-1} /4, 7, 8/ (cf. table 2).

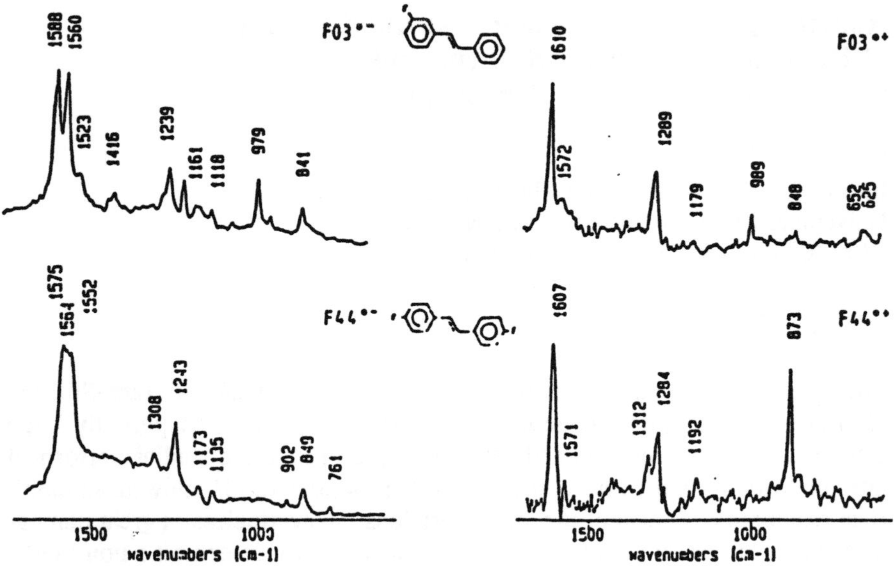

Fig.1 TR^3-spectra of radical ions of ring-fluorinated trans-stilbene

Table 1: Comparison of force constants related to the ethylenic part of the molecule (K = stretching in mdyn/A, H = bending in mdyn A/rad^2)

	K_T	K_R	K_Q	H_β	H_α
TS^0	6.23	4.76	8.37	1.13	1.12
TS^-	6.07	5.88	7.54	1.10	1.03
TS^+	6.37	4.20	7.98	0.90	1.11
F-TS^0	6.04	4.67	8.75	1.51	1.05
F-TS^-	6.03	5.32	7.92	1.03	1.07
F-TS^+	6.34	4.62	7.98	1.36	1.07

DISCUSSION

Both in TS and F-TS, the C=C stretch force constant, K_T, for ring vibrations changes in opposite direction upon oxidation and reduction whereas K_Q (ethylenic stretch) decreases markedly, as is expected from the simple MO picture.

Table 2: Comparison of observed and calculated (in []) C=C stretch mode frequencies (in cm^{-1}) and normal mode analysis. (D00 = TS, D44 = 4,4'-d_2-TS, F44 = 4,4'-F_2-TS, F03 = 3-F-TS).

	Anion		Cation	
	$(C=C)_e$	$(C=C)_{ph}$	$(C=C)_e$	$(C=C)_{ph}$
D00	1553 [1556]	1577 [1576]	1562 [1570]	1605 [1607]
	40Q+16T+23R +32α	60T+17R	39Q+26T+12R +26α	63T+10Q+10α
D44	1553 [1553]	1576 [1572]	1568 [1565]	1602 [1596]
	35Q+29T+20R	20T	29Q+44T	73T
F44	1552 [1557]	1575 [1574]	1571 [1562]	1607 [1608]
	44Q+30R+43α	74T	44Q+22T+23α	68T-10α
F03	1560 [1556]	1588 [1589]	1571 [1562]	1610 [1608]
	45Q+24R	68T	50Q+18T+10R+ 26α	69T+10α

The extent of variation is however different. The force constant K_R (C_e-C_{ph} "single" bond stretching) increases in accordance with expectation only upon reduction, but decreases upon oxidation of TS; the change in force constants from neutral to ionic species of F-TS is similar in size as for TS. Regarding H_α (C_e-H bending) and H_β (C_e=C_e-C_{ph} deformation) only H_β varies significantly upon oxidation and upon transition from TS to F-TS.
From this results one must conclude that the removal of one bonding π-electron effects the electronic structure in the region of the C_e-C_{ph}-bond differently than introduction of an additional electron in the antibonding, equivalent orbital. Fluorine-substitution on the rings changes significantly also the force constants of the ethylenic part of the molecule. Effects on vibrational spectra are therefore larger than expected just as consequence of different masses, and they are again different for cations and anions.
As has been pointed out for TS° /5/ the normal modes of the substituted derivatives can often be expressed as linear combination of the normal modes of the parent molecule (all protonated TS). This is also true for the cations and anions of TS. Several extreme variations in band intensity can be rationalized as being due to constructive (+) and destructive interaction (- linear combination) of intensity carrying modes.
From inspection of table 2 one must furthermore conclude that both the ethylenic and the ring stretching modes observed in the spectra are no pure modes. Remarkable is the sometimes high contribution of the C_e-C_{ph}-stretching coordi-

nate R and the C_e-H bending coordinate α. On the other hand, no normal mode is found in the 1000 - 1300 cm^{-1} range, which is dominated by coordinate R, that is to say, which could be assigned as C_e-C_{ph}-stretching mode.

SUMMARY

With the derived force field, the experimentally observed frequencies can be reproduced within ± 10 cm^{-1}. Due to strong substituent dependent mixing of the local coordinates unusual strong variations in band intensities can be explained without the assumption of a deviation of planar geometry.

Financial support by Fonds der Chemie is gratefully acknowledged.

REFERENCES

/1/ W.Hub, S.Schneider, F.Dörr, J.D.Oxman, F.D.Lewis, J.Am.Chem.Soc. 106, 708 (1984)

/2/ W.Hub, U.Klüter, S.Schneider, J.D.Oxman, F.D.Lewis, J.Phys.Chem. 88, 2308 (1984)

/3/ C.Richter, W.Hub, in Time-resolved Vibrational Spectroscopy, (A.Lauberau, M.Stockburger, Eds.) Springer, Berlin, 1985, p.137

/4/ R.Bug, PhD Thesis, Techn.Univ.München, 1990

/5/ G.Baranovic, Z.Meic, H.Güsten, J.Mink, G.Keresztury, J.Phys.Chem. 94, 2833 (1990)

/6/ J.H.Schachtschneider, Shell Technical Report No.57-65 (1965), modification by G.Baranovic

/7/ S.Schneider, C.Scharnagl, R.Bug, G.Baranovic, Z.Meic, Submitted to J.Phys.Chem.

/8/ The complete set of force constants is available upon request.

Resonance Raman Spectroscopic Studies of Organic Radical Ions

M. Aoyagi, M. Ata, Y. Gondo, H. Kihara, Y. Kubozono, and H. Ujita

Department of Chemistry, Faculty of Science, Kyushu University 33, Hakozaki, Higashiku, Fukuoka 812, Japan

1. Introduction

The highest occupied and lowest unoccupied molecular orbitals (HOMO and LUMO) are important in chemistry. In this context, we have undertaken resonance Raman (RR) spectroscopic studies on organic radical ions of phenanthrenes, phenazine, bipyridyls, pyrazine, tetramethylpyrazine, alkyl-substituted biphenyls, 4-phenylpyridine, and p-terphenyl [1].

2. Experimental

Raman spectra were recorded on a JEOL JRS-400T laser Raman spectrophotometer. An NEC GLG-3200 argon ion laser was used to excite Raman spectra. The radical anions were prepared by reduction with metallic potassium in THF solutions. N,N'-Dihydro-4,4'-bipyridyl radical cation ($PV^{+\cdot}$) was prepared by reduction with zinc dust in aqueous solution containing a small amount of acetic acid; a constant flow of nitrogen gas stirred the solution so that freshly-produced $PV^{+\cdot}$ could always be observed.

3. Results and Discussion

All RR bands observed are totally symmetric on the basis of the polarization measurements. In order to aid the interpretation of the RR spectra, we have also observed the electronic absorption and ESR spectra, and made molecular orbital calculations and normal coordinate analyses. In particular, the RR intensities have been calculated to be compared with the observed intensities. We have discussed the frequency shifts from parent to radical ion, resonance enhancement of the Raman scattering, and ground-state as well as excited-state molecular geometries. These results are significant in radical chemistry to make up for the limited scope of ESR spectroscopy. As typical examples, we describe briefly the results for 4-phenylpyridine radical anion ($PHPY^{-\cdot}$) and $PV^{+\cdot}$ [2,3] in comparison with 4,4'-bipyridyl radical anion ($BPY^{-\cdot}$) [1].

As can be seen from Table 1, on going from parent to radical, $PHPY^{-\cdot}$ shows the largest shift of the inter-ring stretching frequency among biphenyl and its related compounds we have ever studied. As for 4,4'-diethylbiphenyl radical anion, the inter-ring mode is too weak to be observed, probably owing to the mode mixing due to the mass effect of the ethyl groups. The upward shifts of the inter-ring stretching mode indicate the quinoidal ground-state structures of these radicals. For 4,4'-dimethylbiphenyl radical anion, the $C-CH_3$

Table 1. The inter-ring stretching frequencies ν in the radical anions of biphenyl and related compounds

Parent compound	ν/cm^{-1}	$\Delta\nu/cm^{-1}$ [a]	Rel. int.[b]
biphenyl[c]	1326	39	0.26(1587 cm^{-1})
4-methylbiphenyl[d]	1321	46	0.58(1590 cm^{-1})
4-ethylbiphenyl[d]	1300	21	0.10(1604 cm^{-1})
4-phenylpyridine[d]	1364	85	0.53(1586 cm^{-1})
4,4'-bipyridyl[d]	1350	52	0.73(1612 cm^{-1})
p-terphenyl[e]	1315	46	0.50(1595 cm^{-1})

a) Δν = ν(radical anion) - ν(parent molecule).
b) As compared with the mode specified in parentheses.
c) Observed in THF by exciting at 488.0 nm [4].
d) Observed in THF at room temperature by exciting at 488.0 nm.
e) Observed in THF at about -30°C by exciting at 488.0 nm.

Table 2. Raman frequencies observed for 4,4'-bipyridyl (BPY), $BPY^{-\cdot}$, and $PV^{+\cdot}$ (cm^{-1})

BPY	$BPY^{-\cdot}$	$PV^{+\cdot}$	
		This work	Ref. 3
1605	1612	1658	1660
1510	1509	1523	1525
1298	1350	1357	1358
1220	1230	----	1234
----	1043	1042	1045
1000	990	1000	1002
753	742	740	740
312	----	----	----

stretching mode at 1180 cm^{-1} is enhanced, which is interesting from the viewpoint of hyperconjugation.

$PV^{+\cdot}$ has also been studied in glycerol by Barker et al. [3], and their results are quite similar to ours except that we could not identify the 1234 cm^{-1} band, as shown in Table 2. Thus, our preparation procedure is appropriate and may be applicable to other unstable radicals. It may be noted that the 1040 and 1500 cm^{-1} bands are enhanced remarkably both in $BPY^{-\cdot}$ and $PV^{+\cdot}$. Both $BPY^{-\cdot}$ and $PV^{+\cdot}$ are quinoidal in the ground state.

References

1. For example, H. Kihara, Y. Gondo: J. Raman Spectrosc. 17, 263 (1986)
2. M. Aoyagi, Y. Kubozono, M. Ata, Y. Gondo: Chem. Phys. Lett. 131, 201 (1986)
3. D. J. Barker, R. P. Cooney, L. A. Summers: J. Raman Spectrosc. 18, 443 (1987)
4. C. Takahashi, S. Maeda, Chem. Phys. Lett. 24, 584 (1974)

Transient Infrared Spectroscopy of Photochromic Spiropyran-Photomerocyanine Systems in Solution

T. Yuzawa[1], *H. Takahashi*[1], *C. Kato*[2], *and H. Hamaguchi*[2]

[1]Department of Chemistry, School of Science and Engineering, Waseda University, Tokyo 169, Japan
[2]Molecular Spectroscopy Laboratory, Kanagawa Academy of Science and Technology (KAST), KSP East 301, 100-1 Sakato, Takatsuku, Kawasaki 213, Japan

1. Introduction

A number of investigations have been reported on the photochromism of spiropyrans. However, the structure of the colored transient species, the photomerocyanines, has not been well understood. Photomerocyanines are expected to have open-ring structures contributed by two resonance structures, the zwitterion form and the ortho-quinoid form (figure 1). We have previously reported the transient resonance Raman spectra of BIPS (1',3',3' -trimethylspiro[2H-1-benzopyran-2,2'-indoline]) in various solvents [1]. We report here on the transient infrared study of the BIPS-photomerocyanine system in solution.

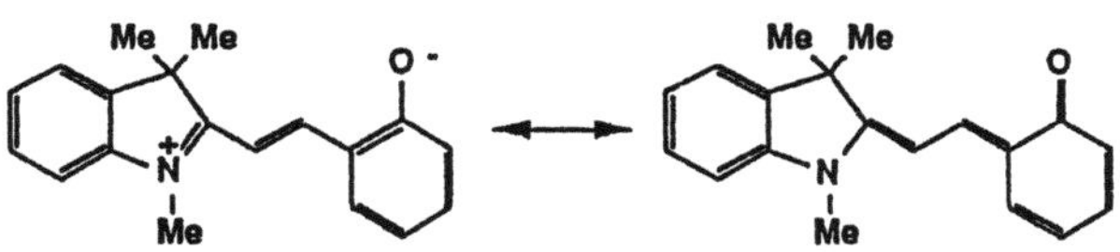

Figure 1. The zwitterion form and the ortho-quinoid form of the photomerocyanine.

2. Experimental

An AC-coupled microsecond time-resolved infrared spectrometer constructed at KAST [2] was used with time resolution of 80 μs. The BIPS molecule was photoexcited by the forth harmonic of Q-switched Nd:YAG laser (266 nm, 50 Hz, 10 mW). The sample solution was circulated in a KBr flow cell. The concentration was 5.0×10^{-2} mol dm^{-3}.

3. Results and discussion

Figure 2a shows the transient infrared difference spectrum of BIPS in cyclohexane. The positive (upward) peaks are ascribed to the photomerocyanine generated from BIPS. The negative (downward) peaks correspond to the depletion of the closed form (ground state) BIPS. It is found that the positive peaks are more intense than the negative peaks. This means that the infrared intensities of the photomerocyanine are much higher than those of closed form BIPS. The C=O stretch of quinoids, which usually appears at around 1650 cm^{-1}, is not found in the spectrum. The absence of

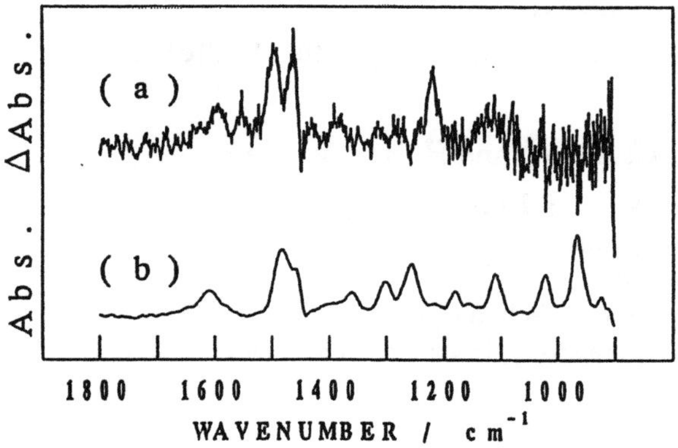

Figure 2. (a) Transient infrared difference spectrum of BIPS in cyclohexane. (b) Infrared spectrum of the closed form BIPS in cyclohexane.

the band in this region may be ascribed to the contribution of the zwitterion form to the resonance hybrid structure of the photomerocyanine. The strong band at 1500 cm^{-1} is also observed in the transient resonance Raman spectrum. This band is assignable to the C=C stretch of the quinoid ring. The combination of assigned infrared and Raman frequencies will give more reliable structural information on the photomerocyanine.

References

[1] H. Takahashi, K. Yoda, H. Isaka, T. Ohzeki, and Sakaino, Chem. Phys. Lett. 140, 90 (1987)
[2] K. Iwata and H. Hamaguchi, Appl. Spectrosc. 44, 1431 (1990)

Structures of Spiropyranic Photomerocyanines Investigated by Time-Resolved Resonance Raman Spectroscopy

Y. Nagasawa, S. Hashimoto, and H. Takahashi

Department of Chemistry, School of Science and Engineering, Waseda University, Tokyo 169, Japan

1. Introduction

Spiropyrans are known to exhibit photochromism in solution, however, the configurations of transient species, photomerocyanines, produced in the photochromic reaction have not been well understood. In this paper, the configurations of the photomerocyanines of 1',3',3'-trimethylspiro[2H-1-benzopyran-2,2'-indoline] (BIPS) and 2,2'-spirobi[2H-1-benzopyran] (SBP) have been investigated by time-resolved resonance Raman spectroscopy with the aid of complexation with metal ions.

2. Results and Discussion

As shown in Fig. 1 the resonance Raman spectrum of photo-produced transient merocyanine of BIPS in acetonitrile is drastically different from that in methanol. However, by adding

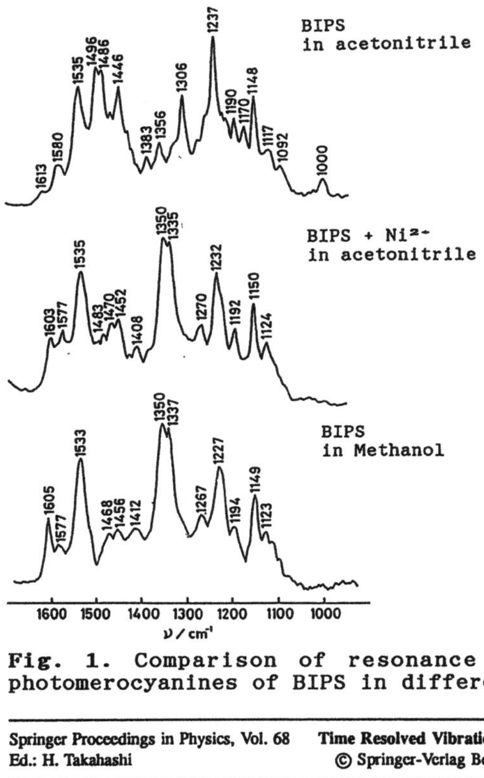

Fig. 1. Comparison of resonance Raman spectra of transient photomerocyanines of BIPS in different solvents.

Springer Proceedings in Physics, Vol. 68 Time Resolved Vibrational Spectroscopy V
Ed.: H. Takahashi

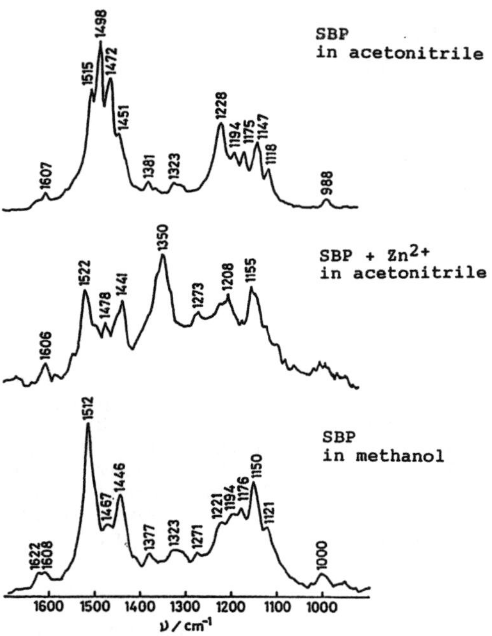

Fig. 2. Comparison of resonance Raman spectra of transient photomerocyanines of SBP in different solvents.

equimolar Ni^{+2} and measuring 800 µs after UV irradiation, acetonitrile solution of BIPS gives the resonance Raman spectrum essentially identical to that of photomerocyanine in methanol. Obviously the spectral change can be attributed to the formation of the Ni^{2+}-photomerocyanine complex. Since in this complex the photomerocyanine is considered to take the zwitterionic form rather than the ortho-quinoidal form, the above results indicate that the photomerocyanine of BIPS takes the zwitterionic form in methanol on account of hydrogen bond formation between phenolate oxygen of the photomerocyanine and hydroxy hydrogen of solvent methanol. The strong Raman band at 1350 cm^{-1} may be regarded as a characteristic band of the zwitterionic form.

In contrast, the resonance Raman spectrum of photomerocyanine of SBP in methanol is quite different from that of its Zn^{2+} complex (Fig. 2). It should be noted that the spectrum of the SBP-photomerocyanine Zn^{+} complex has a close resemblance to that of the BIPS-photomerocyanine Ni^{2+} complex and exhibits a band at 1350 cm^{-1} which is characteristic to the zwitterionic form. These facts indicate that while the photomerocyanine of SBP is in the zwitterionic form in the Zn^{2+} complex, it does not take the zwitterionic form even in strong hydrogen-bond-donor solvents like methanol. This may be attributable to the unstable nature of the zwitterionic form of SBP in which the oxygen atom in the benzopyran ring has to be charged positively.

The resonance Raman spectrum of the photomerocyanine of SBP in acetonitrile excited by 450 nm light closely resembles that of coumarin as shown in Fig. 3. The resonance Raman spectrum of the photomerocyanine of SBP is known to exhibit strong dependence on the excitation wavelength: at wavelength longer than 460 nm, the

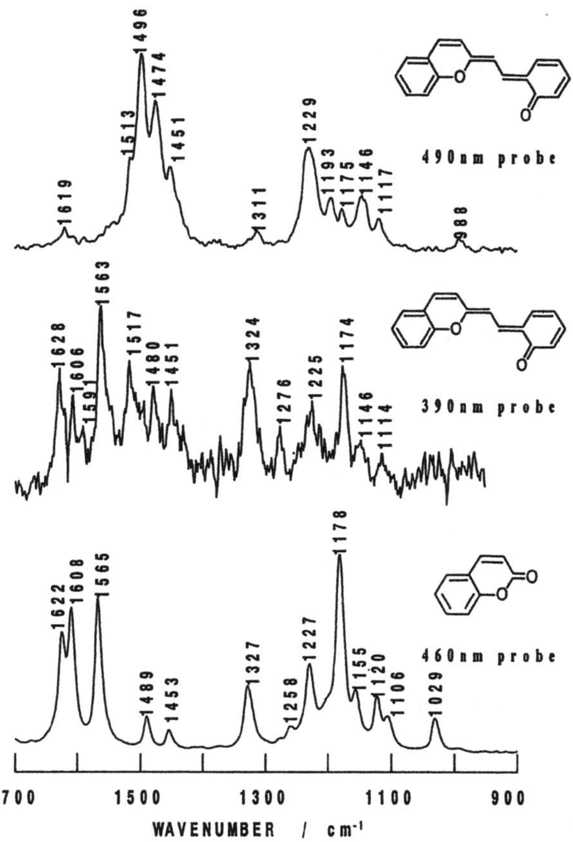

Fig. 3. Comparison of resonance Raman spectra of the photomerocyanines of BIPS in acetonitrile with the Raman spectrum of coumarin.

vibrations of open benzopyran part are strongly enhanced, while at wavelength shorter than 460 nm, only the vibrations of intact (not cleaved) benzopyran part are enhanced [1]. The close resemblance between the Raman spectra of the photomerocyanine of SBP and coumarin suggests that the photomerocyanine of SBP takes the ortho-quinoidal form both in acetonitrile and in methanol; the spectral difference between these two solutions may be attributable to the difference in the relative population of configurational isomers which is known to be very sensitive to the polarity of the solvent [2,3].

References

1. S.Hashimoto, A.Shimojima, T.Yuzawa, H.Hiura, J.Abe and H.Takahashi, J. Mol. Struct. **242**, 1 (1991)
2. H.Takahashi, K.Yoda, H.Isaka, T.Ohzeki, Y.Sakaino, Chem. Phys. Lett. **140**, 90 (1987)
3. H.Takahashi, H.Murakawa, Y.Sakaino, T.Ohzeki, J.Abe, O.Yamada, J. Photochem. Photobiol. **A 45**, 233 (1988)

Raman Spectroscopic Studies of Long-Lived Colored Merocyanine Conformers in the Aggregates: A Poor-Man's Time-Resolved Study

Y. Onai, M. Mamiya, M. Kobayashi, H. Shinohara, and H. Sato

Department of Chemistry for Materials, Faculty of Engineering, Mi'e University, Tsu 514, Japan

Raman spectroscopic studies on long-lived colored merocyanine conformers are made [1]. Colored merocyanine conformers, short-lived ($\lesssim$ seconds) in fluid solutions, survive for days in the large aggregates.

UV irradiation of uncolored (spiropyran) form of 6-nitro-1',3',3'-trimethylspiro[2H-1-benzopyran-2,2'-indoline] (6-nitro BIPS) in nonpolar solvents (e.g. cyclohexane) gave a long-lived (more than days) colored aggregates, which were larger in size (10-200 $\mu m\phi$) compared to "quasicrystals" of Krongauz and Parshutkin [2]. Resonance Raman spectra of the long-lived colored species in the aggregates were measured in KBr discs. As an example, the spectrum of the aggregate formed on UV irradiation (30 min) of 6-nitro BIPS in cyclohexane solution is shown in Fig. 1.

The spectra were very similar to the transient spectra of fluid solutions of 6-nitro BIPS obtained by Takahashi et al.[3] with time-resolved resonance Raman spectroscopy. Additional bands were found near 1190, 1281 and 1532 cm^{-1}. As the structure of merocyanine conformers in the aggregate, the aci-nitro structure coordinated each other through the NO_2 group is proposed. The three bands peculiar to the aggregate are attributed to the unequivalent N-O stretching modes in the coordinated aci-nitro structure of the colored merocyanine conformers in the aggregates.

The aggregate grows during UV irradiation. They remain in solution when they are small in size. On prolonged irradiation

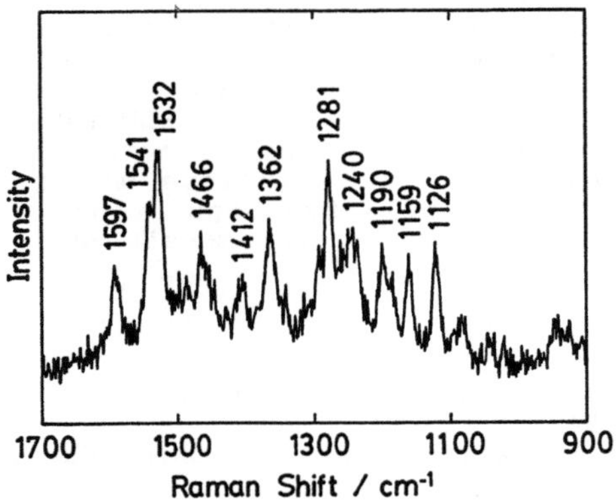

Fig. 1. Resonance Raman spectrum (in a KBr disc) of the long-lived aggregate formed on UV irradiation (30 min) of 6-nitro BIPS in cyclohexane solution.

Springer Proceedings in Physics, Vol. 68 Time Resolved Vibrational Spectroscopy V
Ed.: H. Takahashi

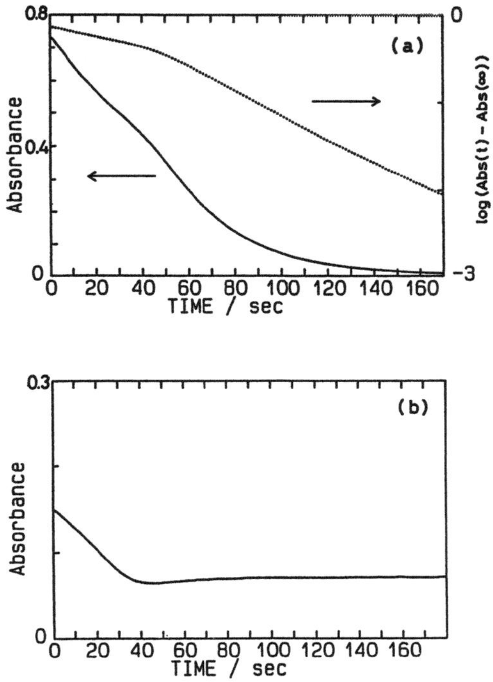

Fig. 2. Temporal variation of absorbance of the UV-irradiated (for 30 s) cyclohexane solution of 6-nitro BIPS (1.0 x 10^{-3} M). Measurement started at 20 s after irradiation was stopped. (a) at 580 nm (merocyanine monomer band), (b) at 700 nm (aggregate). Temperature: 20 ℃

they begin to precipitate. Before the large aggregates as mentioned above precipitate, aggregates of smaller aggregation number are formed in solution. On the UV irradiation of intermediate length (30-60 s) of 6-nitro BIPS in heptane or cyclohexane, the aggregate absorbing around 700 nm was formed. This seems to correspond to the "charge-transfer complex" (CTC) reported by Krongauz and Parshutkin [2]. We have found [4] anomalous non-exponential decay behavior of absorbance at the merocyanine monomer band (Fig. 2) in conditions that the aggregate absorbing near 700 nm is present in solution. This occurs concomitantly with the decay of aggregate absorbing near 700 nm. This can be interpreted by partial decomposition of the aggregates into merocyanine monomers.

In conclusion, in the large aggregates of 6-nitro BIPS the merocyanine conformers are preserved for days. For the aggregates smaller in size (with an absorption band near 700 nm), partial deaggregation to merocyanine monomers is observed.

References

[1] Briefly reported in: Y. Onai, K. Kasatani, M. Kobayashi, H. Shinohara and H. Sato, Chem. Lett., 1809 (1990).

[2] V. A. Krongauz and A. A. Parshutkin, Photochem. Photobiol., 15, 503 (1972); A. A. Parshutkin and V. A. Krongauz, Mol. Photochem., 6, 437 (1974).

[3] H. Takahashi, H. Murakawa, Y. Sakino, T. Ohzeki, J. Abe and O. Yamada, J. Photochem. Photobiol. A. Chemistry, 45, 233 (1988).

[4] H. Sato, H. Shinohara, K. Kobayashi and T. Kiyokawa, Chem. Lett., submitted.

Time-Dependent SERS Study: An Implication of SERS Mechanism

M. Takahashi and M. Ito

Department of Chemistry, Faculty of Science and Technology, Keio University, Hiyoshi 3-14-1, Kohoku-ku, Yokohama 223, Japan

Surface-enhanced Raman scattering (SERS) is highly sensitive to adsorbed species on roughened silver surfaces. It is well-known that two kinds of roughness are necessary for SERS; i.e. large scale roughness of an order of 100 A and small scale roughness of an order of atoms. The former is known to give large electromagnetic field enhancement. We studied recently the contribution of the latter roughness, small scale roughness, formed by the electrochemical treatment and concluded that the latter is originated from atomic scale metal clusters[1]. It is exceedingly important to reveal the role of the clusters.

It has been suggested that the cluster acts as a site for charge-transfer(CT) resonance between adsorbates and metal substrate. To prove the CT resonance, we must determine the Raman excitation profile of an adsorbed molecule. In electrochemical systems, the separation of energy levels of adsorbed molecules and metal can be controlled by changing the electrode potential, accordingly, the excitation profile must be obtained by tuning the electrode voltage. Observed SERS intensity, however, is subjected by following three factors; (1) the scattering cross section of the adsorbate which is directly concerned with the excitation profile, (2) the number of adsorbates and (3) the number of SERS active sites. Therefore, to discriminate these factors, we measured the time response of Raman intensity to stepping electrode potentials, based on the fact that diffusion controlled process is slower than double layer charging[2].

The SERS intensities of pyridine, benzene and 4-cyanopyridine adsorbed on a silver electrode were measured in an I^- (10^{-3} M) solutions. Raman excitation wavelength was 514.5 nm of Ar^+ ion laser or 645 nm of dye laser with the power of 50 mW. Figure 1(a) shows the Raman time response of the intensity due to the ring breathing mode (984 cm^{-1} [3]) of benzene adsorbed in an I^- solution induced by the potential change between -0.95 V and -0.55 V. Within this potential region, SERS intensity changes reversibly, and reached a maximum at -0.95V.

At more negative potential than -0.95V, the occurrence of I^- desorption leads to decrease in the number of SERS active site, resulting in the irreversible decrease in SERS intensity. Figure 1-(b) shows the SERS intensity change caused by the negative potential step to -1.45V from -0.95V. Just after the potential step to negative value, the SERS signal increased in intensity. One second later, it reached the maximum intensity and then, started to decrease in intensity. The gradual decrease is explained by the reduction of SERS active site, which is originated from diffusion

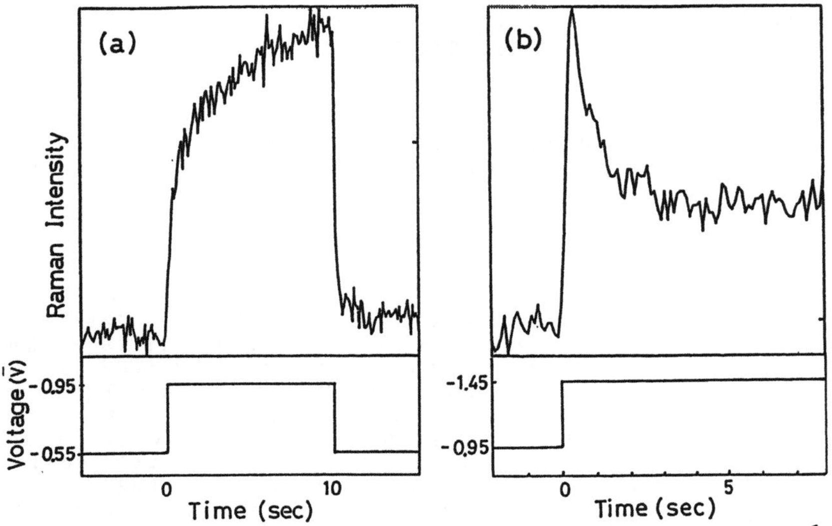

Figure 1. Time response of SERS intensity due to 984 cm^{-1} band of adsorbed benzene in I^- aqueous solution. Potential is stepped (a) from -0.55V to -0.95V and back to -0.55V, and (b) from -0.95V to -1.45V, respectively.

process, while the steep component is largely due to the increase in scattering cross section of adsorbed molecule. Therefore, the step potential change is a useful method to estimate the real Raman cross section of adsorbed species. Such sharp intensity maximum was observed until -1.65V which is the negative potential limit for detection of SERS in this system. Therefore, the scattering cross section increased monotonously up to the detection limit, which indicates that real SERS intensity per one molecule continues to increase up to negative potential limit. No net maximum of the cross section of adsorbed species could be found in present potential range for these molecules, implying that the resonance state for the Raman scattering of adsorbed species is located much higher than 514.5 nm excitation. Therefore, we suggest that, there is no CT level at around 500 nm. In previous SERS result of the small scale silver clusters on atomically smooth Pt(111) surface where large scale roughness is missing [1], clusters must act as antennas for laser light and strengthen remarkably the electromagnetic field resulting in the large enhancement of SERS intensity.

References

[1] M.Takahashi and M.Ito, J.Electron Spectry., 54/55 (1990) 913.
[2] D.L.Jeanmaire and R.P.Van Duyne, J.Electroanal.Chem., 84 (1977) 1.
[3] M.Takahashi, H.Idetsuki and M.Ito, Surf.Sci., 232 (1990) 346.

The Jet-Cooled Fluorescence Excitation Spectra, Potential Energy Functions, and Conformations of 2-Cyclopenten-1-one, Propanal, and Propynal in the $S_1(n,\pi^*)$ Electronic Excited State

C.M. Cheatham and J. Laane

Department of Chemistry, Texas A&M University,
College Station, TX 77843, USA

The jet-cooled fluorescence excitation spectra of 2-cyclopenten-1-one and its 5,5-d_2 isotopomer have been recorded in the 370 to 340 nm region and a portion of the spectrum for the undeuterated species is shown in Figure 1. The electronic origin for this molecule occurs at 27210 cm^{-1} for the $S_1(n,\pi^*)$ electronic excited state. More than fifty separate frequencies were observed and compared to earlier absorption measurements [1]. Many of the transitions involve combinations with the ring-puckering, ring-twisting, or the carbonyl bending modes. The vibrational frequencies for the three carbonyl motions and the nine ring modes were determined for the excited state, and these are compared to ground state values [2] in Table I. Bands at 67, 158, and 256 cm^{-1} for the d_0 species, at 63, 147, and 240 cm^{-1} for the 5-d_1 isotopomer, and at 59, 138, and 227 cm^{-1} for the d_2 species were assigned to the ring-puckering motion in the S_1 state, and the transitions corresponding to these are shown in Figure 2. A single one-dimensional potential energy function,

$$V(\text{cm}^{-1}) = 2.5\times10^6x^4 + 1.3\times10^3x^2, \tag{1}$$

where x is the puckering coordinate, accurately fit the data for all three isotopomers. This function is nearly purely quartic in character and shows the ring to be planar in the electronic excited state. However, it has become less rigid, and this is ascribed to a decrease in initial angle strain within the ring resulting from decreased

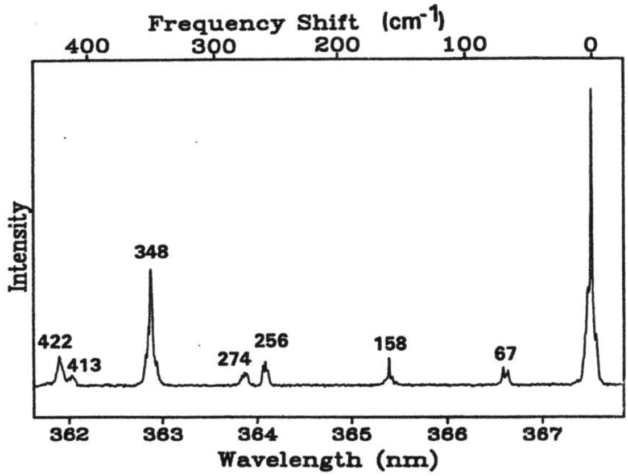

Figure 1. Fluorescence excitation spectrum of 2-cyclopenten-1-one

Table I. Vibrational frequencies (cm^{-1}) for the ground and excited $S_1(n,\pi^*)$ states of 2CP-d_0 and 2CP-d_2.

Approx. Description	2CP-d_0 Ground	2CP-d_0 Excited	2CP-d_2 Ground	2CP-d_2 Excited
ν_5 C=O stretch	1748	1357	1743	1360
ν_6 C=C stretch	1599	1418	1602	1421
ν_{13} Ring mode	1094	1037	1114	1037
ν_{14} Ring mode	999	974	957	960
ν_{15} Ring mode	912	906	851	843
ν_{16} Ring mode	822	849	810	814
ν_{17} Ring mode	753	746	746	727
ν_{18} Ring mode	630	587	630	580
ν_{19} C=O def (‖)	464	348	449	338
ν_{26} α-CH bend	750	768	737	762
ν_{28} C=O def (⊥)	537	422	?	403
ν_{29} C=C twist	287	274	281	267
ν_{30} Ring-puckering	94	67	85	59

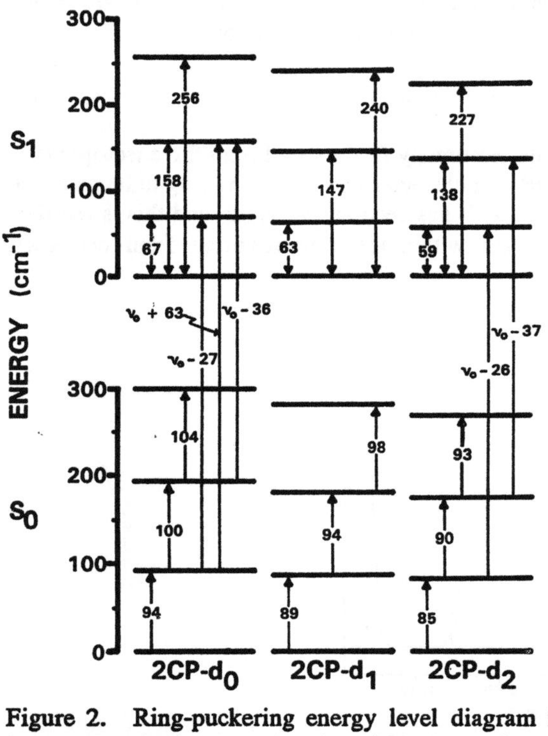

Figure 2. Ring-puckering energy level diagram for 2-cyclopenten-1-one and its isotopomers in the ground and excited electronic states.

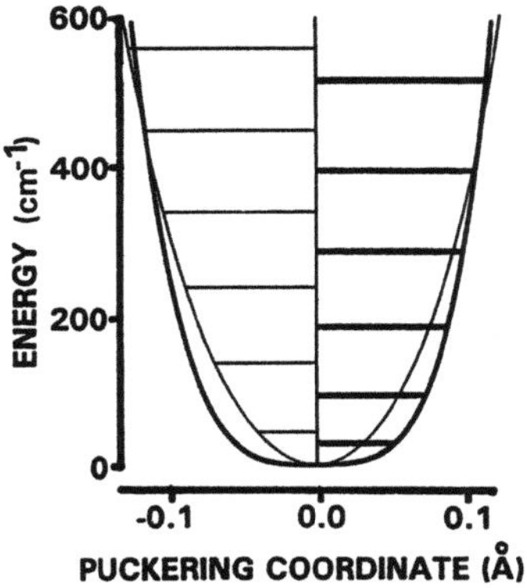

Figure 3. Potential energy functions and energy levels for the ring-puckering vibrations of 2-cyclopenten-1-one in the ground (thin lines) and $S_1(n,\pi)$ excited (thick lines) electronic states.

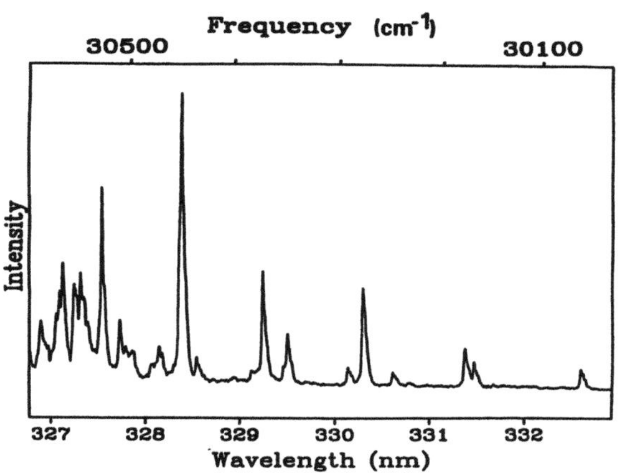

Figure 4. Fluorescence excitation spectrum of propanal

conjugation. Figure 3 compares the ring-puckering vibrations for the ground and excited states of this molecule.

The C=C and C=O stretching frequencies occur at 1418 and 1357 cm^{-1} for the d_0 molecule in the S_1 excited state, and these values have decreased by 181 and 391 cm^{-1}, respectively, as compared to the ground state. Thus, the $n \rightarrow \pi^*$ electronic transition is to an antibonding orbital which is concentrated on the carbonyl group. The ring-twisting frequency for the S_1 state occurs at 274 cm^{-1}. Previous electronic absorption measurements had resulted in a misassignment for this motion [1].

The jet-cooled fluorescence excitation spectra of propanal-d_0, 1-d_1, and 1,1-d_2 and of propynal have also been recorded, and that for the undeuterated popanal is shown in Figure 4. Propanal has two stable conformations, a *cis* and a *gauche,* and two internal rotations which complicate its spectral analysis. In the S_1 electronic excited state the CHO out-of-plane bend has a double-minimum potential function so that the carbonyl group is bent out of the plane of the three carbon atoms. For propynal the electronic band origin shows up as an intense band at 26162 cm^{-1}. A series of bands arising from the carbonyl wag show that the molecule retains its planar skeleton in the S_1 excited state.

REFERENCES

1. R. D. Gordon and D. R. Orr, J. Mol. Spectrosc., **129**, 24 (1988).
2. C. M. Cheatham and J. Laane, J. Chem. Phys., **94**, 5394 (1991).

Part V

Excited State Dynamics

Ultrafast Laser Photolysis Studies on Photoinduced Electron Transfer and Related Relaxation Phenomena

N. Mataga

Department of Chemistry, Faculty of Engineering Science, Osaka University, Toyonaka, Osaka 560, Japan

1. Introduction

Photoinduced electron transfer (ET) is the most important fundamental process in photochemical and photobiological reactions. We have been studying the dynamics and mechanisms of the following ET or charge transfer (CT) reaction processes with femtosecond-picosecond laser photolysis and time-resolved spectroscopy.

(a) The charge separation (CS) at encounter between fluorescer and electron donating or accepting quencher leading to the formation of geminate ion pair (IP) which undergoes charge recombination (CR) and dissociation in strongly polar solutions. In this case the geminate IP seems to have a loose structure (loose IP, LIP) with intervening solvent molecules between the paired ions [1-3].

(b) Intramolecular photoinduced CS and CR of the intramolecular IP or CT state in the D, A system linked by insulating spacer or directly by single bond [4,5].

(c) Excitation of the ground state CT complex, $(A^{-\delta} \cdot D^{+\delta})_S$, and relaxation from the excited Franck-Condon (FC) state, $(A^{-\delta'} \cdot D^{+\delta'})_S^{FC}$, leading to the formation of compact or contact IP (CIP) which undergoes CR and dissociation or transformation to LIP followed by dissociation [6-9].

(d) Slow electron ejection into the surrounding polar solvent or ET to the near-by aggregate of solvent molecules from the fluorescent state of solute molecule with low ionization potential like TMPD(N,N,N',N'-tetramethyl-p-phenylenediamine) [10-13].

(e) Ultrafast electron ejection from solute molecule into surrounding nonpolar solvent by multiphoton excitation with ultrashort laser pulse which leads to the formation of short lived cation-electron pair [14,15].

In the following, we discuss the problems (b), (c) and (d).

2. Results and Discussion

2.1 Photoinduced CS in Linked Systems

We have made femtosecond laser photolysis and time-resolved spectral studies on the systems: p-$(CH_3)_2$N-Ph-$(CH_2)_n$-(1-pyrenyl)(Pn, n=1,2,3), p-$(CH_3)_2$N-Ph-$(CH_2)_n$-(9-anthryl)(An, n=1, 2,3), Ph-N(CH_3)-CH_2-(9-anthryl)(9-AnMe), 9,9'-bianthryl(BIAN) and (9-anthryl)-(N-carbazolyl) (C9A) in alkanenitrile solution [4,5]. In Pn, An and 9-AnMe, we have clearly observed the CS process from the excited state localized in anthracene or pyrene part to the intramolecular IP state. The rise times(τ_{CS}) of the IP state in Pn and An (n=1,2,3) are considerably longer than solvent τ_L and also longer than τ_S except A_1 where τ_{CS} is rather close to τ_S. For example, values of τ_{CS} in acetonitrile, butyronitrile and hexanenitrile were 0.65 ps, 1.0 ps and 1.4 ps for A_1 and 1.7 ps, 2.5 ps and 4.5 ps for P_1, compared with the τ_L values of 0.19 ps, 0.53 ps and 0.98-1.1 ps, respectively.

Therefore, there seems to be intrinsic barrier for the photoinduced ET of Pn and An in alkanenitrile solution other than that for solvent orientation. In this respect we have examined the effect of the excess vibrational energy on the ET process by changing the wavelength of the exciting laser pulse from 355 nm to 295 nm in the case of P_1 and A_1 in hexanenitrile solutions. However, the observed τ_{CS} value was practically the same, indicating that the excess vibrational energy is not effectively used for facilitating the ET and, probably, distributed very rapidly to a multitude of vibrational freedoms before ET, leading to the final dissipation.

On the other hand, 9-AnMe shows a very rapid photoinduced CS, with τ_{CS} value of 0.7 ps in hexanenitrile solution. This value is shorter than τ_L and much shorter than τ_S of hexanenitrile. The mechanism of this result in 9-AnMe is not very clear at present, it may be possible that the ET process is facilitated by coupling with intramolecular vibrations [16].

When the two chromophores are directly combined by single bond as in BIAN and C9A, more or less electronic delocalization interaction between two groups becomes possible. We have observed by femtosecond absorption spectroscopy that the rise curve of the CS state of these systems in alkanenitrile contains slow component with τ_{CS} longer than τ_S in spite of the stronger interaction. This may be due to the structural change necessary to minimize the delocalization interaction between two groups to attain the more complete CS.

2.2 Photoinduced CS in CT Complexes

An extreme case of the photoinduced CS process in the strongly interacting D, A systems is that of the excited CT complexes. We have examined the excited state behaviors of various kinds of CT complexes with femtosecond laser photolysis and time-resolved absorption spectral measurements [6-9].

According to our measurements and the results of the previous MO calculations, the photoinduced CS in the excited state of TCNB (1,2,4,5-tetracyanobenzene) in benzene and methylsubstituted benzene (toluene, mesitylene) solutions seems to take place by two steps [6,7]. First, a slight configuration change within 1:1 complex accompanied with a slight enhancement of CT degree takes place with time constant $\tau_d \sim 1$ ps. This configuration change in those less polar solvents is not sufficient to attain the complete CS, but the formation of 1:2 complex is necessary.

$$(A^{-\delta}\cdot D^{+\delta}) \xrightarrow{h\nu} (A^{-\delta'}\cdot D^{+\delta'})^* \xrightarrow[\text{structural change within 1:1 complex}]{\tau_d}$$

$$(A^{-\delta''}\cdot D^{+\delta''})^* \xrightarrow[D]{\tau_r} (A^{-}\cdot D_2^{+})^* \qquad (1)$$

where τ_d=2 ps, τ_r=20 ps for benzene solution, τ_d=1.5 ps, τ_r=30 ps for toluene solution and τ_d=550 fs, τ_r=40 ps for mesitylene solution. In strongly polar solvent acetonitrile, the photoinduced CS takes place within 1 ps to a considerable extent without the 1:2 complex formation due to the solvent orientation but not completely. For the complete CS, intracomplex structural change within 1:1 complex is necessary. It takes ca. 20 ps for the complete CS to form the CIP in the case of TCNB-toluene complex in acetonitrile solution.

We have examined the TCNB in toluene solution, comparing the result obtained by 295 nm fs pulse excitation with that obtained by 355 nm fs pulse excitation. The observed spectral changes and time constants were the same, which indicated that the excess vibrational energy was not effectively used for the change of the intracomplex configuration.

2.3 Electron Transfer from Excited Solute to the Transient Aggregate of Polar Solvent Molecules

In order to elucidate the mechanism of this ET reaction, we have examined the temperature effect [12] as well as the effect of the excess vibrational energy [13] on this reaction. We have observed clearly that the yield of the IP state formation becomes significantly small and the fluorescence lifetime becomes considerably longer by temperature lowering, leading to the large decrease of CS rate at low temperatures, and the activation energy was obtained to be 7.4 kcal/mol. This result of the temperature effect on the CS process means that the nature of the solvent aggregate or cluster which acts as electron acceptor is not static but very dynamic one formed by rapid fluctuations of surrounding nitrile solvent molecules. When a favorable aggregate of solvent molecules which can accept electron from the relaxed S_1 state is formed near-by the excited solute, the ET will be able to take place.

We have examined also the effect of the excess vibrational energy on this ET process by changing the wavelength of the exciting picosecond dye laser pulse. The excitation of TMPD at 355 nm is very near to the 0-0 transition. The monophotonic ionization yield increases significantly with increasing excitation energy near the origin of the S_1 state and, in the region around 3×10^4 cm^{-1}, it is almost constant. It should be emphasized here that the yield obtained by excitation at 3.39×10^4 cm^{-1} (295 nm) is less than that obtained by excitation around 3.08×10^4 cm^{-1} (325 nm). If ivr (intramolecular vibrational redistribution) is much faster than the ionization from the vibrationally unrelaxed state, the ionization yield should be determined simply by the amount of the excess vibrational energy contrary to the observed result. Observed result indicates that the ionization can compete with ivr. The lower ionization yield when excited at 3.39×10^4 cm^{-1} (295 nm) which is the onset of S_2 state suggests that the ionization yield depends on the optically excited vibrational mode or, if the internal conversion occurs before ionization, the accepting mode of $S_2 \rightarrow S_1$ internal conversion is different from the promoting mode of ionization.

As we have discussed in 2.1 and 2.2, in the case of the D, A systems combined by small spacer, P_1 and A_1, and also in the case of the strongly interacting D, A systems of the charge transfer complex, TCNB-toluene, no effect of the excess vibrational energy was recognized in the photoinduced CS processes. However, its effect was quite remarkable in this special case of very weakly interacting system.

Acknowledgment

The author acknowledges the partial support by a Grant-in-Aid for the Specially Promoted Research (No. 62065006) from the Ministry of Education, Science and Culture.

References

[1] N. Mataga, Y. Kanda, T. Asahi, H. Miyasaka, T. Okada and T. Kakitani, Chem. Phys., 127, 239 (1988).
[2] S. Nishikawa, T. Asahi, T. Okada, N. Mataga and T. Kakitani, Chem. Phys. Lett., submitted.
[3] N. Mataga, T. Asahi, Y. Kanda, T. Okada and T. Kakitani, Chem. Phys., 127, 249 (1988).
[4] N. Mataga, S. Nishikawa, T. Asahi and T. Okada, J. Phys. Chem., 94, 1443 (1990).
[5] S. Nishikawa, T. Okada, N. Mataga and W. Rettig, in preparation.
[6] H. Miyasaka, S. Ojima and N. Mataga, J. Phys. Chem., 93, 3380 (1989).
[7] S. Ojima, H. Miyasaka and N. Mataga, ibid., 94, 4147, 5834, 7534 (1990).
[8] T. Asahi and N. Mataga, ibid., 93, 6575 (1989).
[9] T. Asahi and N. Mataga, ibid., 95, 1956 (1991).
[10] Y. Hirata and N. Mataga, ibid., 87, 1680 (1983).
[11] Y. Hiratga, N. Mataga, Y. Sakata and S. Misumi, ibid. 87, 1493 (1983).
[12] Y. Hirata, Y. Tanaka and N. Mataga, in preparation.
[13] Y. Hirata, M. Ichikawa and N. Mataga, J. Phys. Chem., 94, 3872 (1990).
[14] Y. Hirata and N. Mataga, ibid., 93, 7539 (1989).
[15] Y. Hirata and N. Mataga, ibid., 95, 1640 (1991).
[16] H. Sumi and R. A. Marcus, J. Chem. Phys. 84, 4894 (1986).

Iodine Kinetic Energy: A Monitor for Dynamics of and Structure of Radicals Produced in the Photodissociation of Organic Iodides

H.J. Hwang, J. Freitas, and M.A. El-Sayed

Department of Chemistry and Biochemistry, University of California, Los Angeles, CA 90024-1569, USA

Abstract. The technique of state-selective one dimensional photofragmentation translation energy spectroscopy is developed and the velocity of the one photon photodissociated iodine in the $^2P_{3/2}$ and $^2P_{1/2}$ states is determined for C_2F_5I as well as for axial and equatorial iodocyclohexane.

The velocity, and thus the kinetic energy, of the iodine atoms produced in the ground state from the photodissocition of C_2F_5I is found to be sensitive to the laser polarization direction with respect to the detection axis. This can be explained by proposing that ground state iodine atoms are produced as a result of absorption to two repulsive states with oppositive polarization characteristics. One state gives the iodine atoms directly, while the other crosses to another state that produces the iodine. It is further concluded that energy redistribution occurs during the curve crossing process, leading to less kinetic energy for the dissociated iodine atoms formed from the latter mechanism.

The 304.0 nm photodissociation of the axial iodocyclohexane is found to produce excited iodine atoms with more kinetic energy than those produced from the photodissociation of equatorial iodocyclohexane. This is in spite of the fact that the excited nonbonding electrons on the iodine in the axial position is more strongly coupled to the ring hydrogens than in the equatorial position. This result suggests that the cyclohexyl radical produced from the photodissociation of the axial iodocyclohexane (i.e., the equatorial radical) has less internal energy than that produced from the photodissociation of equatorial iodocyclohexane (i.e., the axial radical). This suggests that the ground state of the cyclohexyl radical has an equatorial configuration.

1. Introduction

Statistical thermal or photo dissociation occurs after the excess energy in the molecule has equilibrated among all the different internal (vibration) degrees of freedom. Due to the rapid exchange of the energy among the different modes, the probability that enough energy accumulates in the dissociation coordinate (which is made of the linear combination of the different orthonormal set of vibration eigen states) is non-vanishing and determines the dissociation probability or the dissociation time of the molecule. For this reason, statistical dissociation of large molecules, having large number of degrees of freedom, usually takes relatively long time but of course the dissociation time decreases as the excess energy in the molecule (i.e., internal temperature) increases. Since such dissociation generally produces fragments with equilibrated internal modes, the kinetic energy released in these types of dissociation is small and is on the order of the energy content in one of the equilibrated vibrational degree of freedom in the dissociating molecule. Since dissociation times are usually much longer than rotational times, the photofragments produced from polarized laser photolysis are produced with equal probability in all directions i.e., polarized photolysis do not produce anisotropic spacial distribution of the photofragments. Thus all information regarding the correlation between the direction of the transition moments in the molecule and the dissociation axis is lost.

Rapid or spontaneous photodissociation on a repulsive energy surface usually occurs on a time scale shorter than molecular rotation times. This produces anisotropic distribution of fragments if the photolysis light source is polarized, e.g., a laser. From this, the angle between the absorbing transition moment and the dissociation axis can be determined. Equally important is the fact that dissociation occurs in a time shorter than or comparable to intramolecular vibration relaxation (IVR) processes. Thus a good fraction of the excess available energy (beyond the bond dissociation energy, D_0) is found in the form of kinetic energy of the fragments, i.e., as the kinetic energy released (KER) in the photodissociation:

$$KER = h\nu - D_0 - \Delta E^{int}$$

where D_0 is the dissociation energy and ΔE^{int} is the amount of internal excitation energy left in the fragments.

Molecules containing carbon iodine (C-I) bonds, e.g., alkyliodides have absorption in the 200-300 nm region.[1] This is due to the absorption to the repulsive potential energy surfaces resulting from the $\sigma^* \leftarrow n$ excitation in this system, where n is the iodine nonbonding electrons and the σ^* involves antibonding molecular orbital which is mostly localized in the C-I bond.[2]

2. Experimental Technique[3]

We have developed[3] a technique of state-selective one dimensional photofragmentation translation energy spectroscopy[4] to determine the kinetic energy (KE) of the iodine atom photoproduced in either the ground I ($^2P_{3/2}$) or excited I* ($^2P_{1/2}$) state by simply using a laser of wavelength of 304.7 or of 304.0 nm, respectively. The technique uses a nanosecond tunable Nd-Yag pumped dye laser with frequency doubling capabilities. The wavelength is selected to allow the one photon spontaneous dissociation of the C-I bond as well as the 2+1 photon ionization of the iodine atoms produced in either the $^2P_{3/2}$ (if λ = 304.7 nm) or the $^2P_{1/2}$ state (if λ = 304.0 nm). After a delay of a few microseconds, to allow the produced ions to travel freely with their recoil velocity and anisotropic distribution in space, a pulsed acceleration electric field is applied and the accelerated ions are detected in a time-of-flight (TOF) mass spectrometer. To make the instrument a one dimensional velocity analyzer, a pinhole is fixed at the detector. Two TOF peaks are observed for the iodine ions produced, one for the iodine ions recoiling towards and the other away from the detector. The TOF of the ions depends on the atom recoil velocity, the acceleration field, the position of the laser focus and the field-free drift distance between the acceleration electrode and the detector. These geometric parameters can be determined from the photolyses of I_2 molecule under the same experimental conditions[3] used for the molecules studied. The kinetic energy released in the photolysis of I_2 is known accurately and thus once one determines the TOF at different delay times, one can solve for these unknown parameters. This allows the conversion of the observed TOF of either peak into velocity and thus into kinetic energy[3].

The relative intensities of TOF peak observed when the laser polarization is parallel and when it is perpendicular to the detection axis give the anisotropy parameter. This is related to the angle between the transition moment of the photolysis absorption and the dissociation axis within the molecular framework (generally the C-I bond for monoiodides) as well as a limit of the dissociation time comparted to molecular rotation time.

3. Results and Discussion

3.1 Vibration Energy Distribution at Electronic Level Crossing in C_2F_5I[3]:

Figure 1a gives the observed TOF spectrum of the iodine ions corresponding to dissociated I from the photodissociation of C_2F_5I. Figure 1b gives the corresponding velocity distribution spectrum. In both figures, the spectra are recorded for parallel (solid lines) and perpendicular (dots) laser polarizations, normalized so that they have similar peak intensities for comparison. It is obvious that both the TOF and thus the velocity distribution for I are sensitive to the laser polarization. This suggests that the iodine atoms are produced from the absorption to more than one state with opposite polarizations and whose dissociation give different

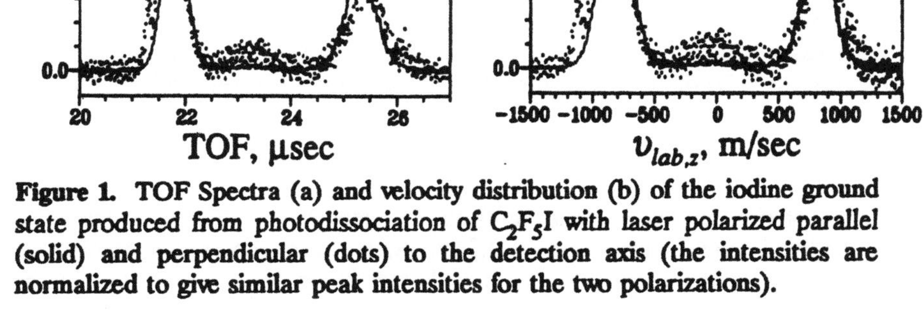

Figure 1. TOF Spectra (a) and velocity distribution (b) of the iodine ground state produced from photodissociation of C_2F_5I with laser polarized parallel (solid) and perpendicular (dots) to the detection axis (the intensities are normalized to give similar peak intensities for the two polarizations).

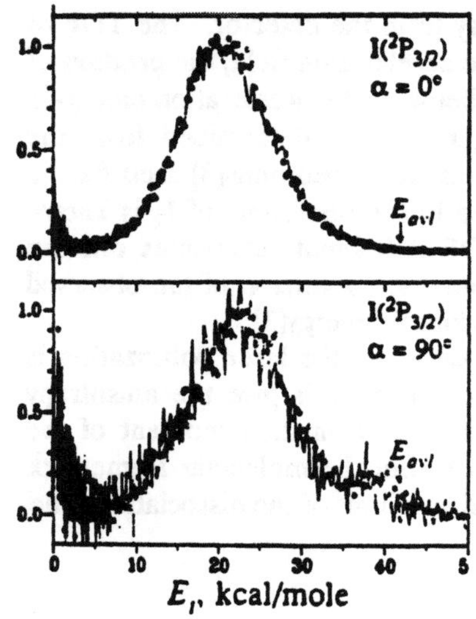

Figure 2. Kinetic energy released in the photodissociation of C_2F_5I to give ground state iodine with laser polarized parallel (top) and perpendicular (bottom) to the detection axis. The difference suggest two mechanisms for the dissociation involving two states of opposite polarization and curve crossing.[3]

values of recoil velocities. This conclusion is supported by the observation that the I* signal is much stronger when the laser polarization is parallel to the detection axis than the I. This is consistent with theroetical predictions[2] which showed that the parallel transition correlates with I* at infinite separation, while the two perpendicular states correlate with the I. From the observed polarization ratio of I, one concludes[3] that 22% of it is produced directly from absorption to a perpendicular state (followed by rapid dissociation) and 78% is produced from absorption to the parallel state followed by curve crossing to another perpendicular state that rapidly dissociates. The fact that the two states have to mix by internal motion that removes the cylinderical symmetry of the C-I electronic system during the curve crossing process, suggests that vibration energy redistribution could occur during the crossing process. This would produce the ground state iodine from the absorption to the parallel state with less kinetic energy than those produced directly from the absorption to the perpendicular state. Thus by increasing the contribution of the iodine atoms produced from parallel absorption (i.e., detecting at parallel polarization) the average kinetic energy observed is expected to decrease. Figure 2 gives the kinetic energy distribution observed for ground state iodine atoms produced with parallel and perpendicular polarizations.

It is clear that there is a difference in the energy of the iodine kinetic energy. One calculates[3] that iodine atoms produced via parallel absorption followed by level crossing lose ~ 3 kcal/mol^{-1} to internal energy of the C_2F_5 radical more than those produced by direct dissociation (via absorption to the perpendicular state.)

3.2 The Ground State Geometry of Cyclohexyl Radical[5]:

Due to steric hindrance resulting from the repulsive coupling between the nonbonding iodine electrons and the 3,5 hydrogens, the energy of iodocyclohexane in the axial-configuration (a-$C_6H_{11}I$) is found to be higher than that in the

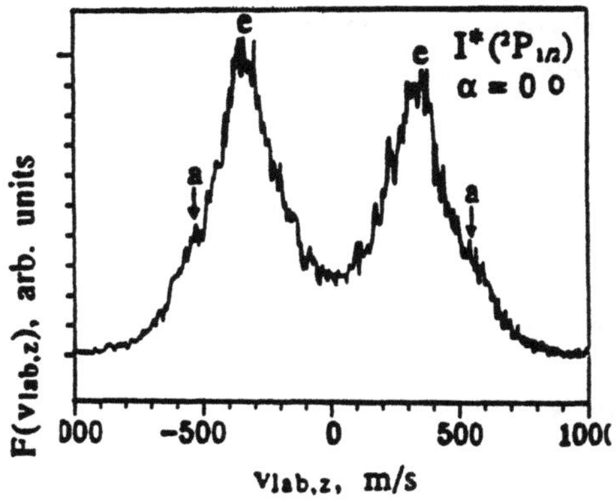

Figure 3. The velocity distribution of I* produced from the photolysis of iodocyclohexane at 298K photolysis of iodocyclohexane at 298K with laser polarized parallel to the detection axis. The velocity of the I* produced from the axial configuration (the shoulder peak) is found to be higher than that from the equatorial, suggesting that the ground state of the radical has the hydrogen on the radical center

equatorial (e-$C_6H_{11}I$) configuration. Because of this, there are 3 times as many[6] molecules in the equatorial as in the axial configuration at room temperature. One should thus expect to see two different kinds of velocity distributions in the photodissociation of iodocyclohexane, one for each of the a- and e-$C_6H_{11}I$, with relative intensities of 1:3 respectively.

Figure 3 shows the velocity distribution of the excited I* produced in the photodissociation-ionization of iodocyclohexane. The distributions are observed with relative intensity of almost 3:1. Interestingly enough, the more intense band resulting from the photodissociation of the e-$C_6H_{11}I$, has a lower average velocity than the a-$C_6H_{11}I$. Thus the molecule with the largest coupling between its excited electrons, the a-$C_6H_{11}I$, produces iodine atoms with high kinetic energy, i.e., it leaves the radical with less internal energy (i.e., close to the ground electronic state of the radical produced). One then concludes that the cyclohexyl radical must have an equatorial configuration. This supports the theoretical predictions of Imam and Allinger[7] but contradicts the ab initio calculations with fixed nuclear configuration of Lloyd et al.[8].

Acknowledgement: The authors wish to thank the National Science Foundation for its financial support.

References

[1] D. Porret and C. F. Goodeve, Proc. R. Soc. London Ser. A **165**, 31 (1938).
[2] R. S. Mulliken, J. Chem. Phys. **8**, 382 (1940).
[3] H. J. Hwang and M. A. El-Sayed, J. Chem. Phys. **94**, 4877, (1991).
[4] R. Ogorzalek Loo, G. E. Hall, H.-P. Haerri, and P. L. Houston, J. Phys. Chem. **92**, 5 (1988).
[5] J. E. Freitas, H. J. Hwang, A. B. Tiknor and M. A. El-Sayed, Chem. Phys. Lett., in press.
[6] J. McMurry, Organic Chemistry, Brooks/Cole, Monterey, California (1984).
[7] M. R. Imam and N. L. Allinger, Molec. Struc. **126**, 345 (1985).
[8] R. V. Lloyd, J.G. Causey and F. A. Momany, J. Am. Chem. Soc. **102**, 2260 (1980.

Vibrational Mode Dynamics of Photocyclization in cis-Stilbene and Its Related Compounds

K. Yoshihara[1], *H. Petek*[1], *Y. Fujiwara*[2], *J.H. Penn*[3], *and J.H. Frederick*[4]

[1]Institute for Molecular Science, Myodaiji, Okazaki 444, Japan
[2]School of Pharmaceutical Sciences, Kanazawa University, Kanazawa 920, Japan
[3]Department of Chemistry, University of West Virginia, Morgantown, WV 26506-6045, USA
[4]Department of Chemistry, University of Nevada, Reno, NV 89557, USA

1. Introduction

The two way photochemical reaction of cis-stilbene was identified by Fischer and his coworkers in solution, namely, cis-trans isomerization leading to trans-stilbene and ring-closure reaction leading to dihydrophenanthrene(DHP)[1]. The quantum yields for Φ (cis $\rightarrow$ trans) = 0.35 and Φ (cis $\rightarrow$DHP) = 0.10 were determined at room temperature as shown in Fig. 1. Recently cis-trans isomerization dynamics have been studied by picosecond time-resolved spectroscopy in several groups[2-5].

The ring-closure reaction has not received much attention. Recently new dimensions in cis-stilbene structure and dynamics were explored by measuring the spectra and dynamics of several cis-stilbene homologs such as 1,2-diphenylcyclohexene (DPC-6), 1,2-diphenylcyclopentene (DPC-5), and 1,2-diphenylcyclobutene (DPC-4) [6]. These molecules contain the same chromophore as cis-stilbene, but are structurally prevented from undergoing cis-trans isomerization. The quantum yield of ring-closure reaction of DPC-5 is 0.5 [1] and of DPC-6 is "high" [7], and DPC-4 is fluorescent because the barrier to photocyclization is high, effectively preventing the formation of the corresponding DPH structure. Nevertheless the vibrationally resolved spectrum of this molecule displays long progressions in the phenyl bending and twisting modes, indicating that the initial motion on the excited surface is along the reaction coordinate for production of DHP.

In the present work, the experimental DPC-4 fluorescence excitation spectrum is modeled in an effort to derive an empirical excited state potential surface for the symmetric bend, α and the symmetric phenyl twist, ϕ, vibrations (see Fig 1).

2. Experimental

Fluorescence excitation spectrum of DPC-4 obtained in a supersonic expansion is shown in Fig. 2. The majority of lines in the DPC-4 spectra can be assigned to fundamentals, overtones, and combination bands of 46.6, 70.5, and 151.7cm^{-1} modes, which are assigned to the out-of-plane, phenyl twisting, and in-plane bending motions, respectively.

Springer Proceedings in Physics, Vol. 68 Time Resolved Vibrational Spectroscopy V
Ed.: H. Takahashi

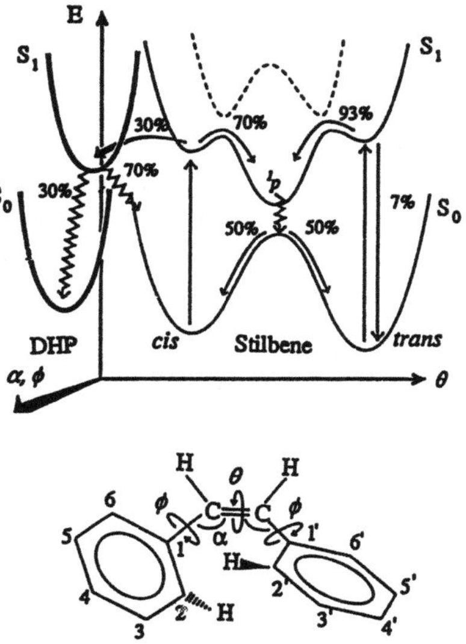

Fig. 1 Schematic potential surfaces of cis-stilbene in α, ϕ, and θ coordinates, which are involved in the photocyclization and cis-trans isomerization reactions. These angles are schematically portrated below the potential energy surfaces.

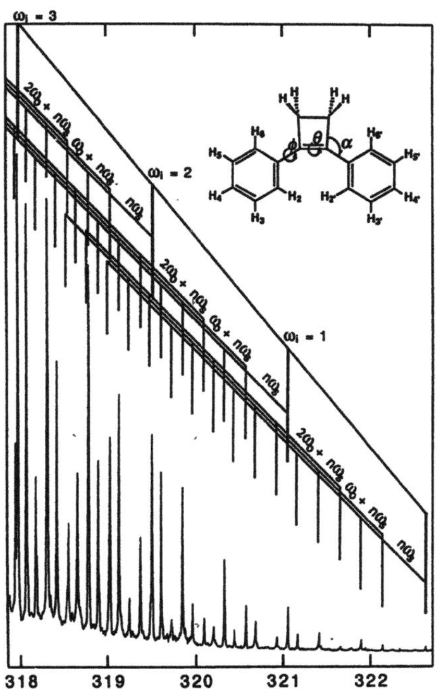

Fig. 2 The low energy region of DPC-4 laser induced fluorescence spectrum under supersonic expansion as cold free molecules.

3. Excited State Dynamics

The structural similarity of the diphenylcycloalkenes and cis-stilbene is exploited to derive excited state surfaces for DPC-5, DPC-6, and cis-stilbene from the surface found for DPC-4. The key to extrapolating the DPC-4 surface to other molecules lies in parametrizing the potential surface in terms of the generalized forces that contribute to the structure and properties of all the molecules. These forces include, for example, the H_2-H_2' repulsive interaction, the conjugative force favoring planarity in the π-system, and the C_2-C_2' attractive interaction on the excited state. Calculations were made by using a time-dependent wave packet method of the sort popularized by Heller and co-workers [8]. The present study enjoys the advantage of having detailed information about the vibrations on the S_1 state of DPC-4 with which to assign the parameters for the potential surface. With a quantitatively accurate DPC-4 surface, it is then much more likely that the extrapolated surfaces for the other molecules will at least be qualitatively representative of their true excited state surfaces in these coordinates.

The empirical surface used in computing the spectrum is shown in Fig. 3 as a contour plot with different contours spaced by 250 cm^{-1} above the potential minimum. This picture suggests that the barrier to the perpendicular twist configuration in DPC-4 is about 1500 cm^{-1}. Also shown in Fig. 3 are a few snapshots from the dynamics of the Franck-Condon wave packet on that surface spanning ~220 fs. Although such a wave packet could only be created in the laboratory using a very short laser pulse, its movement is indicative of the dynamical behavior of the molecule on the excited state.

The surface that results form applying the DPC-4 parameters to DPC-5 is shown in Fig. 4 along with the Franck-Condon wave packet. This molecule allows the two phenyl rings to come close enough to one another in the proper orientation to form DHP-like products. There are actually separate minima on the surface for DPC-5 and DHP with an ~ 800cm^{-1} barrier separating the reactant from product. The second minimum, at α =113.5° and ϕ= 18.5° indicating that the C-C bond has formed.

These features are in accord with experimental observation. No emission is observed for isolated DPC-5 meaning that the molecule decays quickly by

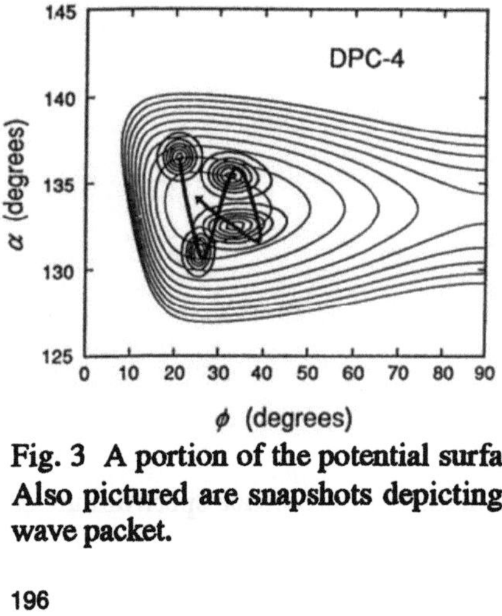

Fig. 3 A portion of the potential surface for S_1 DPC-4 in the α and ϕ coordinates. Also pictured are snapshots depicting the initial dynamics for the Franck-Condon wave packet.

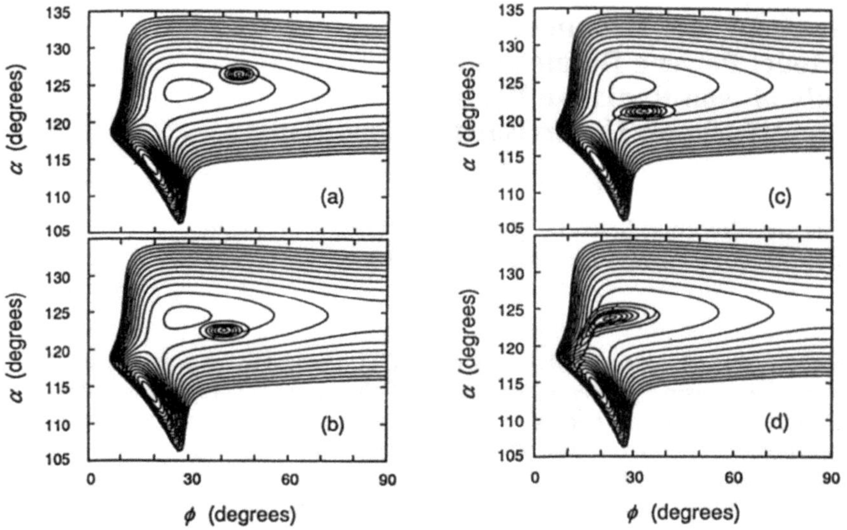

Fig. 4 Snapshots from the early dynamics of the Franck-Condon wave packet on the DPC-5 empirical potential surface. The total time span of the dynamics shown is about 220 fs. The wave packet moves toward the lower left corner of the potential where the photocyclization product (DHP analog) is formed. Since the wave packet is moving parallel to the gap leading to DHP in this case only a small portion "leaks" over into the DHP region of the potential (d).

radiationless processes which include the formation of the photocyclization product. A barrier to DHP formation in DPC-5 of ~2.6 $^{kcal}/_{mole}$ and the quantum yield of this product of ~50%have been measured by Fischer and coworkers[1]. The barrier of 2.3 $^{kcal}/_{mole}$ found for the empirical surface of DPC-5 is in fortuitous agreement with the experimental value.

References

[1] K. A. Muszkat, E. Fischer: J. Chem. Soc. B **1967**, 663; T. Wismonski-Knittel, G. Fischer: E. Fischer, J. Chem. Soc., Perkin Trans. I **1974**, 1930.

[2] M. Sumitani, N. Nakashima, K. Yoshihara: Chem. Phys. Lett., **68**, 255 (1975); M. Sumitani, K. Yoshihara: Bull. Chem. Soc. Jpn., **55**, 85 (1982).

[3] B. I. Greene, R. C. Farrow: J. Chem. Phys., **78** 3336 (1983); F.E. Doany, R. M. Hochstrasser, B. I. Greene, R. R. Millard: Chem. Phys. Lett., **118**, 1 (1985).

[4] D. C. Todd, J. M. Jean, S. J. Rosenthal, A. J. Ruggiero, D. Yang, G. R. Fleming: J. Chem. Phys., **93**, 8658 (1990).

[5] H. Petek, K. Yoshihara, Y. Fujiwara, J. G. Frey: J. Opt. Soc. Am. B, **7**, 1540 (1990).

[6] H. Petek, K. Yoshihara, Y. Fujiwara, Z. Lin, J. H. Penn,J. H. Frederick: J. Phys. Chem., **94**, 7540 (1990); J. H. Frederick, Y. Fujiwara, J. H. Penn, K. Yoshihara, H. Petek: J. Phys. Chem., **95**, 2845 (1991).

[7] K. Ito, S. Tazuke: private communication.

[8] S. Y. Lee and E. J. Heller: J. Chem. Phys. **71**, 4777 (1979); J. H. Frederick, E. J. Heller, J. L. Ozment, D. W. Pratt: J. Chem. Phys., **88**, 2169 (1988).

Spectroscopy and Dynamics of a Model Polyene Decatetraene: A Study of Non-Radiative Pathways in S_1 and S_2 States Under Isolated Conditions

H. Petek[1], *A.J. Bell*[1], *H. Kandori*[1], *K. Yoshihara*[1], *and R.L. Christensen*[2]

[1]Institute for Molecular Science, Myodaiji, Okazaki 444, Japan
[2]Bowdoin College, Brunswick, ME 04011, USA

The spectroscopy and dynamics of decatetraene (DT), a model linear polyene, are studied in order to gain a greater understanding of photophysical properties of this class of molecule, which are pervasive in biological systems and which have interesting electro-optical properties. The photochemistry of a polyene is governed by its two low lying singlet excited electronic states. The optical excitation usually occurs to the S_2 state (1^1B_u), which is strongly allowed. However, due to rapid internal conversion (IC) from the S_2 to S_1 state (1^1A_g), the subsequent photochemistry occurs from the S_1 state. Because the $S_1 \leftarrow S_0$ transition is symmetry forbidden by single photon excitation, it is difficult to study the structure and dynamics of polyenes in their S_1 electronic states. The motivation for this work came from Bouwman *et al.'s* recent demonstration that when intermediate length polyenes (4 - 5 double bonds) are excited to the S_2 state in the gas phase, they emit from both S_2 and S_1 states [1]. This work presents unique spectroscopical and dynamical observations on the S_1 and S_2 states of DT and preliminary results on nonatetraene (NT) under isolated conditions in a molecular beam, where the non-radiative decay processes can be observed as a function of excitation energy without perturbation from a solvent.

The laser excitation spectra of S_1 state of DT are shown in Fig. 1. The S_1 state spectrum consists of a rich progression in low frequency skeletal bending modes,

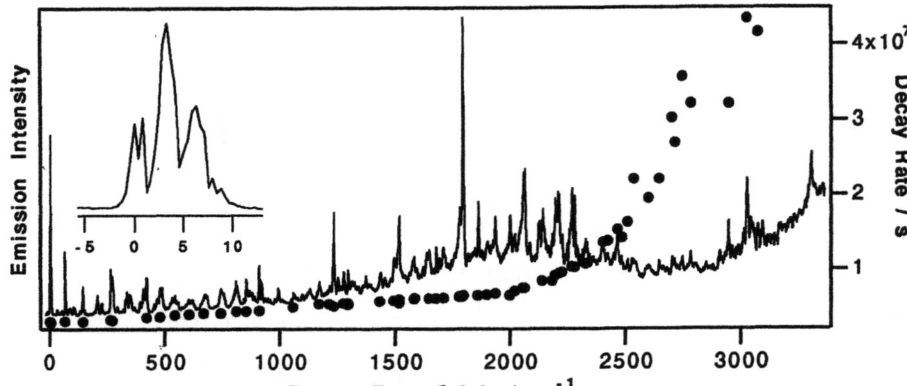

Figure 1. The $S_1 \leftarrow S_0$ fluorescence excitation spectrum of DT. The "false" origin at 344.61 nm in the insert shows triplet structure, which is ascribed to hindered methyl torsions. Decay rates for individual vibronic states are indicated by circles.

Springer Proceedings in Physics, Vol. 68 Time Resolved Vibrational Spectroscopy V
Ed.: H. Takahashi

which are built on false origin at 344.61 nm, and other major features such as C - C and C = C stretching vibrations. This symmetry forbidden $^1A_g \leftarrow {}^1A_g$ transition is made possible by Herzberg-Teller mixing of the b_u vibronic symmetry states in S_1 with the nearby S_2 state. The b_u character can be derived either from in-plane non-totally symmetric skeletal bending modes or from hindered methyl torsions. The triplet structure shown in the inset of Fig. 1 is a signature of modes which are allowed by coupling to the methyl torsion.

The measured decay rates for single vibronic states are also shown in Fig. 1. At the origin the lifetime is ~350 ns. There is a gradual decrease in the lifetime to higher energies due to increased mixing with the S_2 state. At ~2,000 cm^{-1} excess energy a sharp increase in the decay rate in both DT and NT is due to the availability of a non-radiative decay channel, possibly *trans-cis* isomerization. Kohler *et al.* found that *trans-cis* isomerization in the S_1 state of octatetraene (OT) is an activated process with a 1,400 cm^{-1} barrier in solution [2], and they observed the formation of the *cis*-isomer [3]. The non-radiative decay due to *trans-cis* isomerization with a barrier of 1,400 - 2,000 cm^{-1} appears to be common to *all-trans* tetraenes.

The $S_2 \leftarrow S_0$ fluorescence excitation spectrum shows broad vibrational lineshapes due to the rapid IC to the S_1 state. The peaks increase in width from 22 cm^{-1} at the origin to 40 cm^{-1} at 504 cm^{-1} above. Assuming that the widths are due to S_2 - S_1 coupling in the statistical limit, they correspond to an upper limit for the S_2 state lifetime of 0.50 - 0.27 ps. This increase in linewidth may indicate that the low energy skeletal bending vibrations are promoting modes for IC. Similar widths are observed for NT, and Heimbrook *et al.* report a 12 cm^{-1} width for OT [4].

Based on these observations the following mechanism is proposed for the unimolecular decay of tetraenes in the gas phase. Following excitation to the S_2 state, tetraenes internally convert to the S_1 state on the <1 ps time scale. This horizontal radiationless transition leaves tetraenes with >6,000 cm^{-1} of excess energy in the S_1 state. Since this is much greater than the activation barrier to isomerization, tetraenes subsequently decay to the S_0 state in a statistical fashion. The low frequency hindered methyl torsions are an overwhelming factor in determining the density of states; therefore, given that the barriers for isomerization are nearly the same, at a given energy OT will decay the fastest, and DT the slowest.

References

1) W.G. Bouwman, A.C. Jones, D. Phillips, P. Thibodeau, C. Friel, and R.L. Christensen: J. Phys. Chem. *94,* 7429 (1990).

2) J.R. Ackerman, B.E. Kohler, D. Huppert, and P.M. Rentzepis: J. Chem. Phys. *77,* 3967 (1982).

3) L.A. Heimbook, B.E. Kohler, and T.A. Spiglanin: Proc. Nat. Acad. Sci. USA *80,* 4580 (1983).

4) L.A. Heimbook, B.E. Kohler, and I.J. Levy: J. Chem. Phys. *81,* 1592 (1984).

Solvent/Solute Interactions: Mode Specific, Solvent Dependent Dynamics Observed Using Picosecond Transient Raman Spectroscopy

T.L. Gustafson[1], *K. Iwata*[2], *L.A. Huston*[1], and *W.L. Weaver*[1]

[1]Department of Chemistry, The Ohio State University,
120 West 18th Avenue, Columbus, OH 43210-1173, USA
[2]Molecular Spectroscopy Laboratory,
Kanagawa Academy of Science and Technology (KAST),
KSP East 301, 100-1 Sakato, Kawasaki 213, Japan

Abstract. Using picosecond transient Raman spectroscopy, we observe mode specific, solvent dependent variations in the S_1 spectrum of *trans*-stilbene. We attribute the changes in the spectra, in part, to vibrational relaxation via resonance energy exchange, and, in part, to multidimensional isomerization from the excited state. We suggest that the olefin-phenyl torsional motions are strongly coupled to the relaxation pathways.

1. Introduction

While much is understood on a macroscopic scale about the interactions that occur between solute and solvent in solution, the microscopic details, involving the interaction potentials between solvent and solute, transport processes, and energy transfer, are not well understood. The advent of ultrafast laser technology and supercomputers has aided in the advancement of this area over the last decade. But there are still significant deviations between theoretical models, involving bulk solvent properties, and experimental results, particularly at high viscosities.[1]

We are using picosecond transient Raman spectroscopy to probe the mode specific interactions that occur between solute molecules and the surrounding solvent. We have recently shown that certain vibrational bands in the Raman spectrum of S_1 *trans*-stilbene (tS) change peak position and bandwidth with delay.[2,3] We suggest that these data can be explained as a consequence of vibrational relaxation and/or of the multidimensional surface of the isomerization process, both of which involve specific interactions with the solvent. In order to investigate the contribution of vibrational relaxation, in this work we have performed two experiments. The first is to obtain transient Raman spectra at different pump wavelengths, keeping the probe wavelength fixed. This provides a controlled way of introducing excess vibrational energy into the system. The second experiment is to monitor the S_1 anti-Stokes spectrum at various delays. This provides a direct monitor of the molecular "temperature" with delay.

Springer Proceedings in Physics, Vol. 68 Time Resolved Vibrational Spectroscopy V
Ed.: H. Takahashi

2. Experimental

We have previously described the laser system we use for our measurements.[2,3] Briefly, the harmonics of a cw mode-locked Nd:YAG laser are used to pump two independently tunable synchronously pumped, cavity dumped dye lasers. The output from one of the dye lasers is amplified at MHz repetition rates with a six pass amplifier pumped by a cavity dumped argon ion laser.[4] The extraction efficiency of the amplifier is 15-20%, providing ~200 nJ, 3 ps pulses. The amplified output is frequency doubled using a temperature tuned ADA crystal to generate ~20 nJ, 3 ps pulses in the region of *ca.* 280 to 310 nm. The visible probe pulse (*ca.* 577 nm) traverses an optical delay line before being combined colinearly with the UV pump pulse and focussed on the sample with a 5 cm lens. The polarization of the pump and probe beams can be independently controlled; for this work, the polarization is set to the magic angle (54.7°). The pump and probe impinge on the spinning sample cell at a 45° angle and the Raman scattering is collected in a backscattering geometry. The detection system consists of a single spectrograph with a CCD OMA. Data collection times were 20 minutes at each delay.

3. Results and Discussion

We have obtained transient Raman spectra over the range from 170 to 1700 cm^{-1} at various delays for three different excitation wavelengths, 303.2, 294.4 and 285.7 nm, corresponding to ~1800, ~2800, and ~3800 cm^{-1} of excess vibrational energy, respectively, above the origin for tS (*ca.* 31,200 cm^{-1}). For each spectrum we have fit the observed vibrational bands to lorentzian lineshapes and obtained peak positions, bandwidths and integrated intensities. In Figure 1 we show the fit to the data in the region from 1400 to 1700 cm^{-1} obtained at a delay of 20 ps for 294.4 nm excitation. The two peaks at *ca.* 1565 cm^{-1} and 1525 cm^{-1} correspond to motions involving predominately the C=C stretch of the olefin and phenyl portions of the

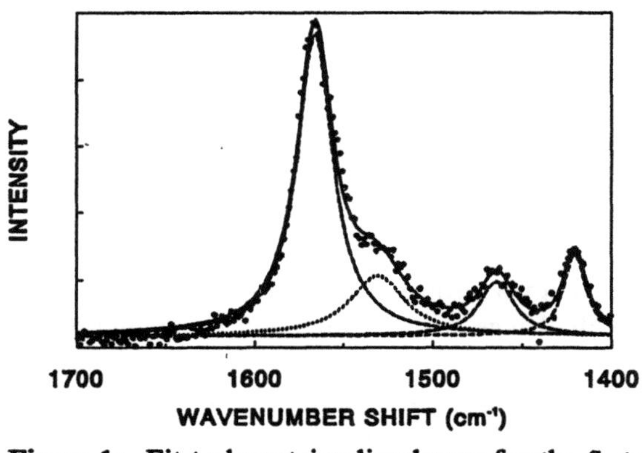

Figure 1: Fit to lorentzian lineshapes for the S_1 transient Raman spectrum of *trans*-stilbene in *n*-hexane: Pump, 294.4 nm; Probe, 577 nm; Repetition rate, 1 MHz.

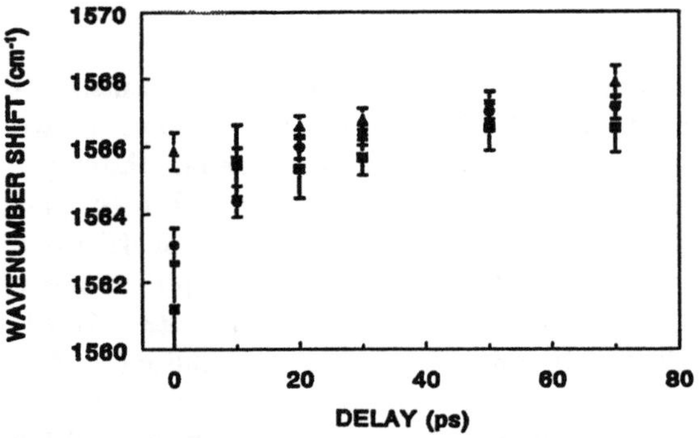

Figure 2: The peak position of the olefin C=C stretching vibration of S_1 *trans*-stilbene in *n*-hexane versus delay for three pump wavelengths: 303.2 nm (▲); 294.4 nm (●); 285.7 nm (■); Probe, 577 nm.

molecule, respectively. In Figure 2 we plot the peak position of the olefin band versus delay for the three pump wavelengths used in this work. There are several important observations from these data. The first is that the peak position shifts to higher frequency with increasing delay for all of the pump wavelengths. In addition, as the amount of excess vibrational energy increases the peak position at the earliest delays decreases. However, the bandwidth change versus delay for the 1565 cm^{-1} band is the same for all three pump wavelengths. We note that the bands at 1525 cm^{-1} and 1420 cm^{-1} do not change peak position or bandwidth for any delay at any pump wavelength. These data would suggest that the dissipation of excess vibrational energy is responsible for the shifts in the peak positions at early times.

In order to probe the nature of the vibrational relaxation process we obtained the anti-Stokes transient Raman spectra of S_1 tS at various delays. The only vibrational band that we could observe on the anti-Stokes side was the 287 cm^{-1} band. No other bands were observable at any delay times. The ratio of the Stokes and anti-Stokes intensities of the 287 cm^{-1} band was constant at all delays, within experimental uncertainty. These data strongly suggest that the vibrational relaxation process does not involve direct coupling of the olefin mode to the solvent. Rather, the vibrational relaxation is likely to involve some other mechanism. One possibility is that of resonance energy exchange via a low frequency vibration, as has been discussed by Harris and co-workers.[5] In their model a "doorway" state, the olefin C=C stretch in this case, couples anharmonically to an "exchange" mode, a low frequency vibration in the molecule, perhaps a torsional motion in tS. The exchange mode exchanges energy with the solvent via resonance energy transfer. The theory predicts that the doorway mode will change both position and bandwidth as the energy is exchanged to the solvent via the low frequency vibration. Such a model can account for the mode specificity because it is possible for specific doorway modes to couple to the low frequency exchange modes.

There are still issues that are not completely addressed by vibrational relaxation. The fact that the bandwidth change is constant regardless of the amount of excess vibrational energy is not anticipated. We suggest that the bandwidth change may reflect the multidimensionality of the isomerization surface, with the torsional motion providing the additional coordinate as modelled theoretically.[1]

We have obtained data for tS in other solvents and for several substituted stilbenes. Similar effects are observed, although different solvents have different rates for various vibrational motions. The rates of change are also affected by the nature of the substituents and their interaction with the solvent and by the relative polarization of the pump and probe.

References

1. N. Agmon and R. Kosloff, *J. Phys. Chem.* 91, 1988 (1987); B. Carmeli and A. Nitzan, *Chem. Phys. Lett.* **106**, 329 (1984).
2. K. Iwata, W. L. Weaver, L. A. Huston, and T. L. Gustafson, *Proc. XIIth International Conference on Raman Spectroscopy*, J. R. Durig, Ed. (John Wiley & Sons, London, 1990), pp. 612-613.
3. T. L. Gustafson, K. Iwata, W. L. Weaver, R. L. Benson, and L. A. Huston, in *Laser Diagnostics of Biological Molecules: Linear and Nonlinear Methods*; S. A. Akhmanov, Ed.; Proc. SPIE 1403, 1990, in press.
4. W. L. Weaver and T. L. Gustafson, *Spectrochimica Acta Rev.*, in preparation.
5. R. M. Shelby, C. B. Harris, and P. A. Cornelius, *J. Chem. Phys.* **70**, 34 (1979); S. Marks, P. A. Cornelius, and C. B. Harris, *J. Chem. Phys.* **73**, 3069 (1980).

Time Dependence of the Depolarization Ratio in Picosecond Time-Resolved Raman Spectroscopy: Effect of Reorientation

K. Iwata[1;2], *L.A. Huston*[1], *W.L. Weaver*[1], *and T.L. Gustafson*[1]

[1]Department of Chemistry, The Ohio State University, 120 West 18th Avenue, Columbus, OH 43210-1173, USA

[2]Molecular Spectroscopy Laboratory, Kanagawa Academy of Science and Technology (KAST), KSP East 301, 100-1 Sakato, Kawasaki 213, Japan

1. Introduction

We usually assume randomly orientated molecules when discussing depolarization ratios in Raman spectroscopy. This assumption does not hold, however, in picosecond time-resolved Raman spectroscopy. Because samples are, in most cases, excited with linearly polarized laser light and the time resolution of the measurement is shorter than the reorientation lifetime of the sample, the probability distribution of the excited molecules is largest along the axis of the polarization of the laser light and smallest for the perpendicular direction. The depolarization ratios at t=0 should be different from the "normal" values because of this anisotropy and should change as the time delay increases. We have measured the time dependence of the depolarization ratio of *trans*-stilbene in the lowest excited singlet (S1) state and observed that the depolarization ratio changes depending on the time delay, which can be explained well by our simple, but realistic model.

2. Model Calculation

We assume that the Raman tensor of S_1 *trans*-stilbene has only one non-zero component, α_{ZZ}, when probed with 600 nm light. This means that the 600 nm probing light is in rigorous resonance with the transition between the two non-degenerate electronic states, S_1 and S_n.

Under this condition, the depolarization ratio, ρ, can be calculated as follows:

$$\rho = \int W(\theta,\phi)\,|(e_Z\cdot m)\,\alpha_{zz}\,(m\cdot e_X)|^2 d\Omega \Big/ \int W(\theta,\phi)\,|(e_X\cdot m)\,\alpha_{zz}\,(m\cdot e_X)|^2 d\Omega, \quad (1)$$

where $W(\theta,\phi)$ is the distribution probability of the direction of S_n-S_1 transition moment(**m**), e_X and e_Z are unit vectors in laboratory frame. The probing light is set to be polarized to Z direction. If linearly polarized light is used to excite *trans*-stilbene whose **m** is known to be parallel to the S_1-S_0 transition moment [1] and the polarization of the pumping light is parallel to that of the probing light,

$$W(\theta,\phi) = \frac{3}{4\pi}\cos^2\theta \quad (t=0) \qquad \text{and} \qquad W(\theta,\phi) = \frac{1}{4\pi} \quad (t=\infty). \quad (2)$$

From equations (1) and (2), we obtain $\rho(t=0) = 1/5$ and $\rho(t=\infty) = 1/3$.

Springer Proceedings in Physics, Vol. 68 Time Resolved Vibrational Spectroscopy V
Ed.: H. Takahashi

3. Results and Discussion

The experimental set up we used has been described earlier [2]. In Figure, we show the observed depolarization ratios for 1180 cm^{-1} and 1565 cm^{-1} bands of S_1 *trans*-stilbene in n-hexane for seven different time delays. The depolarization ratio increased from t=0 ps to t=20 ps and reached a constant value at about t=30 ps. A solid curve in Figure is a simulated change of the depolarization ratio when the reorientation lifetime of the S_1 molecule is 17 ps, which was obtained with a polarized transient absorption set up. Because the observed values and the simulated curve are in good agreement, we conclude that the change of the depolarization ratio is due to the reorientation process of the S_1 molecules. We think that we did not obtain the depolarization ratio of 0.2 at t=0 because the time resolution of our laser system (~ 8 ps) was not short enough to detect the non-relaxed distribution.

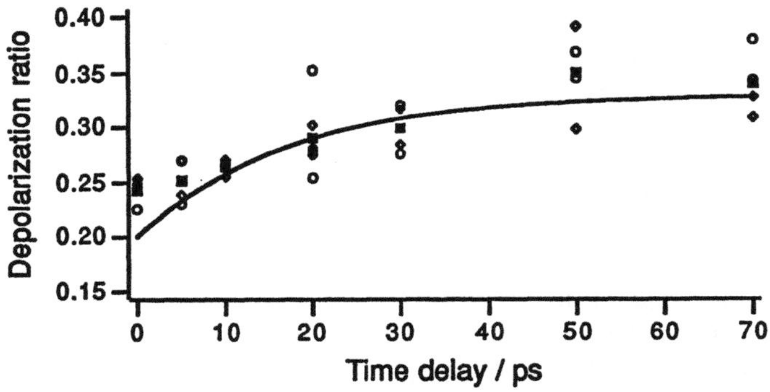

Figure Time dependence of the depolarization ratio of S_1 *trans* -stilbene: observed values (○1565 cm^{-1}, ◇1180 cm^{-1}, ■ average) and the calculated curve.

References

1. S. Abrash, S. Repinec, and R. M. Hochstrasser, J. Chem. Phys. **93**, 1041 (1990).
2. K. Iwata, W. L. Weaver, L. A. Huston, and T. L. Gustafson, *Proc. XIIth International Conference on Raman Spectroscopy*, J. R. Durig, Ed. (John Wiley & Sons, London, 1990), pp. 612-613.

Raman Band Shapes and Vibrational Dynamics of Electronically Excited Molecules in Solution

H. Hamaguchi

Kanagawa Academy of Science and Technology (KAST),
KSP East 301, 100-1 Sakato, Kawasaki 213, Japan

It has been noticed, during the course of transient Raman investigations in the past ten years, that the Raman band widths of electronically excited molecules are always much larger than those of the corresponding ground state molecules [1]. The extra broadenings are most likely to be the manifestations of faster vibrational dephasing times and are expected to carry information on a variety of intra- and intermolecular processes characteristic of electronically excited molecules. Here, we focus on the shape of the ethylenic C=C double bond stretch band of *trans*-stilbene in the lowest excited singlet state (hereafter abbreviated as S_1 tSB).

Probe-laser Power Dependence = Time Dependence

The marked change of the C=C Raman band shape of S_1 tSB with the probe-laser power was already interpreted as due to the competition between the relaxation of the photogenerated "hot" S_1 tSB and the depletion by the $S_n \leftarrow S_1$ photoexcitation by the probe laser [2,3]. By using a simple rate equation, we are able to convert the power dependent spectral changes to the time dependent spectral changes (optical depletion timing). The time dependence of the C=C band shape thus obtained is shown in Fig. 1. The band width decreases as the time goes, showing that the "hot" S_1 tSB molecule exhibits larger C=C band width than the relaxed "cold" molecule. This result was recently confirmed by a real time-resolved measurement by Gustafson, Iwata, and co-workers [4].

Solvent Dependence

The C=C band shape is also very sensitive to the solvent as shown in Fig. 2. On going from methanol to decanol, the width of the C=C band decreases to about one half. It is known that the fluorescence lifetime of S_1 tSB increases from 32 ps in methanol to 121 ps in decanol. Since the only non-radiative decay channel in S_1 tSB is the isomerization over a barrier to the perpendicular form, the increase of the fluorescence lifetime is a direct indication of the increase of the barrier height on going from methanol to decanol. Talking this fact into account, we may be able to relate the observed change of the C=C band width to the change of the barrier height for the isomerization; the band width is larger for the lower barrier height.

Springer Proceedings in Physics, Vol. 68 Time Resolved Vibrational Spectroscopy V
Ed.: H. Takahashi

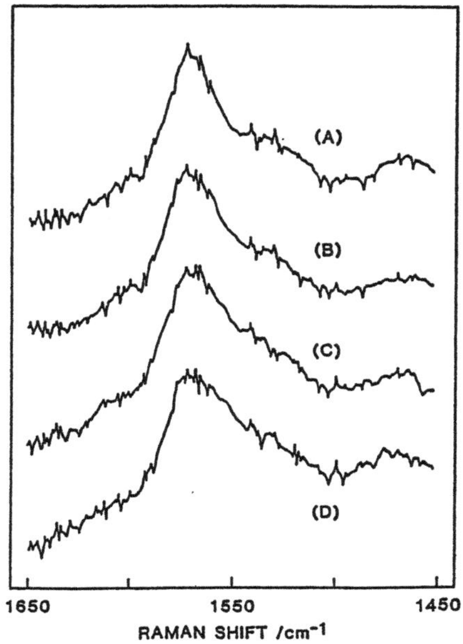

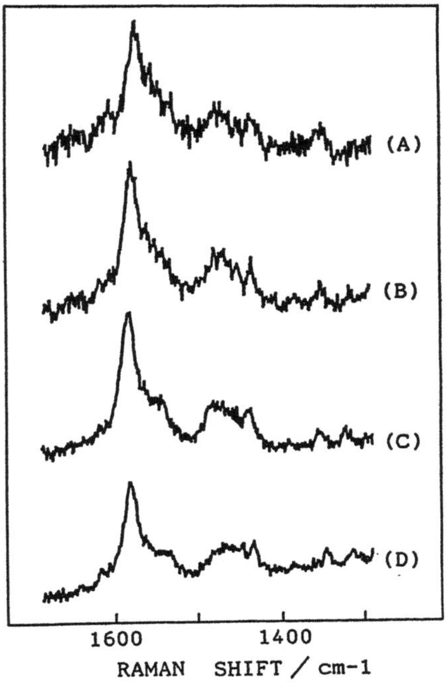

Figure 1. Time dependence of the C=C band shape of S_1 tSB in hexane obtained with the "optical depletion timing" method [2]. (A) 50ps after the photoexcitation, (B) 30ps, (C) 15ps, (D) 7ps. Note that the width in the 7ps spectrum is much larger than that in the 50ps spectrum.

Figure 2. Solvent dependence of the C=C band shape of S_1 tSB. (A) in methanol (flourescence lifetime 32ps), (B) propanol (51ps), (C) hexanol (73ps), (D) decanol (121ps). The band width decreases with increasing fluorescence lifetime.

Increase of the Band Widths with Increasing Vibrational Energies

The width of the C=C stretch band is much larger than those of the other observed Raman bands. The plot of the band widths against the vibrational energy shows a clear trend that they start to increase at around 1000cm^{-1} and rise sharply with further increase of the vibrational energy. This trend is similar to that of the intramolecular vibrational-energy redistribution (IVR) rate in a free S_1 tSB molecule in jet [5].

An IVR-like Vibrational Dephasing Process in S_1 tSB in Solution ?

All the above findings can be explained from a unified viewpoint that the dominant vibrational dephasing process is something like IVR the rate of which is determined by the density of vibrational states. For "hot" S_1 tSB with excess vibrational energy, the density of states is much higher than that of the "cold" molecule. The C=C band width is therefore larger for "hot" S_1 tSB than for "cold" S_1 tSB. The lower barrier to isomerization produces low frequency torsional modes which subsequently increase the density of vibrational states. Thus, the band width is much larger in methanol than in decanol. As the vibrational energy increases, the density of states increases and so does the band width.

This rather native interpretation is yet to be substantiated by further quantitative treatments of the subject. It is worthwhile to note, however, that the band shape analysis has now found a very promising field of application regarding the vibrational dephasings in electronically excited molecules.

References

1. H. Hamaguchi, in "Vibrational Spectra and Structure", J. R. During ed. Elesivier, Vol. 16, pp. 227 (1987).
2. H. Hamaguchi, Chem. Phys. Lett. 126, 185 (1986).
3. H. Hamaguchi, J. Chem. Phys. 89, 2587 (1988).
4. T. L. Gustafson, K. Iwata, W. L. Weaver, L. A. Huston, and R. L. Benson, Proc. SPIE 1403, 1990, in press.
5. J. A. Syage, W. R. Lambert, P. M. Felker, and A. H. Zewail, J. Chem. Phys. 81, 4706 (1984).

Ultraslow and Ultrafast Time-Resolved Resonance Raman Spectra of *cis*-Stilbene Photoproducts

D.L. Phillips, J.-M. Rodier, and A.B. Myers

Department of Chemistry, University of Rochester,
Rochester, NY 14627, USA

1. Introduction

The photochemistry of stilbene has been widely studied as a paradigm for cis-trans isomerization in general and solvation effects on rapid large-amplitude motions in particular. Early time-resolved studies focused mainly on the trans isomer; the excited state dynamics of *cis*-stilbene, with its much shorter lifetime (≤2 ps), have been probed by analyzing ground-state resonance Raman intensities to extract the excited-state dynamic information contained therein [1-3], but have only recently been observed directly on the femtosecond time scale [4-9]. While the time-resolved absorption and fluorescence techniques employed thus far can generate high quality kinetic data, they provide relatively little detailed information about structural relaxation and energy disposal in the photoproducts. To address these questions we are applying time resolved resonance Raman spectroscopy, already used extensively to probe dynamics in the S_1 excited state of *trans*-stilbene [10-12], to the faster photochemistry of the cis isomer.

Excitation of *cis*-stilbene (CS) into its lowest allowed absorption band generates, on the picosecond time scale, a mixture of vibrationally hot ground-state *cis*-stilbene, hot ground state *trans*-stilbene (TS) (Φ=0.35) [13], and the cyclization product, 4a,4b-dihydrophenanthrene (DHP) (Φ=0.10) [14]. The stilbene isomers are both stable molecules whose uv-excited resonance Raman spectra have been well characterized [1,15], although interpretation of the spectra of a mixture of vibrationally hot isomers is challenging. In contrast, DHP is a relatively unstable species which photoconverts readily back to CS, is easily oxidized to phenanthrene, and thermally ring-opens to CS with a rate constant of about 2 hr near room temperature [14]. While its absorption spectrum is known [14], no vibrational spectrum has apparently been reported. For this reason we are carrying out both uv picosecond Raman studies designed to probe the formation and decay of vibrationally hot and/or distorted S_0 *cis*- and *trans*-stilbene, as well as "slow" time-resolved measurements (on the 15-30 min. time scale) to characterize the resonance Raman spectrum of DHP [16].

2. Resonance Raman spectra of 4a,4b-dihydrophenanthrene

Degassed samples of CS in cyclohexane or methanol were photolyzed with 266 nm radiation from a Q-switched Nd:YAG laser for 15 min. The irradiated samples became yellow due to the formation of DHP. The samples were then moved to the Raman spectrometer and the spectra obtained in a backscattering geometry from a stirred cuvette at excitation wavelengths of 436 and 477 nm (Raman-shifted YAG). At these wavelengths DHP absorbs strongly while the stilbene isomers do not absorb. The Raman scattering was collected with a quartz condenser lens system, 0.5-m single spectrograph, and gated intensified diode array detector.

Figure 1 presents difference spectra of the photoproducts formed in both cyclohexane and methanol at 477 nm and in cyclohexane at 436 nm, obtained by subtracting the spectra of the starting material from the spectra of the irradiated samples. Thus they may contain features from not only the resonantly enhanced DHP product but also the TS product, which is expected to be formed in higher yield but is being excited away from resonance. TS Raman lines do appear with moderate intensity in methanol, but are far weaker in cyclohexane. The difference spectra also reveal a number of weak bands which do not belong to TS and are therefore assigned as DHP vibrations. The marked solvent dependence of the TS:DHP Raman intensity ratios indicates that, barring large solvent effects on the Raman cross sections, the ratio of DHP to TS formed through 266 nm irradiation of CS is considerably greater in cyclohexane than in methanol. This conclusion is supported by measurements of the 436 nm absorbance as a function of irradiation time which show that the rate of formation of DHP from CS is two to three times greater in cyclohexane than in methanol under identical irradiation conditions (Figure 2).

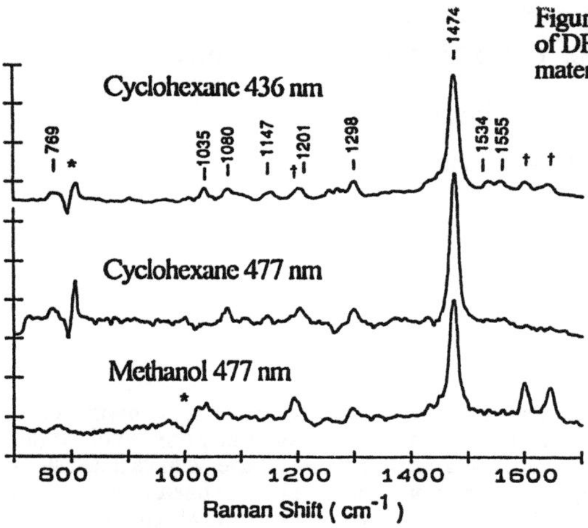

Figure 1. Resonance Raman difference spectra of DHP (uv irradiated sample minus CS starting material). Daggers label *trans*-stilbene lines.

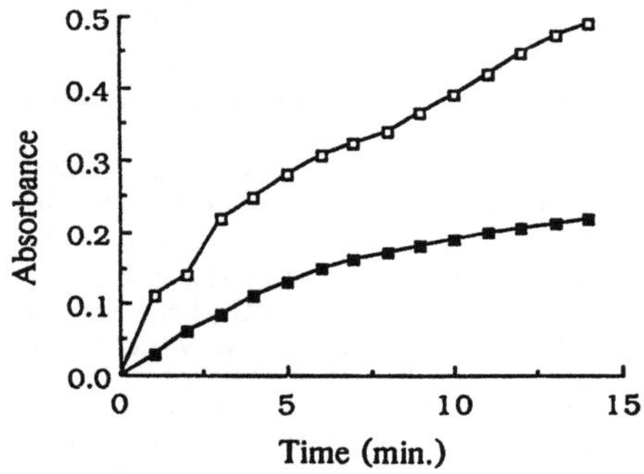

Figure 2. DHP absorbance at 436 nm vs. 266 nm irradiation time for 40 mM CS in cyclohexane (open squares) and methanol (solid).

The rate of DHP destruction under 436 nm irradiation, on the other hand, is found to differ by no more than about 20% between the two solvents. We conclude that while the quantum yield for photochemical ring opening of DHP is not strongly solvent dependent, the yield for photocyclization of CS is at least a factor of two greater in cyclohexane than in methanol. We have not explored the possible solvent dependence of the cis-trans photoisomerization yield, which could also contribute to the observed solvent dependence of the DHP:TS Raman intensity ratios.

Based on the Raman frequencies and intensities predicted through QCFF/PI calculations and by comparison with the Raman spectra of other conjugated linear polyenes we assign the intense 1474 cm^{-1} line as the in-phase combination of the double bond stretches. The unusually low frequency is probably an electronic effect reflecting actual reductions in the force constants for double bond stretching relative to other polyenes in this strained, high-energy species. The two bands at 1534 and 1555 cm^{-1}, which show significant intensity only at the bluer excitation wavelength, are probably the other two double bond stretches having A symmetry, and the 769 cm^{-1} band can be assigned to ring hydrogen out-of-plane wagging. The other observed vibrations in the 1000 to 1300 cm^{-1} range almost certainly involve mixtures of single bond stretching and hydrogen rocking motions as in other polyenes, but isotopic data will be required before specific assignments can be made.

The optical absorption spectrum of DHP is diffuse and similar to that of *cis*-stilbene. In CS the resonance Raman spectrum shows considerable intensity in a number of low-frequency (100-1000 cm^{-1}) modes having predominantly out-of-plane torsional and hydrogen wagging character, and the excited state geometry change along these modes deduced from the Raman intensities has been shown to account for the absence of resolvable stretching progressions in the absorption spectrum [1]. However, DHP seems to have relatively little intensity in low-frequency modes; we have examined the spectrum down to about 100 cm^{-1} and observe no bands with intensities greater than that of the 769 cm^{-1} line shown in Figure 1. Very fast internal conversion of the DHP excited state may make a significant contribution to the electronic spectral breadth; recent femtosecond time-resolved experiments indicate that the ground-state recovery of DHP is complete within ≤0.5 ps [9]. A more quantitative analysis will require higher quality spectra in the low-frequency region as well as accurate absolute Raman cross sections, which are difficult to obtain due to the thermal and photochemical lability of this species.

Our data indicate a two- to three-fold greater rate of photochemical DHP formation in cyclohexane than in methanol. Recent subpicosecond time-resolved studies on CS show nearly a factor of two difference in fluorescence lifetime between alcoholic and hydrocarbon solvents of comparable viscosity [6], and very recent femtosecond pump-probe experiments demonstrate about 2-fold faster formation of DHP from CS in methanol than in hexane, but the yields were not examined quantitatively [9]. These data seem consistent with our results which indicate that the dominance of the isomerization channel over the ring closure pathway is greater in alcohols than in hydrocarbons. We also observe small solvent-dependent differences in the intensities of predominantly out-of-plane vibrations in the ground state *cis*-stilbene resonance Raman spectrum which suggest faster torsional distortion of vertically excited *cis*-stilbene in methanol relative to cyclohexane [16]; however, the analysis is complicated by the fact that these "out-of-plane" modes also have components of in-plane bending that project onto the ring closure pathway [1,17]. Highly accurate ground state normal mode descriptions will be needed to fully interpret the ground-state resonance Raman data.

3. Picosecond antistokes spectra of vibrationally hot *cis*-stilbene photoproducts

Preliminary uv antistokes resonance Raman spectra of the photoproducts formed following excitation of CS have been obtained in a single-pulse configuration. A cw modelocked Nd:YLF laser (Coherent Antares) was used to synchronously pump a home-built dye laser, which was amplified in a two-stage dye amplifier pumped at 500 Hz by the frequency doubled output of a cw-pumped Nd:YLF regenerative amplifier (Quantronix/Medox). The amplified output was frequency doubled to give 5-10 ps, 2 μJ pulses at 289 nm, which were focused with a 50 mm f.l. lens to excite samples of 1 mM CS in cyclohexane in an open flow system.

Figure 3 shows the single-pulse, high-power antistokes spectrum. Reduction of the pulse energy by a factor of five significantly reduces the intensities of all of the stilbene antistokes bands except the 559 cm^{-1} feature, indicating that at the higher pulse energy a population of molecules vibrationally hotter than the initial Boltzmann distribution is being formed within the pulse duration. For comparison, Figure 4 presents the room temperature Stokes spectra of both the cis and trans isomers (excited with nanosecond pulses). The band at 559 cm^{-1} in our picosecond spectrum clearly corresponds to the 560 cm^{-1} band of CS. The other strong antistokes bands at 947, 1172, and 1588 cm^{-1} arise from vibrationally hot molecules, presumably mostly TS (see below), and correspond to the bands observed in

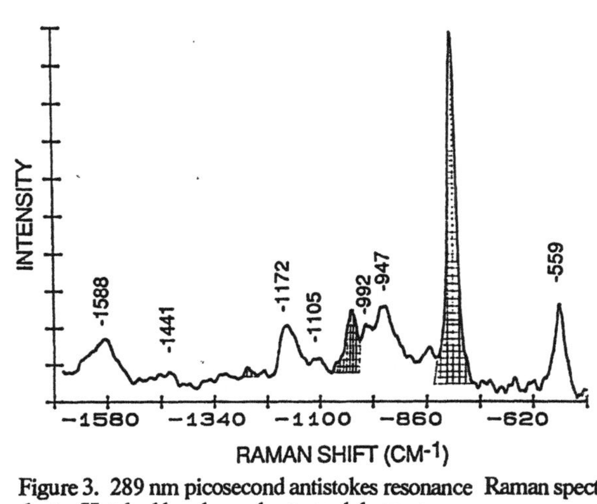

Figure 3. 289 nm picosecond antistokes resonance Raman spectrum of *cis*-stilbene and its photoproducts. Hatched bands are due to cyclohexane.

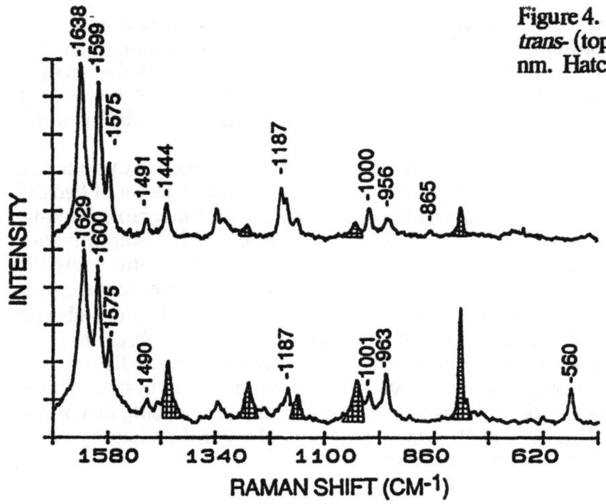

Figure 4. Stokes resonance Raman spectra of *trans*- (top) and *cis*-stilbene (bottom) at 266 nm. Hatched bands are due to cyclohexane.

the room temperature Stokes spectrum at 956 cm^{-1} (hydrogen out-of-plane wag), 1187 cm^{-1} (C-ϕ stretch and H rock), and 1638/1599 cm^{-1} (ethylenic and phenyl double bond stretch) [15,18].

The dependence of the picosecond antistokes spectrum on pulse energy implies that most of the antistokes intensity arises from CS molecules that have absorbed a photon, internally converted back to vibrationally excited levels of the ground state, and then been probed by the trailing edge of the same pulse. Both the formation of S_0 TS and the decay of S_1 CS are reported to occur in less than 2 ps [7-9], and femtosecond pump-probe measurements of TS fluorescence following excitation of CS at 312 nm have been analyzed to infer a vibrational cooling time for S_0 TS of 14 ps [8]. We therefore expect that vibrationally hot populations of both isomers will contribute to our antistokes spectrum. The observed frequencies alone cannot be used to distinguish between the isomers, as the antistokes bands we observe are quite broad and are shifted down in frequency by 10-40 cm^{-1} from the corresponding room temperature Stokes bands of either isomer. Broadening of the bands and shifting to lower frequencies is expected for a distribution of vibrationally excited and perhaps conformationally distorted molecules [10,11]. Although definite assignments cannot yet be made, we believe that our picosecond antistokes spectra arise predominantly from vibrationally hot S_0 TS molecules formed through photoisomerization of CS within the 5-10 ps pulse duration. While the published quantum yields [13,14] imply a partitioning of excited CS among ground-state CS, TS, and DHP in yields of 55, 35, and 10%, respectively, the TS is expected to dominate the 289 nm excited Raman spectrum because it is the strongest absorber; the absorption coefficient of TS at 289 nm is about 2.6 times that of CS, and the Stokes scattering from room-temperature TS is roughly a factor of 7 stronger than that of CS near this wavelength [19]. (This does not take into consideration the absorption red-shift of the vibrationally hot products [8], which will alter the resonance Raman enhancements.) Since no strong antistokes bands appear near 1474 cm^{-1}, DHP apparently does not make a significant contribution to our 289 nm excited antistokes spectrum, consistent with its lower quantum yield for formation and its relatively weak absorption at this wavelength [14]. Electronically excited *cis*-stilbene is unlikely to contribute much to the spectrum in view of its very short lifetime [6-9], although it cannot be ruled out without knowledge of its full uv absorption spectrum; data obtained down to 320 nm suggest that the 289 nm absorbance may be strong [7].

We are presently extending this work to two-pulse, one- and two-color pump-probe measurements which will enable us to time resolve the antistokes spectrum and distinguish among the various contributing species. This, combined with computational modeling of the frequencies and intensities based on our considerable knowledge of the ground-state vibrations and excited-state potential energy surfaces of *cis*- and *trans*-stilbene [1,15,20], should provide a detailed picture of the energetic and conformational relaxation pathways following photoexcitation of *cis*-stilbene in solution.

References

1. A.B. Myers and R.A. Mathies, J. Chem. Phys. 81 (1984) 1552.
2. A.B. Myers and R.A. Mathies, in: Biological Applications of Raman Spectroscopy, ed. T.G. Spiro (Wiley, New York, 1987) p. 1.
3. A.B. Myers, J. Opt. Soc. Am. B 7 (1990) 1665.
4. B.I. Greene and T.W. Scott, Chem. Phys. Lett. 106 (1984) 399.
5. F.E. Doany, R.M. Hochstrasser, B.I. Greene and R.R. Millard, Chem. Phys. Lett. 118 (1985) 1.
6. D.C. Todd, J.M. Jean, S.J. Rosenthal, A.J. Ruggiero, D. Yang and G.R. Fleming, J. Chem Phys. 93 (1990) 8658.
7. S. Abrash, S. Repinec and R.M. Hochstrasser, J. Chem. Phys. 93 (1990) 1041.
8. R.J. Sension, S.T. Repinec and R.M. Hochstrasser, J. Chem. Phys. 93 (1990) 9185.
9. S.T. Repinec, R.J. Sension, A.Z. Szarka and R.M. Hochstrasser, J. Phys. Chem., submitted.
10. T.L. Gustafson, D.M. Roberts and D.A. Chernoff, J. Chem. Phys. 81 (1984) 3438.
11. T.L. Gustafson, K. Iwata, W.L. Weaver, L.A. Huston and R.L. Benson, in: Laser diagnostics of biological molecules: Linear and nonlinear methods. Proc. SPIE 1403, ed. S.A. Akmanov (1990).
12. H. Hamaguchi, Chem. Phys. Lett. 126 (1986) 185.
13. S. Malkin and E. Fischer, J. Phys. Chem. 68 (1964) 1153.
14. K.A. Muszkat and E. Fischer, J. Chem. Soc. B (1967) 662.
15. A.B. Myers, M.O. Trulson and R.A. Mathies, J. Chem. Phys. 83 (1985) 5000.
16. J.-M. Rodier, X. Ci and A.B. Myers, Chem. Phys. Lett., submitted.
17. H. Petek, K. Yoshihara, Y. Fujiwara, Z. Lin, J.H. Penn and J.H. Frederick, J. Phys. Chem. 94 (1990) 7539.
18. G. Baranović, Z. Meić, H. Güsten, J. Mink and G. Keresztury, J. Phys. Chem. 94 (1990) 2833.
19. A.B. Myers, Ph.D. dissertation, University of California, Berkeley (1984).
20. X. Ci and A.B. Myers, Chem. Phys. Lett. 158 (1989) 263.

Resonance Raman Studies of CH_3I Sub-Picosecond Mode-Specific Photodissociation and Radiation-Induced Pure Dephasing Effects

L.D. Ziegler

Department of Chemistry, Northeastern University, Boston, MA, USA

1. Introduction

The polarization characteristics of spontaneous resonance Raman (RR) scattering of free rotors are sensitively dependent on the dephasing time of the resonant optical transition. [1] When this time falls in the time regime between a vibrational and rotational period this frequency domain based polarization technique can determine, with considerable precision, vibronically mode specific lifetimes due to ultrafast dynamical processes in excited electronic states. [2] The Raman scattering of methyl iodide vapor excited by radiation resonant with the predissociated Rydberg B-state (201 nm-190 nm) provides an excellent example of this method and reveals rather unexpected quantum specific predissociation effects. [3] Due to the rapid photochemistry all B-state vibronic absorption bands are rotationally diffuse. However, the methyl iodide bond breaking is not so fast as to obscure most vibrational resolution. The goal of these studies is to uncover, on this rapid time scale corresponding to a few vibrational periods, how the photodissociation rate of this isolated molecule depends on excess total vibrational energy above the electronic origin, vibrational symmetry and type of initial nuclear motion excited on the multidimensional electronic surface. It may be anticipated that when statistical descriptions of energy randomization do not apply, unimolecular reaction rates may be enhanced or appear mode selective if the initial excitation can be localized in the apparent reaction coordinate. Such expectations are tested by these studies.

When the Raman excitation bandwidth approaches the homogeneous linewidth of the resonant electronic transition then the polarization values, as well as the spectral redistribution of the resonant secondary radiation, will change due to this radiation based source of pure-dephasing. [4] Thus, even for isolated molecules, resonance fluorescence features may be observed due to the finite bandwidth of the incident driving fields. These effects have been observed for the resonance emission of molecular oxygen when monochromatic excitation is replaced by partially-coherent laser excitation from an ArF excimer laser. [5]

2. Theoretical Considerations

The theoretical description of the resonance emission of isolated molecules used here is based on a sum-over-all rovibronic states Kramers-Heisenberg (KH) tensoral approach. [1] Only when the angular dependence of all molecular states, i.e. initial, resonant and final, are explicitly considered in a KH framework (or time domain equivalent) are the intensities and hence polarization properties of the RR scattering of free rotors properly captured. For the experiments of interest here all excited state dephasing is due to predissociation in the limit of monochromatic excitation at the experimental pressures employed here. When the excited state dephasing constant corresponds to times on the order of a vibrational period, i.e. much shorter than a rotational period, off resonance or classical values of the depolarization ratio are obtained. However, when the excited state lifetime is between a rotational and vibrational period totally

Springer Proceedings in Physics, Vol. 68 Time Resolved Vibrational Spectroscopy V
Ed.: H. Takahashi

symmetric vibrational Raman bands have depolarization ratios larger than their off-resonance values. In our frequency domain approach this phenomenon can be seen to arise from the contribution of all three scattering invariants (isotropic, anisotropic and antisymmetric) to the RR scattering cross-section regardless of the symmetry of the Raman vibrational transition. [1] The dispersion of the Raman depolarization ratio as the excitation is tuned through an absorption system is consequently a sensitive probe of the resonant excited state lifetime with vibronic specificity.

In the absence of material pure-dephasing, the resonant secondary radiation spectrum excited by incoherent radiation is just the convolution of the incident spectral density with the scattering cross-section due to monochromatic radiation. [4] When this spectral density is larger than the resonant homogeneous linewidth the resonant emission is most properly described as resonance fluorescence (RF). When these widths are comparable complicated lineshapes including interferences between RR and RF components may be observed. Furthermore, off-resonant Raman-like (RL) contributions may be enhanced relative to FL components in a RSR spectrum excited by resonant incoherent excitation. The resulting bandshape may not be clearly assignable under these circumstances as FL or RR. [5]

3. Experimental

The experimental set-up has been previously reported. [1-3,5] Tunable radiation in the 202 to 190 nm region is produced by the fourth anti-Stokes Raman shift in H2 of doubled Nd:YAG pumped red dye output. The scattering of CH_3I in a flowing stream or in a He seeded supersonic jet was observed. O_2 in a static cell (~1 atm) was used in the pure-radiation induced dephasing experiments. Polarization analysis of the spontaneous scattered radiation is accomplished with a stacked plate polarizer.

4. Results

4.1. Polarization Analysis of Methyl Iodide Photodissociation

Vibronically specific lifetimes of the predissociated bands of the B-state of methyl iodide vapor are determined by fits of theory to the observed depolarization ratio dispersion as Raman excitation is tuned through vibronic absorption regions. Observed polarization values and the corresponding best-fits to theory for resonance with two vibronic B-state levels are shown in Figure 1. The vibronic lifetimes for the B-state of both CH_3I and CD_3I determined by this technique are summarized in Table 1.

As seen in Table 1, the only initial nuclear motion found to accelerate this photodissociation upon excitation to this B-state surface is the totally symmetric C-H stretching mode. In contrast to physically intuitive expectations, excitation of one quantum of C-I stretch decreases the photodissociation rate by a factor of three. The v_3 dependence may be understood in terms of a simple Fermi-Golden rule picture. Due to the location of the curve crossing seam between the bound B-state and the unbound surface responsible for the predissociation Franck-Condon overlaps of the corresponding vibrational wavefunctions will have a strong quantum dependence. For the case of this B-state photodissociation the continuum wavefunction overlap is decreased for the $v_2 = 1$ level as compared to the origin level due to the nodal pattern of the $v = 1$ vibrational wavefunction. In addition, these predissociation rates do not appear to be directly dependent on total energy or vibrational mode symmetry. These data determine the location of the curve crossing seam along the multidimen-

Predissociation Dynamics and Structure of the Higher Vibronic Levels in the Methyl Radical Rydberg 3s State

S.G. Westre[1], *L.D. Ziegler*[2], *and P.B. Kelly*[1]

[1]Department of Chemistry, University of California, Davis, CA 95616, USA
[2]Department of Chemistry, Northeastern University, Boston, MA 02115, USA

The structure and dynamics of the [0100] level in the $\tilde{B}$ $^2A_1'$ Rydberg 3 s state were investigated by examining the Raman spectrum of CD_3 in resonance with the [0100]-[0100] $\tilde{B}$-$\tilde{X}$ transition. Figure 1 shows Raman excitation of the totally symmetric C-D stretch which exhibits a strong Q branch and resolved R and S rovibrational features. Raman excitation profiles for the S(2) through S(9) Raman lines were obtained by measuring the intensities of the S lines as a function of excitation wavelength. The J′ specific excited state dephasing constants, Γ, were extracted by modelling the Raman excitation profiles with the intensity expressions for resonance Raman scattering developed by Ziegler et al. [1]. For the methyl radical dissociation, the excited state mean lifetime, τ, is related to the dephasing constant by $\tau=(2\pi c\Gamma)^{-1}$. The CD_3 lifetimes decrease from 350 fs for J′ = 3 to 150 fs for the J′ = 10 level. The lower rotational levels exhibit a slight alternation in lifetime as a function of rotational level suggesting the predissociation may be enhanced via Coriolis coupling of the ν_2 and ν_3 motions. The predissociation rates (given as $1/\tau$) are approximately two times faster than those in the corresponding

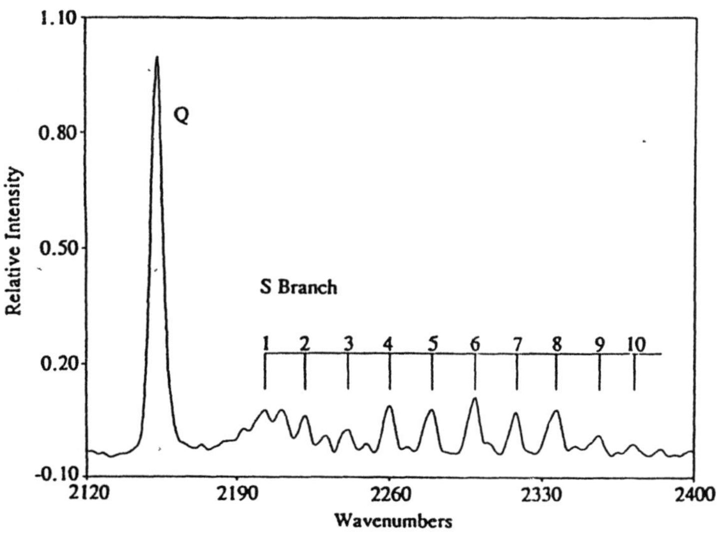

Figure 1. Raman spectrum of CD_3 at 211.5 nm in resonance with the [0100]-[0100] absorption.

Springer Proceedings in Physics, Vol. 68 Ed.: H. Takahashi
Time Resolved Vibrational Spectroscopy V

[0000] level [2]. This is in marked contrast to the Rydberg 3 s state of ND_3 where the dissociation rate from $\nu_2=1$ is estimated to be three times slower than in the [0000] level [3]. The difference in relative dissociation rates for the methyl radical and ammonia can be attiributed to an exit channel that is wider with respect to the out of plane coordinate for the methyl radical. The modelling of the excitation profile data yields a band origin for the [0100]-[0100] transition of 47271 cm^{-1}. Our previous resonance Raman study determined the CD_3 [0000]-[0000] band origin to be 46635 cm^{-1}. Using the ground state ν_2 value of 457.8 cm^{-1} from Frye et al.[4], the ν_2 frequency in the Rydberg 3 s state of CD_3 is estimated at 1094 cm^{-1}. This is consistent with the absorption spectra of Callear and Metcalfe [5] and agrees with Hirota's analysis of the pseudo Jahn-Teller coupling of the ground $\tilde{X}\,^2A_2'$ state to the $\tilde{B}\,^2A_2'$ state by a vibronic interaction involving ν_2 [6].

We have examined the CD_3 and CH_3 absorption features centered at 207 nm. The only features present in the Raman spectrum are ν_1 and $2\nu_1$. These stretching vibrations appear in the Raman spectrum with equal intensity indicating resonance with the $\nu_1=1$ level of the $\tilde{B}$ state. No vibrations were observed in the anti-Stokes region suggesting that this transition originates in the v=0 level of the $\tilde{X}$ state. These two observations lead us to confirm Callear and Metcalfe's identification of this vibronic transition as [1000]-[0000]. We obtained vibrational Raman excitation profiles of this absorption feature by following the intensities of ν_1 and $2\nu_1$ Q branches as a function of excitation wavelength. The Raman excitation profiles were modelled to yield the center and bandwidth of the feature. For CH_3 this yields a band origin of 206.85 nm. Combining this with the [0000]-[0000] band origin yields a frequency for ν_1 in the $\tilde{B}$ state of 2040 cm^{-1}. The dephasing constant for the CH_3 [1000] level is modelled to be 400 cm^{-1} corresponding to an excited state lifetime of 13 fs.

References

1. L.D. Ziegler, Y.C. Chung, P. Wang, and Y.P. Zhang, in *Time Resolved Spectroscopy*, edited by R.J.H Clark and R.E. Hester (Wiley, New York, 1989) and references therein.
2. S.G. Westre, P.B. Kelly, Y.P. Zhang, and L.D. Ziegler, J. Chem. Phys. **94**, 270 (1991).
3. J. Biesner, L. Schnieder, G. Ahlers, X. Xie, K.H. Welge, M.N.R. Ashfold, and R.N. Dixon, J. Chem. Phys. **91**, 2901 (1989).
4. J.M. Frye, T.J. Sears, and D. Leitner, J. Chem. Phys. **88**, 5300 (1988).
5. A.B. Callear and M.P. Metcalfe, Chem. Phys. **14**, 275 (1976).
6. E. Hirota, *High-Resolution Spectroscopy of Transient Molecules*, (Springer-Verlag, Berlin, New York, 1985), p.175.

Part VI

Vibrational Dynamics

Intramolecular Vibrational Redistribution (IVR) of Jet-Cooled Alkylphenols and Alkylanilines as Revealed by Stimulated Emission Ion Dip Spectroscopy

M. Ito and T. Ebata

Department of Chemistry, Faculty of Science,
Tohoku University, Sendai 980, Japan

Intramolecular vibrational redistribution (IVR) is an important nonradiative process in an isolated large molecule, and it is being extensively studied experimentally and theoretically. Especially, IVR in an electronically excited state has been studied by various experimental means such as fluorescence excitation, dispersed fluorescence spectroscopies and the measurement of fluorescence life-time [1]. As a result, our information on IVR for electronically excited states is now considerably accumulated. On the other hand, the study on IVR in electronically ground-state is very few. This is due to lack of suitable experimental means. The study by infrared radiation is an orthodox way. However, because of poor time response of the infrared detection and of severe selection rule for infrared absorption, the IVR study of a ground state molecule by infrared light is greatly restricted. Instead of direct vibrational excitation by infrared light, the ground-state vibrational level of an isolated molecule can be populated by stimulated emission from an electronically excited state, say, S_1 state with UV/visible laser light (ν_2). One of the stimulated emission methods is "stimulated emission pumping" in which the stimulated emission from S_1 to a ground-state vibrational level is monitored by the dip of the fluorescence from the S_1 level pumped with ν_1[2,3]. The depth of the fluorescence dip represents the decrease of the S_1 state population which is determined by the loss of the S_1 state population by the stimulated emission with ν_2 and the gain of the population by the reabsorption with same ν_2 from the ground-state vibrational level. When the life-time of the ground-state vibrational level to which the stimulated emission occurs is long, the molecules in the vibrational level have a great chance to come back to S_1, resulting in a small fluorescence dip. Conversely, when the life-time (decay rate) is short (large), we have a large dip. Since the life-time of the ground-state vibrational level of an isolated molecule is in most cases determined by IVR process, we can obtain the IVR rate from the observed depth of the fluorescence dip. Therefore, the depth of the fluorescence dip measured in percentage relative to the fluorescence signal in absence of the stimulated emission provides us with very useful information on the IVR rate of ground-state vibrational level. However, the quantitative determination of the percentage fluorescence dip depth is very difficult because it requires complete spatial matching of the two laser beams ν_1 and ν_2. The observed percentage dip depth sensitively varies by mismatching of the two beams whose elimination is practically impossible.

Recently, Suzuki et al. developed another version of stimulated emission method [4-6]. In this method, the first laser light ν_1 pumps a jet-cooled molecule to a particular level in S_1, and the second laser beam ν_2 is introduced to induce the stimulated emission from S_1 to the vibrational level in S_0 (See Fig. 1). ν_2 is also used for the ionization of the S_1 state molecule. The ν_1 laser power is suppressed as low as possible, therefore, the ionization by ν_1 alone does not occur. Under this condition, all the ion signals occur only in the sample space where both ν_1 and ν_2 coexist. When the ν_2 frequency coincides with the energy difference between S_1 and a ground-state vibrational level, the stimulated emission occurs and the population of the S_1 state decreases. The decrease of the population is detected as a dip of the ion signal due to $\nu_1 + \nu_2$ ionization. This method my be called "stimulated emission ion dip (SEID) spectroscopy". The method has a great advantage to obtain quantitative data of the ion dip intensity. If we measure the dip intensity by the percentage depth of the ion dip relative to the ion signal in the off-resonant condition, this percentage dip intensity is not seriously influenced

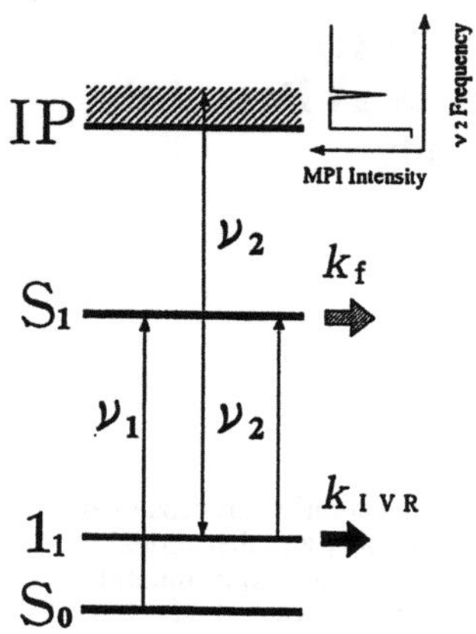

Fig. 1
Energy level and schematic diagram of stimulated emission ion dip (SEID) spectroscopy.

by mismatching of the two laser beams ν_1 and ν_2 because both the ionization and stimulated emission occur only in the space where the two laser beams overlap. The quantitative ion dip intensity leads us to the determination of reliable IVR rate of the ground-state vibrational level which is very difficult by other means. Therefore, we use here SEID spectroscopy for the study of the IVR rate of ground-state vibrational levels.

We applied SEID spectroscopy for the determination of the IVR rate of the ground-state vibrational levels of o-alkylphenols and p-alkylanilines having the side carbon chain of n = 1-6. It was shown in a previous paper [7] that the IVR rate of the ground-state vibrational level of phenol is very small. We are interested in how the IVR rate of phenol changes when an alkyl chain is put at p-position of the molecule. The alkyl chain has low frequency vibrations such as C-C torsions. The density of the vibrational state of phenol increases by the addition of such low frequency modes and it increases with the chain length. It is expected therefore that the IVR rate will increase with the increase of the chain length.

Fig. 2 shows the stimulated emission ion dip spectra of p-methylphenol (p-cresol), p-ethylphenol, p-propylphenol and p-hexylphenol in supersonic jets. All the spectra were obtained by exciting the molecules to their zero-point levels in S_1 with the first laser light of ν_1 and by scanning ν_2 which is used for the stimulated emission as well as for the ionization. In the case of methylphenol, the dip intensity is very weak similar to the case of phenol. In this molecule the low frequency modes due to internal rotation of the CH_3 group are added to the phenol molecule. However, the observed spectrum indicates that these internal rotational modes is not effective to increase the IVR rate. In p-alkylphenol, the dips appear with appreciable intensities. The appearance of the dips suggests effective contribution of the C-C torsion of the ethyl group to the IVR rate. The dip intensity becomes larger in p-propylphenol, and increases further in p-hexylphenol. However, the increase seems to be saturated as the chain becomes longer.

For each molecule, we observed the laser power dependence of the dip intensities and the IVR rates of several vibrational levels were estimated using the rate equations. The results are summarized in Table 1. The mode 1 is one essentially localized in the phenol moiety. Therefore, the rate obtained here gives a measure of vibrational flow of the localized mode into isoenergetic vibrational states which are mainly contributed from the low-frequency chain modes. As seen from the table, the rate is very small for p-methylphenol although only its upper limit was estimated. The rate increases rapidly with an increase of the chain length. But the increase tends to become slow for the molecules of $n \geq 3$ or 4. The results shows that the IVR rate is primarily determined by the density of vibrational state which rapidly increases with an increase of the chain length. However, the trend of the saturation suggests that the vibrational modes effectively coupled with the ring mode are restricted to ones in which the motions are localized in the chain near the ring and the modes involving the motions far from the ring are not effective in the coupling. In the same table, the IVR rate of the mode 1 in the S_1 state is given for three alkylphenols. The rates were determined from the dispersed fluorescence spectra from the vibrational levels of the mode 1 in S_1. The rates in the S_1 state

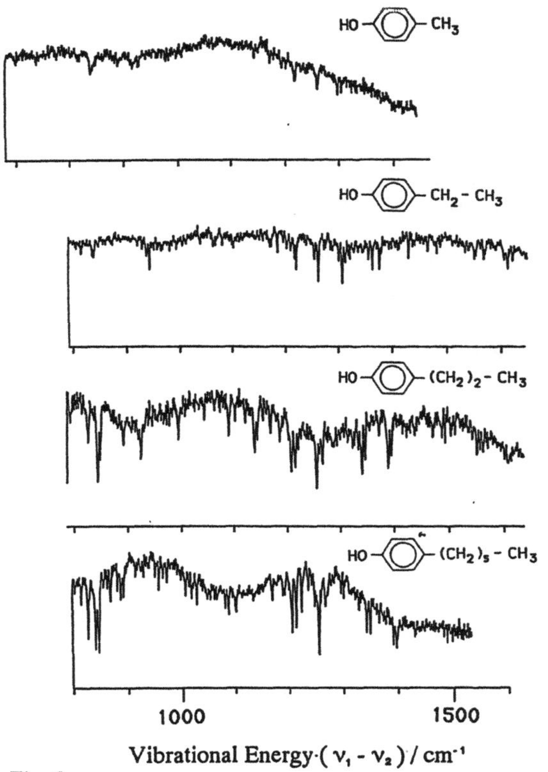

Fig. 2

SEID spectra of jet-cooled p-alkylphenols. All the spectra were obtained by exciting the molecules to their zero-point levels in S_1 with ν_1 and by scanning ν_2 which is used for the stimulated emission as well as for the ionization.

Table 1. IVR rate constant of mode 1 of p-alkyl phenols determined by SEID spectroscopy

k / x 10^9 sec^{-1}

	p-cresol 1_1	p-ethyl phenol 1_1	p-propyl phenol 1_1	p-butyl phenol 1_1	p-hexyl phenol 1_1
S_1	0.25	3.3	5.1		
	(822 cm^{-1})	(817 cm^{-1})	(815 cm^{-1})	(809 cm^{-1})	(816 cm^{-1})
S_0	< 0.1	2.0	3.3	5.0	5.0
	(839 cm^{-1})	(836 cm^{-1})	(844 cm^{-1})	(849 cm^{-1})	(843 cm^{-1})

are comparable in the order of magnitude with those in the ground state, suggesting that the IVR rate is roughly independent of the nature of electronic state.

We also determined the IVR rates of the ground-state vibrational levels of p-alkylanilines by SEID spectroscopy. The results obtained are similar to those for p-alkylphenols. The IVR rate progressively increases with increase of the chain length and the IVR rate in the ground-state is also comparable with that in the excited state.

In conclusion, stimulated ion dip spectroscopy is useful for the study of the IVR rates of large polyatomic molecules.

References

[1] See, for example, (a) C. S. Parmenter, J. Phys. Chem. **86**, 1736 (1982), (b) R. E. Smalley, J. Phys. Chem. **86**, 3504 (1982), (c) R. E. Smalley, Ann. Rev. Phys. Chem. **34**, 129 (1983).
[2] C. E. Hamilton, J. L. Kinsey and R. W. Field, Ann. Rev. Phys. Chem. **37**, 493 (1986).
[3] H. L. Kim, S. Reid and J. D. McDonald, Chem. Phys. Lett. **132**, 361 (1986).
[4] T. Suzuki, N. Mikami and M. Ito, Chem. Phys. Lett. **120**, 333 (1985).
[5] T. Suzuki, N. Mikami and M. Ito, J. Phys. Chem. **90**, 6431 (1986).
[6] T. Suzuki, M. Hiroi and M. Ito, J. Phys. Chem. **92**, 3774 (1988).
[7] T. Ebata, M. Furukawa, T. Suzuki and M. Ito, J. Opt. Spc. **Am. B**, 1890 (1990).

The Influence of a Nearly Free Methyl Rotor on the Dynamics of Vibrational Predissociation and IVR

Z.-Q. Zhao and C.S. Parmenter

Department of Chemistry, Indiana University, Bloomington, IN 47405, USA

1. Introduction

Intramolecular vibrational redistribution (IVR) rates can be profoundly changed by simple chemical modifications. An example occurs in the molecular pair p-difluorobenzene (pDFB) and p-fluorotoluene (pFT). These molecules are related by the replacement of a pDFB fluorine with a methyl group to produce pFT. IVR in 300 K gas phase molecules has been characterized for regions of the S_1 manifold where the ring mode density is 100-2000 states per cm^{-1} [1,2]. The methyl substitution increases IVR rates by about a hundred-fold [2]. The time scales change from 15-300 psec to sub-psec.

A different modification of pDFB is equally interesting. If we expand pDFB with Ar in a supersonic jet, pDFB-Ar van der Waals complexes are produced in which a new IVR process occurs [3]. This IVR involves coupling of the ring modes with the van der Waals modes created by Ar motion against the ring. The time scale of this IVR on the order of 5000 psec. The rate is slow on account of the frequency mismatch between ring modes and the van der Waals modes. Once again, however, methyl substitution to make the pFT-Ar complex increases the rate of this new IVR by at least an order of magnitude.

The interesting question is why the methyl group has such a big effect on IVR. We know that the eight methyl vibrational modes are not involved. The effect is due to the methyl internal rotation. Since the S_1 rotation barrier is only 34 cm^{-1} [4], the rotor is nearly free. Hence we ask how a nearly free internal rotation can have such a prominent IVR influence?

2. Theoretical and Experimental Approaches.

Moss et al. [5] provided a theoretical interpretation. They showed that "internal collisions" between the methyl hydrogens and adjacent ring atoms provides a natural coupling between internal rotation (i) and ring vibrations (v). This i•v interaction opens new paths for coupling states with different ring mode identities and does so with large (1-10 cm^{-1}) coupling matrix elements.

We explore the methyl interactions experimentally by using cold jet S_1-S_0 fluorescence spectroscopy to establish the positions and zero order assignments of levels on the first 410 cm^{-1} of the pFT S_1 manifold. We then hunt for details of specific interactions between these levels. We have explored pFT with a light ($-CH_3$) or heavy ($-CD_3$) rotor. Our fluorescence study extends that of Okuyama et al. [4] who used the spectroscopy to establish the pFT-h_3 S_1 internal rotation energy barrier and energy levels.

3. Results and Discussion

3.1 Methyl Interactions in pFT

The S_1-S_0 fluorescence excitation spectra show the positions of some S_1 levels [4,6]. The dispersed fluorescence spectra after pumping S_1-S_0 absorption bands establish the zero order assignments. These data not only reveal the internal rotation levels that Okuyama et al. assigned but also vibrational levels and levels involving combination of vibration and internal rotation. Comparisons of spectra from cold and warmer molecules and from pFT-h_3 vs. pFT-d_3 help secure band assignments.

The full level structure in the range 0-410 cm^{-1} of pFT-h_3 encompasses 117 states. Twelve are levels of the internal rotation alone. Twenty are levels of ring vibrations, involving eight modes. The remaining 85 levels are combinations involving both ring modes and internal rotation. The pFT level count in this region exceeds that of pDFB by about a factor of six.

There is abundant evidence in the fluorescence excitation spectrum for strong interaction among degrees of freedom. About half of the bands are zero-order symmetry forbidden. Additionally the rotational contour of almost every band is unique. Simulated contour analysis [6] shows that the variety is established by the internal rotation-overall rotation interaction (i•r) that is described by the microwave spectroscopists [7]. Such an interaction can also turn on the forbidden rotor bands. Other bands involving ring vibrations gain intensity from vibronic interaction.

Many specific couplings between zero-order states are established from analysis of dispersed fluorescence. For example, we have recently described [6] a pFT-d_3 interaction between a pumped rotor state m^6 (the rotor quantum number $m' = 6$) and two other states X^1m^0 and 30^2m^0 lying 33 cm^{-1} and 112 cm^{-1} away, respectively. ($\nu_x' = 123$ cm^{-1} is a ring mode of unknown identity.) The coupling matrix elements are large: 10 cm^{-1} and 18 cm^{-1}, respectively. Note that the rotor has provided a path to couple states with different vibrational identity. Such interactions are precisely those that underlie IVR.

The interactions prove ubiquitous. In pFT-h_3 for example, we were able to examine dispersed fluorescence from 36 levels in the 0-410 cm^{-1} region. Fifteen levels were seen to be mixed with coupling matrix elements exceeding 1 cm^{-1}.

We can now use symmetry considerations to define the simplest mechanism that can produce these interactions. The i•v coupling described by Moss et al. [5] accounts for some but not all. This i•v interaction occurs naturally in pFT but so does the i•r interaction. It is therefore likely that all three degrees of freedom interact to form the wonderful acronym i•v•r. This interaction satisfies the symmetry requirements of the remaining couplings.

3.2 IVR in pDFB-Ar and pFT-Ar.

Our view of IVR in the van der Waals complex derives from modelling of the dissociation [3]. Early attempts [8,9,10] to explain the specific vibrational states in which the pDFB dissociation product is formed were based on the assumption of direct coupling between the initially pumped ring level and the dissociation continuum. While these standard approaches to vibrational predissociation (VP) replicated the phenomenon of mode selectivity in the pDFB dissociation product, they failed completely to account for the states actually observed.

We have followed the example of those who model VP of the s-tetrazine-Ar complex [11], particularly Kelley and Bernstein [12], by including the possibility of VP from states in the complex reached by IVR. For the pDFB-Ar complex, we have assumed that the IVR occurs as a discrete kinetic step that always precedes dissociation [3]. This modelling then leads naturally to predictions of the correct dissociation channels. While the modelling is qualitative, the correct predictions of VP channels occurs for so many of the initially pumped complex states that the IVR assumption appears valid.

IVR within the pDFB-Ar complex involves coupling of an initially pumped ring level that has no excitation of the three van der Waals modes to isoenergetic states that are built from combination of some lower energy ring level with excited van der Waals modes. IVR is thus a "flow" of vibrational energy from ring levels into the van der Waals modes. The levels reached by IVR occur in sets containing a common ring level component. In pDFB-Ar, IVR from an initially pumped ring level is highly selective in that the coupling to one of these sets usually exceeds greatly the coupling to others. The IVR selectivity occurs because of a propensity to minimize quantum number changes in ring modes.

The IVR selectivity plays a dominant role in establishing mode selectivity in dissociation of the pDFB-Ar complex. Such a role is ironic. IVR usually destroys mode selective chemistry. Here it creates it.

We can perhaps anticipate what might change upon methyl substitution to make pFT-Ar. Since methyl substitution to make free pFT introduces new paths to couple states involving different ring modes, we might expect the same in pFT-Ar. Hence IVR might be faster, as is observed. Another result would be a diminished mode selectivity in pFT-Ar dissociation on account of much reduced selectivity in the IVR.

The dissociation is in accord with such expectations. For example dissociation of pDFB-Ar pumped to the complex S_1 state 5^1 (819 cm^{-1}) produces the free pDFB molecule in two dominant vibrational states from a choice of more than 20 possibilities. Dissociation of pFT-Ar from the same 5^1 state (803 cm^{-1}) produces free pFT in at least 12 states. Such a qualitative difference between the two is quite in accord with the anticipated methyl effect.

Another observation concerning the pFT-Ar dissociation is also consistent with enhanced IVR occurring within the complex. The initially pumped S_1 complex state has no excitation of the methyl rotor. The pFT molecules formed by dissociation, on the other hand, are born with the rotor excited! These rotor quantum changes are just those associated with IVR, as typified by the specific example of state coupling given in the preceding section.

4. Summary

S_1-S_0 cold jet fluorescence excitation and dispersed fluorescence spectroscopy in pFT has been used to characterize the interactions between the methyl internal rotation and the ring vibrations that accelerate IVR in pFT. The zero order assignments and positions of all levels in the first 410 cm^{-1} of the pFT-h_3 and pFT-d_3 S_1 manifolds have been established. Over 100 levels occur and most are combinations of ring mode vibrations with internal rotation. Clear evidence is seen for interaction between internal rotation and over all rotation. Many specific couplings with large matrix elements between individual levels are identified. The simplest mechanisms that can generate the interactions are i•v as described by Moss et al. [5] and i•v•r.

It is seen in these results that a single added degree of freedom generated by the transformation of pDFB to pFT, namely the internal rotation, not only introduces a big increase in level density, but generates effective mechanisms for coupling many of these levels. If we extrapolate to S_1 regions of high density where IVR resides, we can see that the i•v and i•v•r interactions would persist to generate the big effect on IVR.

One would presume that these couplings would occur within the van der Waals complex pFT-Ar. We would predict that they would have a prominent influence on IVR involving coupling of ring modes with van der Waals modes. Comparisons of pFT-Ar dissociation with that observed for pDFB-Ar complexes is consistent with this idea. First, the IVR is faster in pFT-Ar. Second, the dissociation of pFT-Ar produces the free pFT molecule in many vibrational states as opposed to the high selectivity observed for pDFB-Ar. Finally, the pFT-Ar dissociation produces free pFT molecules with excited internal rotation. This transfer of energy to the rotor is almost certainly a consequence of fast IVR within the undissociated complex.

References

[1] K.W. Holtzclaw and C.S. Parmenter, J. Chem. Phys. 84, 1099 (1986).
[2] C.S. Parmenter and B.M. Stone, J. Chem. Phys. 84, 4710 (1986).
[3] H.-K. O, Ph.D. Thesis, Indiana University, 1990.
[4] K. Okuyama, N. Mikami and M. Ito, J. Phys. Chem. 89, 5617 (1985).
[5] D.B. Moss, C.S. Parmenter, and G.E. Ewing, J. Chem. Phys. 86, 51 (1987).
[6] Z.-Q. Zhao and C.S. Parmenter, "Proceedings of the 24th Jerusalem Conference in Quantum Chemistry and Biochemistry", B. Pullman and J. Jortner, Editors. (Kluwer Academic Publishers, Dordrecht, 1991), (in press).
[7] H.D. Rudolph and H. Seiler, Z. Naturforsch. A20, 1682 (1965); P.N. Ghosh, J. Mol. Spectrosc. 138, 505 (1989); *ibid.*, 142, 295 (1990).
[8] K.W. Butz, D.L. Catlett, Jr., G.E. Ewing, D. Krajnovich, and C.S. Parmenter, J. Phys. Chem., 90, 3522 (1986).
[9] H.-K. O, C.S. Parmenter, and M.-C. Su, Ber. Bunsenges. Phys. Chem. 92, 253 (1988).
[10] A.R. Tiller, A.C. Peet and D.C. Clary, Chem. Phys. 129, 125 (1989).
[11] P.M. Weber and S.A. Rice, J. Chem. Phys. 92, 5470 (1988).
[12] D.F. Kelley and E.R. Bernstein, J. Phys. Chem. 90, 5164 (1986).

Femtosecond Time-Resolved CARS: Time Domain Observation of Molecular Vibrational Dynamics

H. Okamoto[1] *and K. Yoshihara*[2]

[1]Department of Chemistry, Faculty of Science,
The University of Tokyo, 7-3-1 Hongo, Bunkyo-ku, Tokyo 113, Japan
[2]Institute for Molecular Science, Myodiji, Okazaki 444, Japan

1. Introduction

Time-domain coherent light scattering techniques such as time-resolved coherent anti-Stokes Raman scattering (TRCARS) have provided important information on dynamics of molecular vibrations and rotations [1]. Many vibrational dephasing in liquids and solids are observed by TRCARS technique. Beating phenomena caused by neighboring rovibrational levels are sometimes observed. A number of theoretical studies have also been devoted to these fields.

Recent technical developments in ultrashort pulse lasers enable us to obtain light pulses shorter than 100 fs. This time scale corresponds to the period of molecular vibrational motions (*e.g.*, 100 fs = 333 cm^{-1}). One can coherently excite more than two vibrational modes at a time by femtosecond pulses, and monitor relaxation processes in the time domain. TRCARS is applicable to a variety of vibrational modes since any Raman active vibrational transitions can be selected if the Stokes laser frequency is tunable. Femtosecond TRCARS experiments, which were first reported by Zinth and co-workers [2,3], showed ultrafast decays and beating patterns of the coherent signal for several molecular liquids. We describe here observations of (polarized) TRCARS from molecules in the liquid phase [4~7].

2. Experimental

The experimental setup is shown schematically in Fig. 1. The setup consists of a fixed-frequency femtosecond laser, a tunable picosecond laser, and a detection system [4]. The polarization of the radiation incident on samples can be arbitrarily chosen.

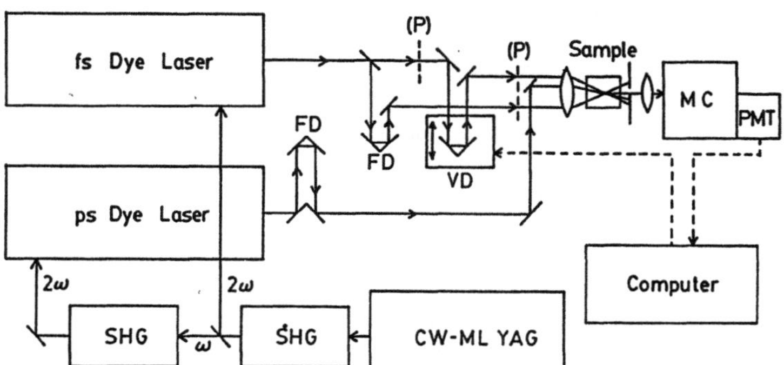

Figure 1 Block diagram of the experimental setup. FD: fixed optical delay, VD: variable optical delay, P: half-wave plate, MC:monochromator.

3. Results and discussion
3.1 Benzene --- Polarized TRCARS

Vibrational dephasing of the benzene 992 cm^{-1} band is examined under various polarization conditions [4]. It is found that two relaxation processes with different time constants are observed depending on the polarizations. The slower one ($\tau = 2.4$ ps), which is a dominant component at the parallel polarizations of the pump and Stokes light, corresponds to the vibrational dephasing. The faster one ($\tau = 0.39$ ps), which dominates at the pump and Stokes polarizations perpendicular, is considered to originate from decay of the intermolecular rotational coherence [8].

3.2 DQOCI --- Rigorous resonant TRCARS

DQOCI (1,3'-diethyl-4,2'-quinolyloxacarbocyanine iodide) has the 0-0 absorption maximum at ~590 nm in a methanol solution, with a very large extinction coefficient. Our pumping laser wavelength is almost exactly tuned to the 0-0 transition of DQOCI. A decaying signal with a time constant of 0.31 ps has been observed when the frequency difference between the pump and Stokes beams are tuned at 1300 cm^{-1} [4]. In general rise and decay signals are expected in the rigorous resonance case [9]. We did not observe the rise of the anti-Stokes signal, probably because of the large inhomogeneous broadening of the electronic transition.

3.3 Carotenoids *in vitro* and *in vivo*

Carotenoids are very important not only as model compounds for the conjugated polyene molecules with mid-range chain lengths but also as biological pigments related to photosynthesis. They have very strong electronic absorption bands which are attributed to the $^1B_u^+ \leftarrow {}^1A_g^-$ transitions around 500 nm, and the Raman intensities are pre-resonantly enhanced. We describe here the observation of the vibrational relaxation of the C=C stretching fundamental vibration (~1520 cm^{-1}) of carotenoids *in vitro* and *in vivo* [5,6].

TRCARS for the in-phase C=C stretching band of β-carotene shows a nearly single exponential decay. The decay time constants (= $T_2/2$) are: 0.30 ps in cyclohexane, 0.28 ps in benzene, 0.29 ps in carbon tetrachloride, 0.28 ps in carbon disulfide, 0.29 ps in chloroform, and 0.30 ps in a cyclohexane/acetone mixed solvent (~1:1 v/v). It may be concluded that the vibrational relaxation rate is almost independent of the solvent. This observation suggests that the vibrational relaxation rate is determined mainly by the intramolecular processes. The solute-solvent interactions seem to have the least bearing upon the relaxation rate. β-Carotene consists of 96 atoms and the density of background (optically forbidden) vibrational states is very high (~10^{12} states/cm^{-1}) at the energy of the C=C stretching band. It is reasonable that the intramolecular relaxation process has a time constant of sub-picosecond regime, considering this extremely high density of vibrational states.

Next, we report observation of femtosecond TRCARS from *in vivo* carotenoids which are contained in intracytoplasmic membranes (chromatophores) of *Chromatium vinosum*, a species of photosynthetic bacteria.

The dominant carotenoids in chromatophores of *Ch. vinosum* are rhodopin and spirilloxanthin. These two carotenoids and β-carotene have nearly equal molecular weights so that the vibrational relaxation rates due to intramolecular processes may be similar among the three. The TRCARS from these carotenoids has been observed

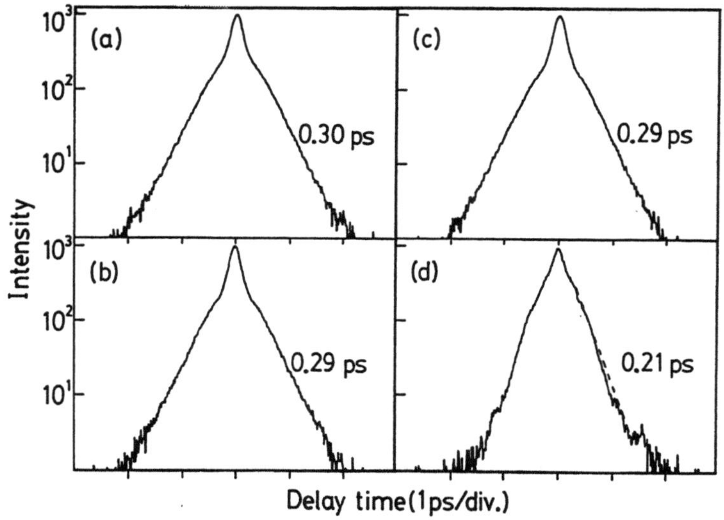

Figure 2 TRCARS from rhodopin and spirilloxanthin in (a) benzene, (b) chloroform, (c) chloroform/acetone mixture, and (d) chromatophores. The dashed line in (d) was used to determine the decay time constant.

in solution of benzene, chloroform, and chloroform/acetone mixture. The decay time constants are found in the range between 0.29 and 0.31 ps, which are solvent independent and almost the same as those of β-carotene. Although rhodopin and spirilloxanthin have polar functional groups at the ends of the chains, intramolecular dissipation processes seem to be more important in the vibrational relaxation of the in-phase C=C stretching than intermolecular interactions between the polar functional group and the solvent molecules.

On the other hand, the decay rate in chromatophores is different. Figure 2 shows TRCARS from rhodopin and spirilloxanthin in solvents (a~c) and in chromatophores (d). A mixture of the two carotenoids (~1:1 in weight) is used in (a~c), where the decay time constants are between 0.29 and 0.30 ps. The decay in chromatophores (~ 0.21 ps) is definitely faster than those in the solvents.

The result obtained for the carotenoids in chromatophores is interesting in view of the fact that the vibrational relaxation rate *in vitro* is solvent independent even in a highly polar environment. The carotenoids may be so tightly bound to the protein that the vibrational motion of the in-phase C=C stretching is coupled to that of the environmental system, causing its faster relaxation. Very recently, Hayashi *et al.* have observed vibrational population relaxation of the in-phase C=C stretching of spirilloxanthin in benzene and in chromatophores by picosecond time-resolved Raman spectroscopy [10]. They have found that the population relaxation rates are nearly equal in benzene and in chromatophores. This is different from the present results on the phase relaxation, although the two kinds of results do not contradict each other. The environment of the pigment-protein complexes of *Ch. vinosum* is different from that *in vitro*, but this difference seems to have little effects on vibrational population relaxation.

3.4 Acetonitrile --- Effects of hydrogen bonding environment

Acetonitrile has a strongly polar group (C≡N) which easily forms a hydrogen bonding with other molecules. Phase relaxation rate of the C≡N stretching vibration is expected to depend on the hydrogen bonding abilities of the environmental molecules. We measured TRCARS for mixtures of acetonitrile and nine kinds of liquid molecules and observed vibrational relaxation of the C≡N stretching [7]. The dephasing rate has been found to be faster with increasing proton-donating ability of the mixed liquid molecule.

References

[1] A. Laubereau, W. Kaiser: *Rev. Mod. Phys.*, **50**, 607 (1978).

[2] R. Leonhardt, W. Holzapfel, W. Zinth, W. Kaiser: *Chem. Phys. Lett.*, **133**, 373 (1987).

[3] W. Zinth, R. Leonhardt, W. Holzapfel, W. Kaiser: *IEEE J. Quantum Electron.*, **24**, 455 (1988).

[4] H. Okamoto, K. Yoshihara: *J. Opt. Soc. Am. B*, **7**, 1702 (1990).

[5] H. Okamoto, K. Yoshihara: *Chem. Phys. Lett.*, **177**, 568 (1991).

[6] H. Okamoto, H. Hayashi, K. Yoshihara, M. Tasumi: *Chem. Phys. Lett.* (in press)

[7] R. Inaba, H. Okamoto, K. Yoshihara, M. Tasumi: *Chem. Phys. Lett.* (submitted)

[8] M. Hayashi, Y. Fujimura, H. Okamoto, K. Yoshihara: *J. Chem. Phys.* (submitted)

[9] H. Okamoto: *Chem. Phys.* (in press)

[10] H. Hayashi, T. Brack, T. Noguchi, M. Tasumi, G. H. Atkinson: *J. Phys. Chem.* (submitted)

Femtosecond Time-Resolved CARS from Acetonitrile and Solvent Effects on the Vibrational Dephasing of the C≡N Stretching Band

R. Inaba[1,2], *H. Okamoto*[1;2], *K. Yoshihara*[2], and *M. Tasumi*[1]

[1]Department of Chemistry, Faculty of Science, The University of Tokyo, 7-3-1 Hongo, Bunkyo-ku, Tokyo 113, Japan
[2]Institute for Molecular Science, Myodaiji, Okazaki 444, Japan

Vibrational phase relaxation in liquids has attracted considerable attention in recent years. It has been demonstrated that the vibrational phase relaxation reflects liquid state structures and dynamics. We report here time-resolved coherent anti-Stokes Raman scattering (time-resolved CARS) for the C≡N stretching of acetonitrile in mixtures with other solvents. The experimental setup for polarized time-resolved CARS was already described [1]. Duration of the excitation laser pulse was 85 fs. Polarization condition was so chosen that the time-resolved CARS signal was free from the molecular rotational contribution.

Typical example of time-resolved CARS for the C≡N stretching band of acetonitrile are shown in fig.1(A). It is evident that the dephasing rate is influenced by the solvent mixed. Apart from the beating structure found in the region < 3 ps, the CARS signal for the C≡N stretching shows a non-exponential decay. This behaviour is ascribed to interference between fundamental and hot band transitions. This interference structures prevent us to estimate exact vibrational dephasing time constants from the time-resolved CARS. In order to discuss the solvent effect, we define an "apparent decay time constant" in the 0.72-2.32 ps delay time region, where each signal seems to decay almost exponentially. This apparent decay time constant may differ slightly from the genuine vibrational dephasing time. However, this parameter is useful for the discussion of solvent effects as is shown in the following.

In fig. 1(B) the apparent decay time constants are plotted against a parameter α, which represents proton donating ability of a solvent and is determined from NMR and electronic absorption data [2]. The apparent decay time constant is clearly correlated to α in fig. 1(B). A larger value of α gives a faster decay of the CARS signal. This result can be qualitatively explained as follows. The hydrogen bond combines the acetonitrile molecule and a solvent molecule, and couples the C≡N stretching mode of acetonitrile with intramolecular vibrational modes of the solvent molecule and/or with intermolecular modes. As the hydrogen bond increases in strength, the coupling becomes stronger, and the relaxation of the C≡N stretching vibrational motion becomes more rapid. This tendency does not hold in the acetic acid solution. The apparent decay time of the C≡N stretching in acetic acid solution corresponds to that in a non-hydrogen-bonding solvent like CCl_4. It is probably because the acetic acid molecules form dimers, and the carboxyl group of

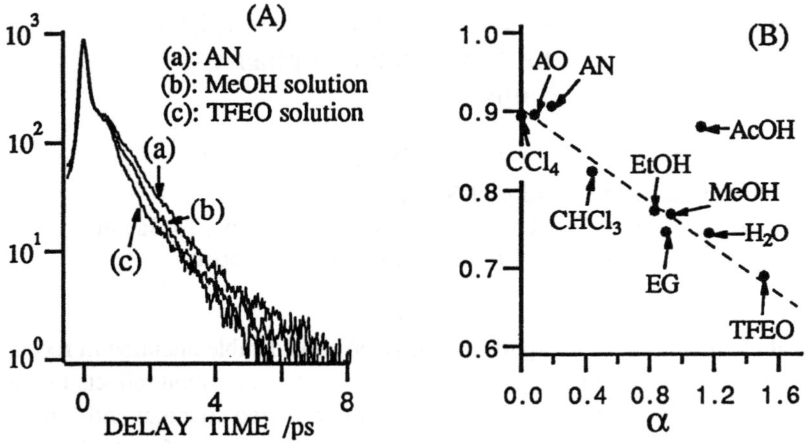

Figure 1 (A) Time-resolved CARS for the C≡N stretching (2252 cm^{-1}) of neat acetonitrile and its solution. Mole fraction of acetonitrile is 75% in each solution. (B) Relation between the apparent decay time-constant and the hydrogen bond ability of the solvent (α). The following abbreviations are used: AO, acetone; AN, neat acetonitrile; EtOH, ethanol; MeOH, methanol; EG, ethylene glycol; AcOH, acetic acid; TFEO, trifluoroethanol. The dashed line is a guide to the eye.

acetic acid has almost no interaction with the C≡N group of acetonitrile. This result on the acetic acid is important since it implies that vibrational dephasing rate provide a new kind of information on solution structure.

References

[1] H. Okamoto, K. Yoshihara: J. Opt. Soc. Am. B7, 1702 (1990).
[2] M. J. Kamlet, J. -L. M. Abboud, M. H. Abraham, R. W. Taft: J. Org. Chem. **48**, 2877 (1983).

Vibrational Relaxation of Very Highly Excited Azulene: Collisional Transition Probabilities in the Quasi-Continuous Energy Range

U. Hold, T. Lenzer, K. Luther, and A. Symonds

Institut für Physikalische Chemie, Universität Göttingen, Tammannstr. 6, W-3400 Göttingen, Fed. Rep. of Germany

1. Introduction

Very highly vibrationally excited molecules are involved in various molecular processes. In particular they play a crucial role in unimolecular and recombination reactions, in which collisional deactivation or activation are essential for a complete dynamic description. Hence there is a strong motivation for direct measurements of collisional deactivation in the energy range of most interest for chemical reactions. However in the larger polyatomic molecules the densities of states quickly rise to extremely high values (e.g. > 10^{12} cm^{-1} in toluene at E = 20000 cm^{-1}). In this quasi-continuous range of rovibrational states various spectrocopic techniques have been used for time resolved observation of intermolecular vibrational relaxation. All of them can give access to only one characteristic parameter of collisional energy transfer, $\langle \Delta E \rangle$, the average amount of energy transferred per collision.

Much more detailed observations of vibrational deactivation at very high densities of states are now possible with help of the KCSI-method (Kinetically Controlled Selective Ionization) (1). It allows to monitor directly the evolution of the molecular populations in fixed energy "windows" during the collisional deactivation following an initial, monochromatic excitation of a sample. A detailed master equation analysis of such data allows to extract the complete information on the time-resolved vibrational energy distributions in terms of energy dependent collisional transition probabilities, and includes the determination of the first and higher moments of energy transfer. The first of such KCSI-studies has recently be performed on highly excited toluene (2). In that work i. a. a case of very small fractions of very efficient ("strong") collisions could be directly shown and precisely quantified. Pronounced distortions in the energy distributions during the deactivation due to such events of transfer of large amounts of energy were found. In this work we report results of our present KCSI-study on azulene deactivation.

2. Experimental and Results

In our experiments vibrationally highly excited azulene is produced by UV absorption of a pump pulse (532, 337 or 266 nm) followed by fast internal conversion to S_0. The deactivation is probed by time delayed KCSI-pulses. This process uses an intermediate electronic state (S_2 in azulene) with strongly energy dependent lifetime for a two photon (2 color) ionization. Kinetic competition between intramolecular transitions and further laser pumping (2nd photon) limits efficient ionization

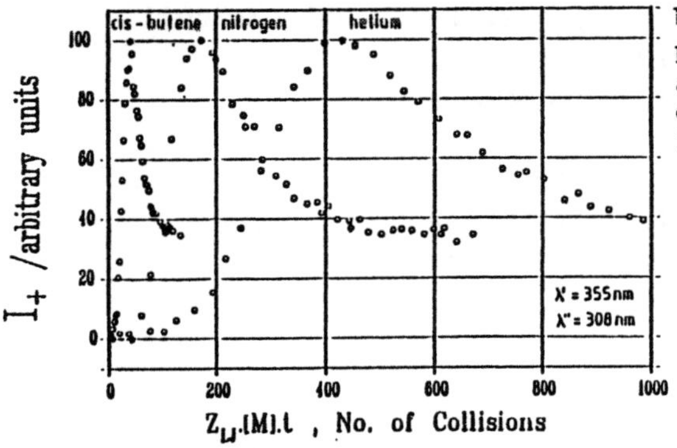

Fig. 1
KCSI-ion signals of azulene (S_o^*) in various colliders. Probe: 355 + 308 nm

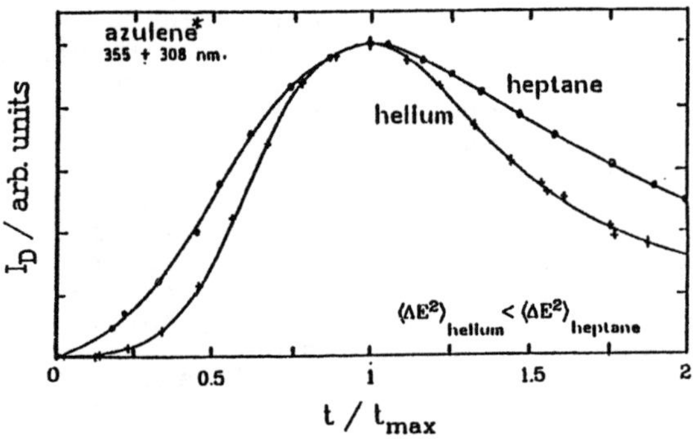

Fig. 2
Azulene deactivation in He and C_7H_{16}. Normalized representation

to an energy interval in the intermediate state. A corresponding energy "window of KCSI-ionization exists on the S_o energy scale and its position can be shifted by variation of the wavelength of the first photon λ'. See (1).

Examples of KCSI-signals are shown in Fig. 1 for three colliders on a timse-scale in "No. of collisions". Azulene was excited at 532 nm to about 20000 cm^{-1} vibrational energy. The curves show how the relaxing excited azulene arrives at the energy "window" of observations and passes through. Approximate values of average <ΔE> can be read directly from the "arrival times" of the KCSI-signals.

In a normalized graph of KCSI-curves (example in Fig. 2) the width (slope) of curves from various colliders follows the respective $\langle \Delta E^2 \rangle$. Distinct contributions from "strong" collisions can be identified directly from irregularities in these KCSI-curves (2). Indications for "strong" collision effects are also found in azulene. Details and results of a master equation analysis will be given in a forthcoming publication (3).

(1) H. G. Löhmannsröben and K. Luther, Chem. Phys. Lett. 144, 473 (1988); K. Luther and K. Reihs, Ber. Bunsenges. Phys. Chem. 92, 442 (1988).
(2) K. Luther, K. Reihs and A. Symonds, J. Chem. Phys., to be published.
(3) U. Hold, T. Lenzer, K. Luther, A. Symonds, to be published.

Picosecond Infrared Studies of Intramolecular Vibrational Energy Transfer

S.A. Angel, J.C. Stephenson, and E.J. Heilweil

Molecular Physics Division, Room B268, Building 221,
National Institute of Standards and Technology,
Gaithersburg, MD 20899, USA

Abstract. CO and NO stretching modes of $Co(CO)_3NO$ at 295 K in the gas phase and in dilute CCl_3F solution are investigated using picosecond infrared (IR) pump/probe transient spectroscopy. Intra- and inter-molecular vibrational energy transfer mechanisms are inferred from these results.

1. Introduction

Picosecond IR spectroscopy has been used to study excited vibrational state dynamics of ground electronic state polyatomic molecules in solution [1-5] and CO ligands bound to metals both on surfaces and in solution for many years [1]. To answer questions raised by these previous studies, we have done experiments to assess rates and mechanisms of intramolecular energy redistribution (IVR) and solvent-induced intermolecular energy transfer (VET). This paper presents evidence for collision free IVR within a metal carbonyl-nitrosyl system as studied by picosecond IR spectroscopy. The effect of solvation on vibrational lifetimes (T_1) is also discussed.

2. Experimental

Broadband and narrowband IR detection were employed in this work. Broadband experiments used a narrowband pump with a multichannel analyzer to detect broadband probe signals (Figure 1) as described previously [6]. Independently tunable narrowband pump and probe IR pulses for kinetic measurements (Figures 2-5) were generated by frequency down conversion in $LiIO_3$ between the output of two tunable red dye lasers and the Nd:YAG second harmonic (532 nm).

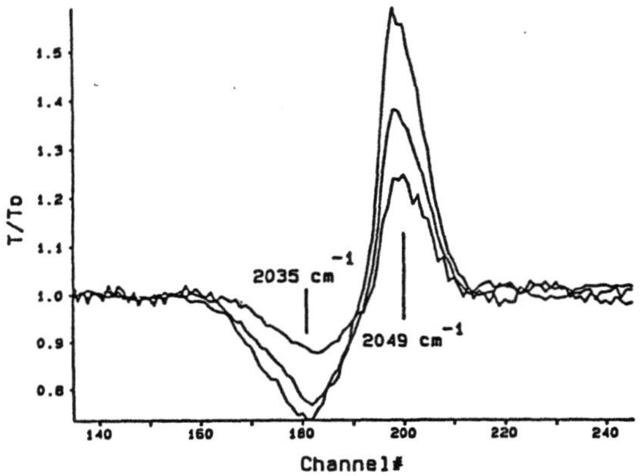

Figure 1: Transient broadband IR spectra of $Co(CO)_3NO$ in the gas phase at 40 Torr, 295 K, following excitation at 2052cm^{-1}. The 3 spectral traces, in order of decreasing intensity, correspond to delay times of 40ps, 440ps and 680ps.

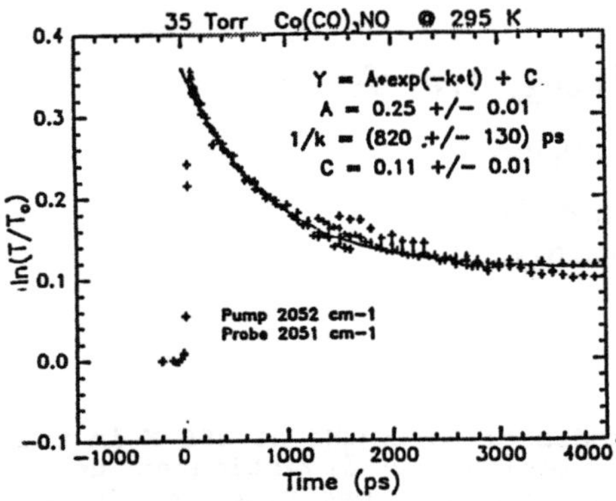

Figure 2: Time dependence of the bleach signal when probing at the v=0→1 transition energy of the CO asymmetric stretching mode (ν_8) with IR excitation of ν_8. (+): Experimental data; (-): Non-linear least squares fit.

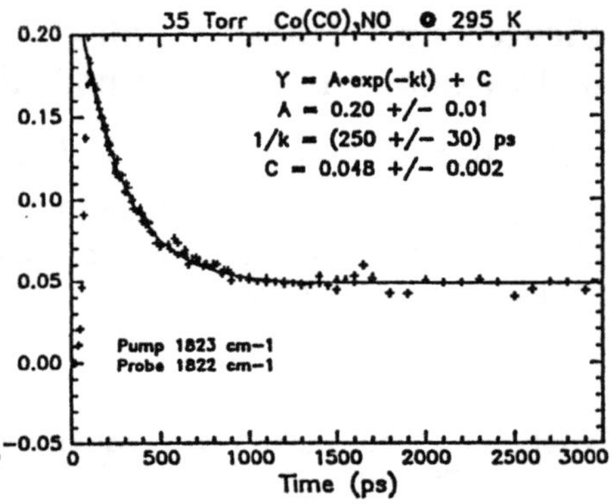

Figure 3: Time dependence of the bleach signal when probing at the v=0→1 transition energy of the NO stretching mode (ν_2) with IR excitation of ν_2. (+): Experimental data; (-): Non-linear least squares fit.

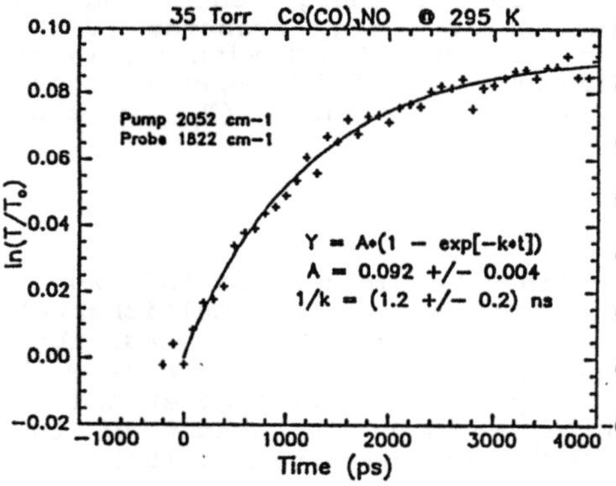

Figure 4: Time dependence of the bleach signal when probing at the v=0→1 transition energy of the NO stretching mode (ν_2) with IR excitation of the CO asymmetric stretching mode (ν_8). (+): Experimental data; (-): Non-linear least squares fit.

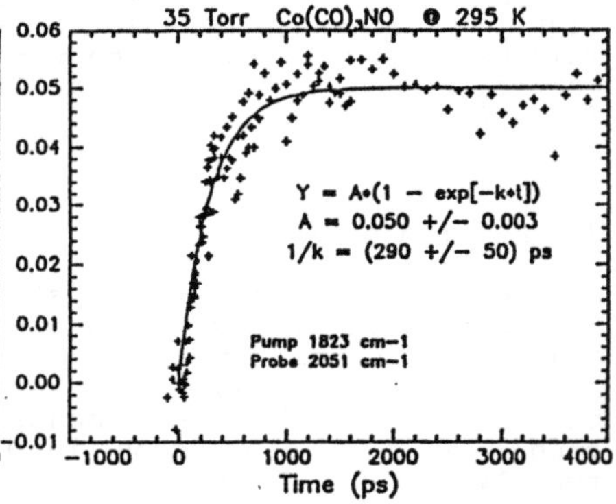

Figure 5: Time dependence of the bleach signal when probing at the v=0→1 transition energy of the CO asymmetric stretching mode (ν_8) with IR excitation of the NO stretching mode (ν_2). (+): Experimental data; (-): Non-linear least squares fit.

The pump pulse (20 ps FWHM, 4 cm^{-1} FWHM, 200 μ diameter) energy ranged from 4 μJ at 1800 cm^{-1} to 7 μJ at 2000 cm^{-1}. Two HgCdTe detectors with boxcar integrators and a computer were used to measure and discriminate shot-to-shot transmitted probe and reference energies. Two hundred shots were typically averaged with the pump pulse on (to obtain the excited sample transmission T) and off (T_o) to obtain the value $\ln(T/T_o)$ at each time delay. One millimeter pathlength sample and reference beam cells were opened to a stainless steel vacuum manifold to administer and continuously monitor gas pressure. The $Co(CO)_3NO$ pressure was varied from 15 to 45 Torr at room temperature with no kinetic effect. Gas phase kinetic data were taken at 35 Torr (ca. 3 ns hard sphere collision time) while $CFCl_3$ solutions with the same solute density (ca. 0.1 mM) were prepared by exposing a known volume of gas to solvent in the same sample cells.

3. Results and Discussion

The time-dependent broadband spectra shown in Figure 1 exhibit bleaching ($\ln(T/T_0) > 1$) and transient absorption ($\ln(T/T_0) < 1$) in the 2000 cm^{-1} region when the gaseous sample is excited at 2052 cm^{-1}. Absorption of the gas phase v=1 CO asymmetric stretching mode (ν_8) of $Co(CO)_3NO$ is centered at 2048cm^{-1}. The bleach signals in Figure 1 (ca. 2048 cm^{-1}) are attributed to transient population in v=1 of ν8 and/or the appearance of population in other modes anharmonically coupled to ν8. Populated modes that are not anharmonically coupled to ν8 will not contribute to a bleach signal at this frequency. The red-shifted absorption at 2035 cm^{-1} results from the 12 cm^{-1} anharmonic shift [7] of the v=1-2 transition of ν_8. Note that shifts of these spectral features do not occur with increasing delay time.

The gas phase T_1 decay kinetics of ν_8 and the NO-stretching mode (ν_2 = 1822 cm^{-1}) are presented in Figures 2 and 3, respectively. As has been suggested for similar systems [1,2], these modes are likely to directly relax to combinations and/or overtones of anharmonically coupled modes such as the M-C, M-N stretches and/or M-CO, M-NO bends with v=1 at 300-600 cm^{-1} energy. The data are well fit to a single exponential which decays to a constant bleaching level. If the exponential decays arise from v=1 population relaxation of the excited states, then T_1 for ν_8 and ν_2 are 820 ± 130 ps and 250 ± 30 ps, respectively. The difference in the apparent rate constants for the relaxation of these two modes suggests that population does not rapidly equilibrate between v=1 of ν_8 and ν_2. The constant bleach level at long delay must arise from population in modes which are anharmonically coupled to the pumped vibrational states. Energy appears to remain equilibrated in these lower energy states out to 4 ns.

The probe pulse transmission kinetics of gas phase $Co(CO)_3NO$ at 1822 cm^{-1} subsequent to ν_8 excitation at 2052 cm^{-1} are shown in Figure 4, while Figure 5 depicts the bleach kinetics at 2051 cm^{-1} after exciting ν_2 at 1823 cm^{-1}. The data are well fit to single exponential risetimes of 1.2 ± 0.2 ns and 290 ± 50 ps respectively. If the risetime of the bleach probed at 1823 cm^{-1} is attributed to direct transfer of population from v=1 of ν_8 to v=1 of ν_2, then the kinetics of Figure 4 should reflect T_1 of ν_8 as the rise of the bleach convoluted with a decay given by T_1 of ν_2. The measured risetime in Figure 4, when compared to the decay time of Figure 2, indicates that v=1 of ν_2 is not responsible for the bleach signal in Figure 4. Likewise, there is no indication of signal decay in Figure 5 which would occur if v=1 of ν_8 became populated.

Solution phase decay kinetics of ν_8 and ν_2 were fit to single exponential decays that reach the baseline ($\ln(T/T_0) = 0$). If these data represent v=1 state relaxation, T_1 for ν_8 and ν_2 in CCl_3F are 410 ± 20 ps and 54 ± 3 ps, respectively. Within the temporal resolution and noise limits of these experiments, no bleach was detected either at the ν_8 transition when ν_2 was pumped or at ν_2 when ν_8 was excited - indicating that energy rapidly leaves the anharmonically coupled lower energy modes when $Co(CO)_3NO$ is solvated by CCl_3F.

4. Summary

The gas phase data in Figures 2-5 suggest that energy supplied to ν_8 decays into modes that are anharmonically coupled to both ν_8 and ν_2. Likewise, energy supplied to ν_2 decays to modes which are anharmonically coupled to both ν_2 and ν_8. However, within noise and time limitations, there is no evidence for vibrational energy transfer between ν_8 and ν_2. These results are contrary to an earlier report and are qualitatively different from the double exponential decays and IVR mechanism reported for $Co_4(CO)_{12}$ in hexane solution [5]. In CCl_3F solution, ν_8 and ν_2 appear to relax back to the ground state 2-5 times faster than in the gas phase.

In a future communication we will discuss specific energy relaxation pathways for this system and report effects on T_1 when the collision rate is changed by varying buffer gas and pressure.

References

[1] E.J. Heilweil, M.P. Casassa, R.R. Cavanagh, J.C. Stephenson, Ann. Rev. Phys. Chem., 40, 143 (1989) and references therein.

[2] E.J. Heilweil, R.R. Cavanagh, J.C. Stephenson, J. Chem. Phys., 89, 230 (1988).

[3] H. Graener, Chem. Phys. Lett., 165, 110 (1990); H. Graener, R. Dohlus, A. Laubereau, Chem. Phys. Lett., 140, 306 (1987).

[4] H.J. Bakker, P.C.M. Planken, A. Lagendijk, J. Chem. Phys., 94, 6007 (1991).

[5] S.A. Angel, P.A. Hansen, E.J. Heilweil, J.C. Stephenson, "Ultrafast Phenomena VII", Springer-Verlag, Berlin, Vol. 53, p. 480, 1990.

[6] E.J. Heilweil, Opt. Lett., 14, 551 (1989).

[7] L.H. Jones, R.S. McDowell, B.I. Swanson, J. Chem. Phys., 58, 3757 (1973).

Time-Resolved Infrared Spectroscopy of Organic Molecules

H.J. Bakker and A. Lagendijk

FOM-Institute for Atomic and Molecular Physics,
Kruislaan 407, NL-1098 SJ Amsterdam, The Netherlands

We present a time-resolved study of the vibrational relaxation after excitation of the C-H vibration for dilute solutions of $CHBr_3$ and $C_2H_2Br_4$ and the pure liquid $CHBr_3$. The experiments are performed with a single-color infrared pump-probe technique using intense tunable (2200-7200 cm^{-1}) infrared pulses with a pulse duration of 20 ps.

For all these liquids, the vibrational energy relaxation takes place via two consecutive relaxation processes. In the first process a large part of the energy of the excited C-H stretch vibration is transferred to other vibrations within the molecule (IVR), leading to an anharmonic shift of 10-15 cm^{-1} of the absorption band to lower frequencies.[1] If the excitation frequency (ν_{exc}) of the laser pulses is significant below the maximum of the absorption band (ν_{max}), the absorption band of the excited molecules will shift better into resonance leading to a strong decrease of the transmission of the probe pulse. In the second process the energy is equilibrated over all degrees of freedom (IET).

The IVR is only possible when the energy mismatch between the excited C-H vibration and other vibrations is compensated by the interactions with the solvent. In a polar solvent the electrostatic interactions are much stronger than in a nonpolar solvent so that the intramolecular energy transfer is much faster (fig.1).[2]

We observe that for $C_2H_2Br_4$ the rate of IVR does not depend on the solvent (fig 2).[3] This molecule possesses many vibrations so that the energy mismatch will be small. Therefore even the weak interactions of the excited molecule with nonpolar solvent molecules are sufficient to compensate for the energy mismatch and make energy transfer possible.

For pure $CHBr_3$ we observe that the IET process is very slow which indicates that it takes a long time before all molecular vibrations that get excited in the IVR process and

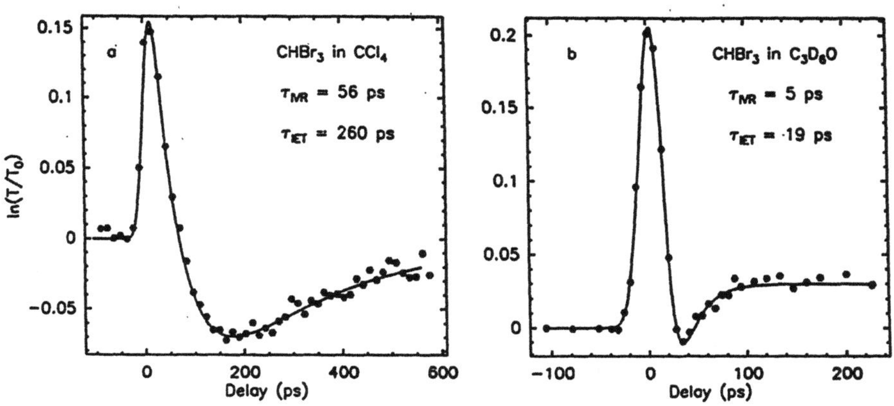

Fig. 1 : Relative transmission change of the probe for $CHBr_3$ in solution : a) in CCl_4 $\nu_{max} = 3032$ cm^{-1}, $\nu_{exc} = 3025$ cm^{-1}; b) in C_3D_6O $\nu_{max} = 3028$ cm^{-1}, $\nu_{exc} = 3016$ cm^{-1}

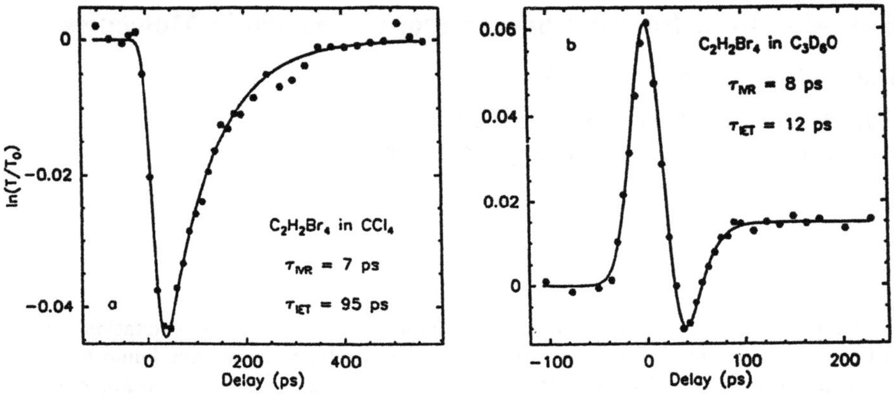

Fig. 2 : Relative transmission change of the probe for $C_2H_2Br_4$ in solution : a) in CCl_4 $\nu_{max} = 2993\ cm^{-1}$, $\nu_{exc} = 2985\ cm^{-1}$; b) in C_3D_6O $\nu_{max} = 2987\ cm^{-1}$, $\nu_{exc} = 2976\ cm^{-1}$

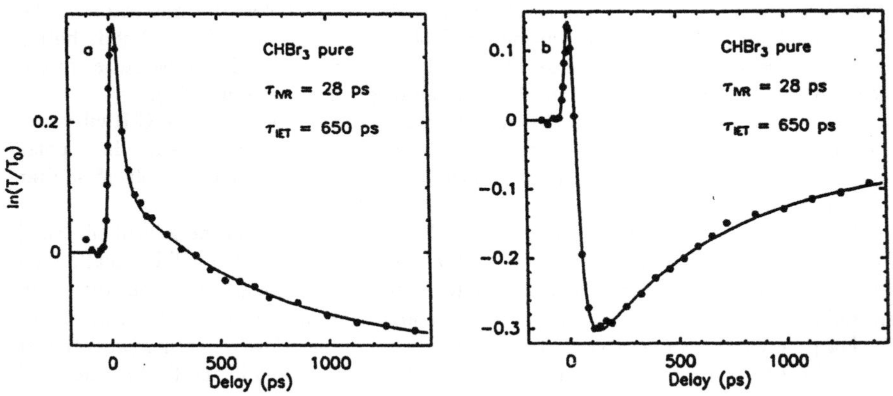

Fig. 3 : Relative transmission change of the probe for pure $CHBr_3$ ($\nu_{max} = 3022\ cm^{-1}$). a) $\nu_{exc} = 3030\ cm^{-1}$; b) $\nu_{exc} = 3016\ cm^{-1}$

induce shifts of the C-H absorption band have transferred their energy to other degrees of freedom. We also observe that for large delays the transmission of the probe is at a strongly decreased level. Due to the high concentration of absorped energy there will still be a lot of energy in the lowest-energy vibrations of the $CHBr_3$ molecule after the IET. This results in an increase of the absorption coefficient of the C-H stretch vibration (fig.3).

[1] H. Graener, Chem. Phys. Lett. **165**, 110 (1990)
[2] H.J. Bakker, P.C.M. Planken and A. Lagendijk, Nature **347**, 745 (1990)
[3] H.J. Bakker and A. Lagendijk, J. Chem. Phys. **94**, 6007 (1991)

SEP-LIF Spectra of Some Molecules and van der Waals Molecules in a Molecular Beam

M. Takayanagi and I. Hanazaki

Institute for Molecular Science, Myodaiji, Okazaki 444, Japan

SEP-LIF(stimulated emission pumping-laser induced fluorescence) technique was applied to examine the dynamical behavior of vibrationally excited molecules and van der Waals complexes in a pulsed molecular beam. In the SEP-LIF scheme, a specific species in the molecular beam is excited to a specific vibrational state in the ground electronic state by SEP, and then by the LIF measurement the vibrationally excited species and/or the species which are vibrationally relaxed or predissociated from the initially excited state are probed.

For the SEP excitation, frequency-doubled outputs of two dye lasers pumped simultaneously by the second harmonics of a single Nd:YAG laser were used, while the frequency-doubled output of a dye laser pumped by an excimer laser was used for the LIF measurement. The decay of the SEP-LIF signal intensity, which corresponds to the relaxation or predissociation of the vibrationally excited molecules, was measured by varying the delay (300 ns - 1.4 μs) between the SEP and LIF laser pulses.

Figure 1 shows the LIF spectra for benzonitrile(BN) seeded in Ar (a) with and (b) without SEP excitation to the 460 cm^{-1} vibrational state. With the SEP excitation, two bands, marked with arrows appear near 460 and 48 cm^{-1}, which correspond to the 0←1 and 1←1 transitions respectively of vibrationally excited BN. The same measurements were performed for BN monomer, dimer, BN·Ar and BN·H_2O with the SEP excitation to the 460, 1000 and 1460 cm^{-1} vibrational states. The SEP-LIF

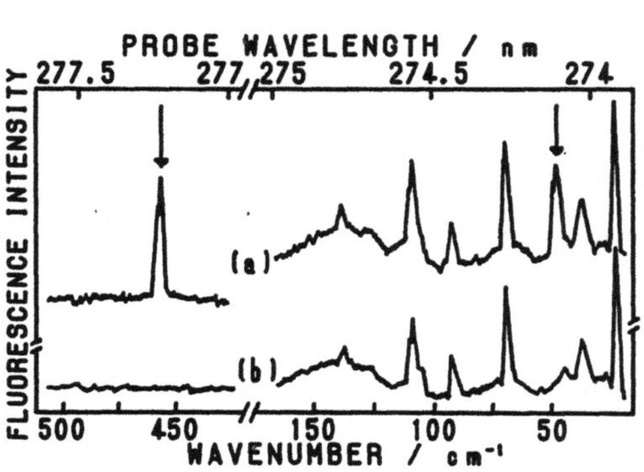

Fig. 1 LIF spectra of benzonitrile (a) with and (b) without SEP excitation to the 460 cm^{-1} vibrational state. The delay between the SEP and LIF pulses is 400 ns. The frequency difference between the probe light and 0-0 transition is shown on the lower abscissa.

signal was observed for the monomer with the excitation to each vibrational state, while no SEP-LIF signal was observed for BN·Ar and BN·H_2O with the 1460 cm^{-1} excitation and for the dimer even with the shortest delay between the SEP and LIF pulses. This experimental result indicates the fast relaxation of the prepared vibrationally excited state for the complexes. These complexes have low frequency intermolecular vibrational modes, which make the state density higher and would accelerate the vibrational relaxation. The Ar complex and H_2O complex with the 460 or 1000 cm^{-1} vibrational excitation showed SEP-LIF signals. The state densities of the complexes in the energy regions are estimated to be high enough for the fast relaxation if the relaxation rate is determined by the golden rule. The present experimental results show that the relaxation rate is not governed by the statistical process.

Some information on the rate of relaxation or predissociation of the vibrationally excited species could also be obtained from semiquantitative analyses of the SEP-LIF signal intensity [1]. On this basis, the relaxation rate of the state with the higher vibrational energy was found to be higher for each species. The relaxation rate of the H_2O complex was found to be higher than that of the Ar complex.

The monomer showed a slow decay, which is independent of the vibrational state. The decay could be due to the escape of the vibrationally excited species out of the probe region. Finite decay rates were obtained for BN·Ar and BN·H_2O, which were higher than that for the monomer. The rates were not sensitive to the species and the vibrational energy. The cause of the insensitivity is not known, although the decay rates seem to reflect the relaxation (or predissociation) rates of the vibrationally excited species.

We measured also the SEP-LIF spectra of anisole. Anisole has the low frequency O-Me torsional mode and its vibrational state density is higher than that of BN. In the LIF spectrum measured with the SEP excitation to the 788 cm^{-1} vibrational state, many hot bands which were not observed without the SEP excitation appeared. The hot bands are considered to be due to the species in plural vibrational states because the relative intensities of the bands depend strongly on the nozzle condition and the delay between the SEP and LIF laser pulses. Vibrational relaxation of the excited anisole seems to occur.

We have shown that the applicability of the SEP-LIF technique to the study of the dynamics of vibrationally excited molecules and van der Waals complexes. It is not clear for both of anisole and the complexes containing benzonitrile whether the relaxation or the predissociation is caused by the soft collision in the molecular beam or by the coupling of vibrational levels. A further study of the relaxation (or predissociation) mechanism seems to be needed.

Reference

1. M. Takayanagi and I. Hanazaki: J. Opt. Soc. Am. B 7, 1898 (1990)

Femtosecond Optical Kerr Study of β-Carotene Solutions

S. Califano[1], *P. Foggi*[1], *V.F. Kamalov*[2], *R. Righini*[1], *and R. Torre*[1]

[1]European Laboratory for Nonlinear Spectroscopy,
University of Florence, Largo E. Fermi, 2, I-50125 Florence, Italy
[2]Photochemistry Department, Institute of Chemical Physics,
Russian Academy of Sciences, Kosygin str. 4, 117977 Moscow, Russia

Temporal evolution of low frequency intramolecular vibrational coordinates and ultrafast decay component attributed to librational motion were studied for simple liquids and solutions recently by optical Kerr effect (OKE) [1]. The purpose of this study is the creation of anisotropy of β-carotene solutions by optical Kerr effect to study dynamical behavior of β-carotene at the femto and picosecond time scales.

1. Experimental

The anisotropy created in β-carotene solutions by powerful pump pulse were measured by transmittance of probe pulse passed through the crossed polarizers. The angle between linearly polarized pump and probe beams was 45°. Durations of pump and probe pulses were about 200 fs. Wavelength of pump was 605 nm, probe was tuned close to β-carotene electronic transition about 500 nm. The dependence of anisotropy created on the time delay between pump and probe pulses were measured for β-carotene in dodecane solutions with concentration 100 μM.

2. Results and Discussion

At low pump intensity the anisotropy created during the laser pulses decays exponentially with the time 8±1 ps. This reflects the creation of a deep in the orientational distribution of molecules in the ground electronic state due to one-photon absorption of pump beam, and subsequent decay of anisotropy due to repopulation of the ground state with the time 8 ps.

The contributions from nonlinear optical processes increases with the pump intensity. The signal of anisotropy detected at high pump intensities is caused by birefringence generated in the β-carotene solutions due to OKE. The high level of the signal observed was due to intrinsic anisotropy of β-carotene, high value of β-carotene hyperpolarizability [2], and its resonant enhancement due to resonance of the probe beam with the electronic transition.

Fast and slow components are observed in the decay of OKE induced anisotropy (Fig.1). Dashed line in Fig.1 is the numerical fit obtained by convolution of bi-exponential response function with the time profile of the pump and probe pulses. The slow component (215 ps for dodecane) in the decay was attributed to the rotational diffusion process, and the dependence of rotational diffusion rate on the viscosity was measured. The interpretation of 250 fs relaxation is less straightforward. Fast dynamics in molecular liquids is generally associated with the interaction-induced (I-I) contribution into the

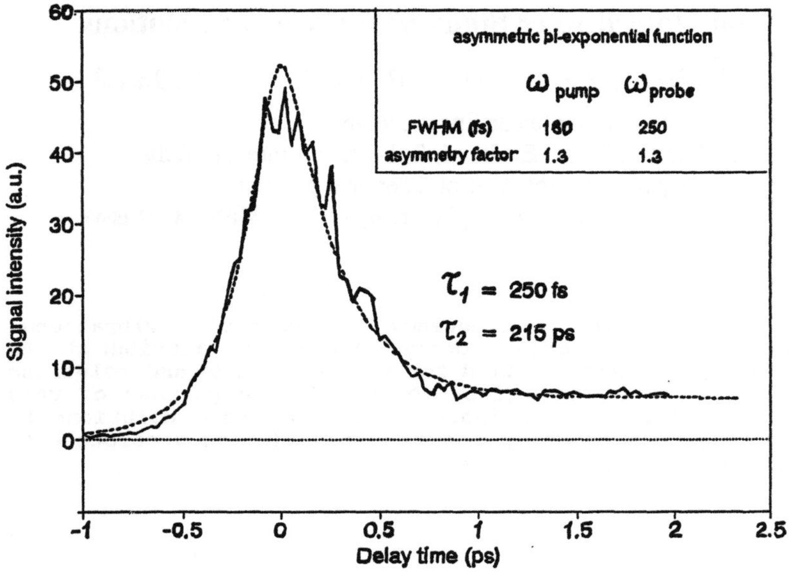

Fig.1. Transient optical Kerr response of β-carotene in dodecane solution. $C=10^{-4}$M. T=293 K.

susceptibility. The I-I contribution originated from overdamped librations of the solute molecule within the local solvatation structure. This fast decay of the anisotropy can be also caused by electronic dephasing at ^1B_u-^1A_g transition of β-carotene.

References

1.D.McMorrow, W.T.Lotshaw, G.A.Kenney-Wallace, IEEE J.Quantum Electron. QE-24, 443 (1988).
2.J.P.Hermann, D.Ricard, J.Ducuing, Appl.Phys.Lett.23,178 (1973).

Part VII

Dynamics in Solids, Aggregates, and Liquid Crystals

The Effect of Diagonal and Off-Diagonal Disorder on Optical Properties of Molecular Aggregates

D.A. Wiersma, H. Fidder, J. Terpstra, and J. Knoester

University of Groningen, Department of Chemistry,
Ultrafast Laser and Spectroscopy Laboratory,
Nijenborgh 16, NL-9747 AG Groningen, The Netherlands

Recent results of a study of the absorption lineshape, superradiant emission, and resonant light scattering measurements on aggregates of PIC are discussed in light of numerical simulations on linear molecular aggregates with diagonal and off-diagonal disorder.

1. Introduction

Molecular aggregates are abundantly observed in photo-biological systems where they act as light-harvesting entities to excite a chromophore or group of chromophores. This triggers, for instance, a structural change of the surrounding protein [1] or initiates a chemical reaction [2]. The high efficiency of the underlying energy transfer process results from the fact that it is completed in a few picoseconds. The potential of aggregates for technological use was long ago recognized in the photographic industry [3] where aggregates have found widespread use as sensitizers for films and other photosensitive devices. Recently, it has also been suggested [4] that aggregates exhibiting high optical nonlinearities have potential for optical communication applications. The idea is that the molecular nonlinearity is greatly amplified in the aggregate [5] such that a low threshold exists for nonlinear optical processes as frequency doubling and mixing. Activity in this field was further pushed by the commercial availability of Langmuir-Blodgett (LB) troughs, which enables easy fabrication of mono- or multilayers of many aggregate-forming dye molecules.

Scheibe [6,7] and Jelly [8] were the first to report on the change in optical spectrum that occurred upon aggregation of the dye molecule pseudoisocyanine (PIC) in solution. Since these first experiments much work has been done concerning the understanding of the optical dynamics of this dye aggregate. A detailed paper on the optical dynamics of aggregates of PIC in solid solution was recently published [9].

In this Letter we wish to address the question of the effect of diagonal and off-diagonal disorder on the lineshape and optical dynamics of molecular aggregates. While we will often refer to PIC, to compare results of numerical simulations with experimental findings in this system, many of our conclusions apply to other aggregates as well.

2. Theory

The Hamiltonian that governs the exciton dynamics in a linear chain of molecular aggregates, in the absence of phonons, has the following form [10]:

$$\mathbf{H} = \sum_{n} (\langle\varepsilon\rangle + D_n)|n\rangle\langle n| + \sum_{n,m\,,\,n\neq m}\sum J_{nm}|m\rangle\langle n| \equiv \sum_{n}\sum_{m} H_{nm}|m\rangle\langle n| \qquad (1)$$

Here $|n\rangle$ denotes the state in which molecule n (n=1,...,N) is excited and all other molecules are in the ground state. $\langle\varepsilon\rangle$ is the average molecular excitation energy and D_n the inhomogeneous offset energy of molecule n (diagonal disorder). J_{nm} in Eq. (1) is the transfer integral between molecule n and m, which is assumed to be of dipolar origin. In our calculations all dipolar interactions on the chain are taken into account. Disorder in the interactions J_{nm} (off-diagonal disorder) arises from configurational disorder. We assume that a Gaussian distribution exists in lattice positions of molecules around an average nearest neighbour distance. The relevant optical observables were calculated from the wave functions obtained from diagonalizing Hamiltonian (1) for different sets of diagonal and off-diagonal energies, which were chosen randomly from prescribed energy distribution functions.

3. Results and Discussion

Figure 1 displays a calculation of the lineshape of the J-aggregate of PIC-Br in an ethylene glycol/water glass at 1.5 K for a linear chain of 250 molecules and a D/J ratio of 0.106. D is the standard deviation of the Gaussian distribution of the offset frequencies D_n and $J \equiv J_{12}$ was taken to be $-600\ cm^{-1}$[9]. The solid bars in Fig. 1 present the results of a calculation for a homogeneous linear chain of the same length. These results clearly show that for self-assembled aggregates of PIC the exciton lineshape is dominated by diagonal disorder and that disorder shifts the exciton absorption to lower energy. Fig. 2 displays the results of a calculation of the degree of localization $\mathcal{L}(\varepsilon)$ for this system with the same set of parameters. $\mathcal{L}(\varepsilon)$ is 1 in case a state is completely localized at a single molecule and 3/2(N+1) in case of complete delocalization of the wave function over the N molecules of the chain [10]. The most noteworthy feature of Fig.2 is that it shows that disorder strongly localizes the excitons near the top and bottom of the band. The calculations further show that excitons near the bottom of the band have

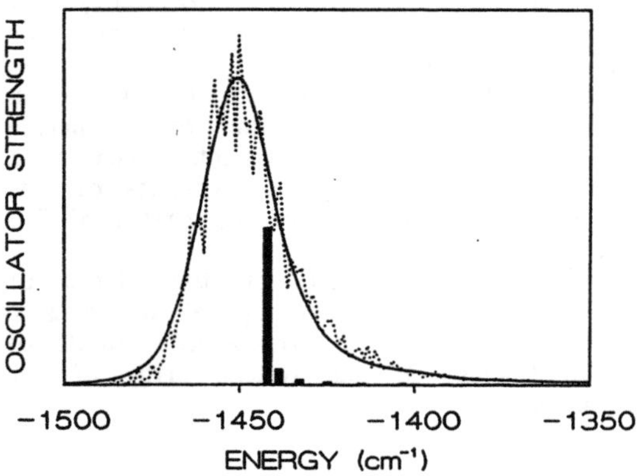

Fig. 1 (a) Oscillator strength (in units of μ_{mon}^2) against energy for the eigenstates at the bottom of the exciton band, found by diagonalizing one chain of 250 molecules with random site energies (D/J=0.106). The dotted line is the absorption spectrum, at a resolution of $1\ cm^{-1}$, obtained by averaging over 500 chains. The solid line is the experimental spectrum. The solid bars represent the simulated optical spectrum of the same 250 molecules in the absence of disorder. The width of the bars were chosen arbitrarily.

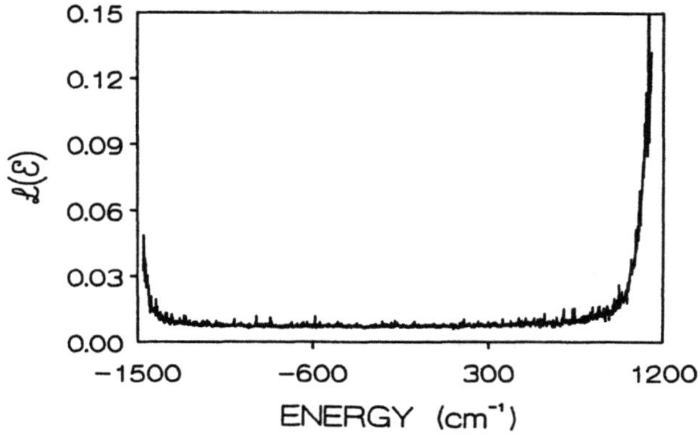

Fig.2 Degree of localization $\mathcal{L}(\mathcal{E})$ for diagonal disorder with D/J=0.106 for chains of 250 molecules with a resolution of 1 cm^{-1}.

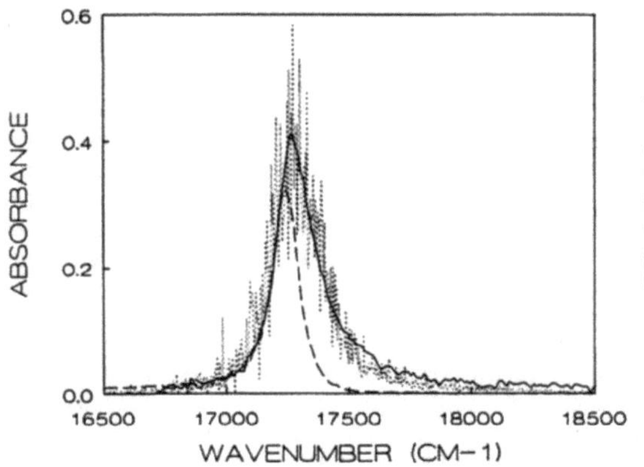

Fig.3 Absorption spectrum (solid line), emission spectrum (dashed line) and computer simulation (dotted curve) of a bilayer of PIC at 77 K.
The numerical calculations were performed on 120 chains of 250 molecules. D/J=0.353; σ=0.083 (Ref.10) and the resolution is 3 cm^{-1}.

large oscillator strength leading to ultrashort *radiative* lifetimes of these states [9]. This superradiant effect is a striking example of many molecules emitting in phase. This cooperative effect, however, is counteracted by optical dephasing processes like exciton-phonon scattering. For J-aggregates the relation between superradiance and optical dephasing has been thoroughly investigated but a quantitative model is still lacking [9].

We now turn to a discussion of the dynamics of J-aggregates in LB films. Figure 3 shows the optical spectra of a bilayer of PIC at 77 K. The "noisy" curve in this figure presents the simulated lineshape with both diagonal and off-diagonal disorder included. Use of only diagonal or off-diagonal disorder in the simulation leads to a worse fit. We therefore surmise that for PIC in LB films, in contrast to a glass, off-diagonal disorder is an important factor in determining the exciton dynamics. For further details we refer to Ref. 10. The dotted curve in Fig. 3 presents the fluorescence spectrum of the bilayer at 77 K. It shows that emission comes from states below the top of

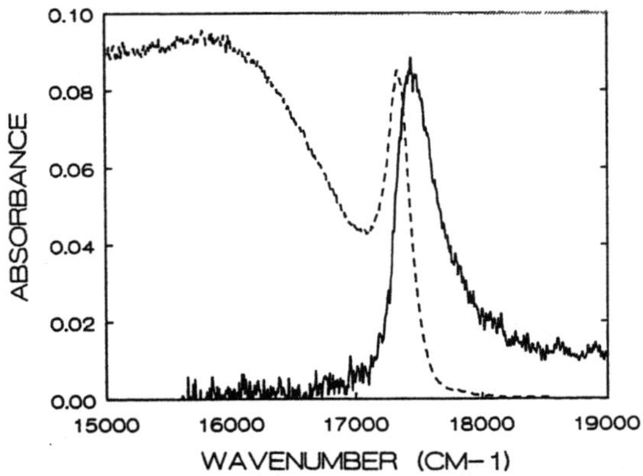

Fig.4 Absorption (solid line) and emission (dashed line) spectra of a monolayer of PIC at 77 K.

the disordered exciton band. This indicates that the mobility of excitons above the "renormalized bandedge" is much higher than below this edge. Resonance Rayleigh scattering experiments on this bilayer show [11] that the frequency dependent optical dephasing time constant T_2 shortens drastically above the bandedge, in accordance with our interpretation of the emission spectrum. Regarding the optical dynamics in this bilayer we therefore conclude that a *mobility edge* exists at the renormalized bandedge. While the absorption lineshapes of a mono- and bilayer are quite similar, the exciton dynamics, however, are found to be rather different. Figure 4 shows the absorption and emission spectrum of a monolayer of PIC. The most striking observation is that at 77 K a large fraction of the emission comes from states *below* the exciton band. The emission from these states is very broad and reminiscent of excimer emission in molecular solids. Our calculations show that off-diagonal disorder indeed creates dimer-like states below the band, which can act as deep traps for excitons. A similar effect has recently been observed in LB layers of anthracene [12]. When the temperature of the PIC-monolayer is further lowered to 1.5 K, the excitonic emission remains barely visible and the spectrum is dominated by excimer-like emission.

We conclude that off-diagonal disorder is an important factor in determining the linear and nonlinear optical properties of molecules in LB films. Future research in our group will be directed at exploring, in a more quantitative way, the effects of disorder on the optical nonlinearities in molecular aggregates.

4. References.

1. R. R. Birge, Ann Rev Biophys Bioeng **10** 315 (1981).
2. G. Feher and M.Y. Okamura,in: *The photosynthetic bacteria*,eds.R.K. Clayton and W. F. Sistrom (Plenum Press, New York, 1978) pp. 349-387.
3. P.B. Gilman. Phot. Sci. Eng. **18** 418 (1974)
4. Y. Wang, Chem.Phys.Lett.**126**, 209 (1986).
5. E. Hanamura, Phys.Rev.B **37**, 1273 (1988).
6. G. Scheibe, Angew.Chem.**49**, 563 (1936); Angew.Chem.**50**, 212 (1937).
7 G. Scheibe in *Optische Anregungen organischer Systeme*, p.109ff., ed. W. Först (Verlag Chemie, Weinheim, 1966).

8 E.E. Jelly, Nature (Lond.) **138**, 1009 (1936); Nature (Lond.) **139**, 631 (1937).
9. H. Fidder, J. Terpstra and D.A. Wiersma, J.Chem.Phys. **94**, 6895 (1991).
10 H. Fidder, J. Knoester and D.A. Wiersma, submitted to J.Chem.Phys.
11 J. Terpstra, H. Fidder and D. A. Wiersma, Chem.Phys. Lett.**179**,349(1991).
12 D. Weiss, R. Kietzmann, W. Storck, J. Lehnert and F. Willig, Macromolecular Chemistry, in press.

Ultrafast Geometrical Relaxation Induced by Electron-Phonon Interaction in Conjugated Polymers

T. Kobayashi

Department of Physics, Faculty of Science,
University of Tokyo, 7-3-1 Hongo, Bunkyo-ku, Tokyo 113, Japan

Conjugated polymers are known to have large third-order optical nonlinearity. In order to clarify the mechanism of the optical nonlinearity, we studied the ultrafast relaxation in the polymers such as polydiacetylenes (PDA-3BCMU (blue phase) and PDA-4BCMU(red and blue phases)) and polythiophenes (P3MT(poly 3-methylthiophene) and P3DT(poly3-dodecylthiophene)) by pump-probe experiment using a CPM laser of about 100 fs pulse width. [1-3]

We observed various ultrafast nonlinear optical processes, which appear very rapidly just at the pump pulse. In our previous paper we reported the perturbed free induction decay, AC-Stark shift, resonant Kerr effect, hole burning, and Raman gain [3]. Other phenomena such as induced phase modulation and perturbed free Raman decay are also observed in the present study. The time-resolved spectra of PDA-3BCMU are shown in Fig. 1. At slightly longer delay time, we observe the formation of self-trapped exciton (STE) from free exciton (Table 1).

Table 1. The formation time (τ_f) and the decay time (τ_d) of self-trapped excitons in several polymers and the relaxation mechanisms

	τ_f(fs) (290K)	relaxation mechanisms	τ_d(ps) (10K)	fluorescent (f) or nonfluorescent (n)
PDA-3BCMU (blue)	150±40	(1)(2)(3)	1.9±0.2	n
PDA-4BCMU (blue)	140±50	(1)(2)(3)	2.1±0.1	n
PDA-4BCMU (red)	<200	(1)(2)(3)	0.96±0.09*	f
P3MT	70±50	(1)(3)	0.62±0.07*	f
P3DT	100±50	(1)(3)	0.45±0.06*	f

*The decay kinetics of STE in the fluorescent polymers cannot be represented by a single exponential. The decay becomes slower and slower at longer delay time. This can be explained by slowing down of the tunneling from STE to the ground state associated with thermalization and cooling down as will be discussed later.[3,4] The decay of fluorescent polymer cannot be described by a single exponential function. Therefore τ_d is obtained form the slope of the semilogarithmic plot of the absorbance change near the zero delay time.

There is no barrier between the bottoms of the adiabatic potential curves of the free exciton and the ST exciton in a 1D system. Hence it is sometimes expected that the formation time of the ST exciton corresponds to the oscillation period of the strongly coupled vibrational mode. The experimentally determined formation time constants are all several times longer than the oscillation period of the strongly coupled c=c and/or c≡c stretching modes. This may be explained by the following three mechanisms (Table 1).

(1) There is no slope at the minimum of the adiabatic potential of the free exciton. Hence it takes finite time for the wave packet of free exciton photogenerated to start sliding along the adiabatic potential of the ST exciton. [5]

Springer Proceedings in Physics, Vol. 68 Time Resolved Vibrational Spectroscopy V
Ed.: H. Takahashi

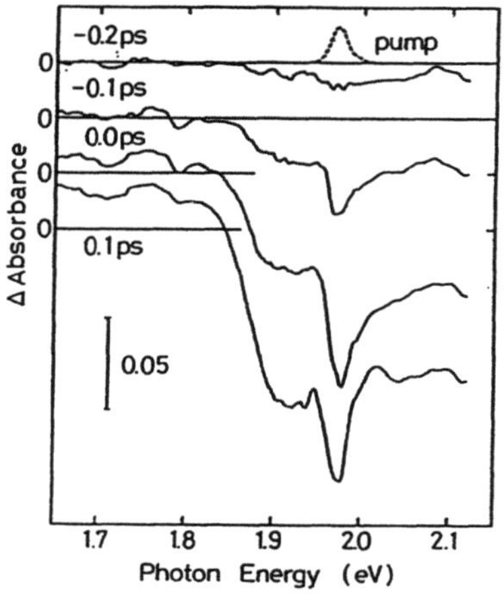

Fig. 1 Transient absorption spectra of PDA-3BCMU induced by 1.97 eV pump pulse at 290 K. The excitation photon density is $2.0x10^{15}$ photons/cm^2.

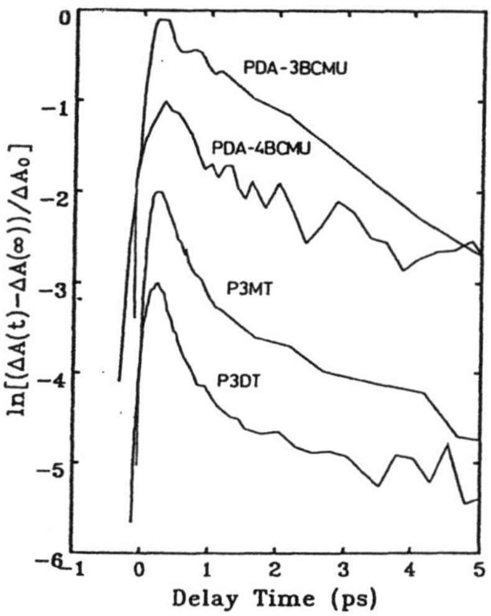

Fig.2 The time dependence of the absorbance change in PDA-3BCMU in the blue phase in PDA-4BCMU in the red phase and P3MT and P3DT at 10K.

(2) The formation of the ST exciton is associated with the geometrical relaxation including extending side chains out form the main chains.

(3) The adiabatic potential represented as a function of the reaction coordinate of the exciton is extremely flat in one-dimensional system because of the continuous reduction of the size of the ST exciton from infinity. [6]

The main mechanisms deducted from comparison of experimental results among several samples are listed in Table 1.

Figure 2 represents the time dependences of the absorbance change due to the ST excitons in PDAs and PTs, after the very long lived component being subtracted. The decay of nonfluorescent PDA-3BCMU and PDA-4BCMU in the blue phase are approximately represented by an exponential function except the early stage within about 500 fs.

The other fluorescent polymers have nonexponential decay from very early delay time until about 30 ps in the absorption decay curves. The initial decay times defined by the slope of the semilogarithmic plot of ΔA vs time just after excitation are shown in Table 1. The decay curve of $\Delta A(t)$ for P3DT may be represented by the power law $\Delta A(t)=\alpha t^{-\beta}+\gamma$, with constants α and γ which represents long lived species. The power β varied from 0.6 to 1.1 depending the probe wavelength. Since in the case of the geminate recombination in a 1D chain, β is expected to be 0.5, the wavelength dependence of β indicates that the polaron is not a main contribution. From this and the consideration of luminescence spectra as in the case of poly(p-phenylenevinylene),[7] we prefer to explain the decay in terms of the ST exciton. The decay becomes slower at longer delay times. The differences in the decay kinetics of excitons in the nonfluorescent PDA-3BCMU (blue phase) and PDA-4BCMU (blue phase) and the fluorescent PDA-4BCMU (red phase) and P3MT are discussed using the potential curves (Fig. 3) representing the free exciton, self-trapped exciton, and the ground state.

The relaxation of photoexcitations in the polymers can be described as follows. A free exciton(FE) shown in the potential curve (Fig. 3) is self

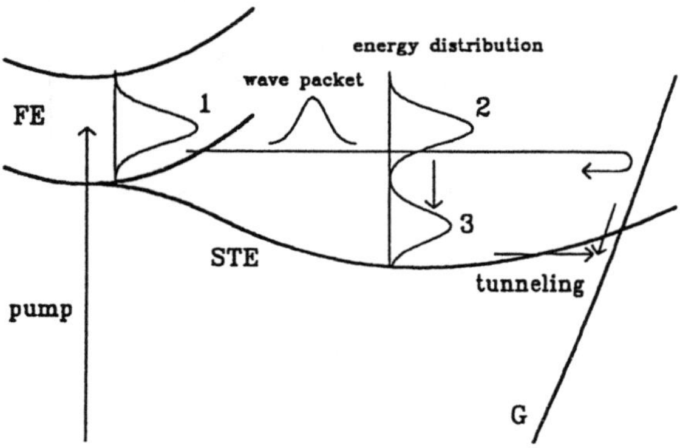

Fig. 3 Adiabatiac potential curves of the ground state(G), free exciton(FE), and self-trapped exciton(STE) of the nonfluorescent PDAs. There is no barrier between the FE and STE potential curves.

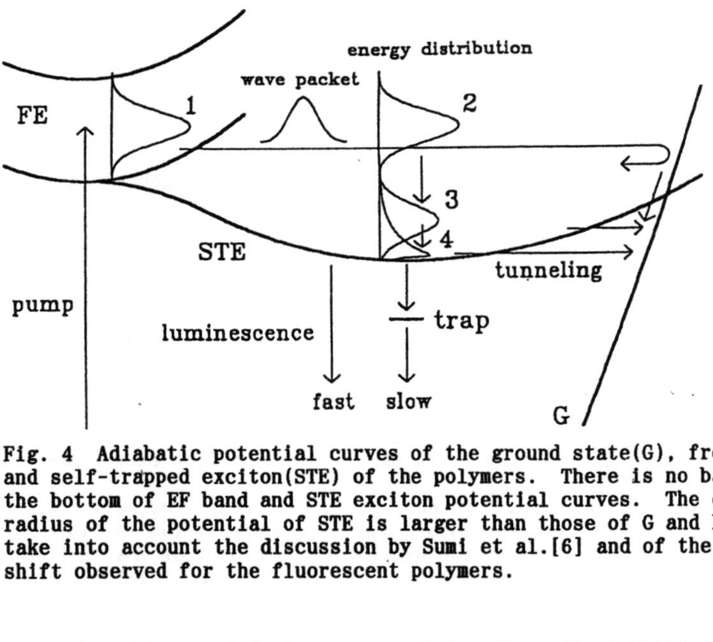

Fig. 4 Adiabatic potential curves of the ground state(G), free exciton(FE), and self-trapped exciton(STE) of the polymers. There is no barrier between the bottom of EF band and STE exciton potential curves. The curvature radius of the potential of STE is larger than those of G and FE states, to take into account the discussion by Sumi et al.[6] and of the small Stokes shift observed for the fluorescent polymers.

trapped within 10-20 fs to create a hot unthermalized STE (process 1-2). Then the oscillation of the strongly coupled mode is activated and the excited electron wave-packet passes or comes close to the crossing point between the ground-state and ST potential curves in the nonfluorescent PDAs. Then the distribution decays with a lifetime (τ_f) of about 100 fs by phonon emission (process 2-3)(Table 1). In the process (process 2-3) of the phonon emission, the major fraction is relaxed over the crossing point or by tunneling through the barrier. The residual unthermalized STE near the bottom of the STE potential then relaxes by tunneling through the barrier with about 2 ps lifetime at 10 K. It is estimated that only a small fraction (say a few percent) of the free excitons remains 2 ps after excitation in nonfluorescent PDA-3BCMU and PDA-4BCMU both in the blue phase, while the other residual excitons relax directly to the ground state before finishing cooling down.

While in the fluorescent PDA-4BCMU (red phase) and P3MT, the crossing points are higher than in the nonfluorescent polymers exciton. After a few tens picoseconds, the excitons in the fluorescent polymers are thermalized to the state (process 3-4), where the temperature can be defined among the exchange modes connected to the bath mode. The temperature defined is still higher than the experimental temperature, which is equal to the bath temperature. Then cooling down process of the exchange mode temperature to the experimental temperature by the vibrational energy flow to the bath mode (process 4-5 not indicated in Fig.4). During this cooling down process, the tunnelling rate becomes smaller and smaller because of the increase in the barrier thickness and height. After cooled down to the bottom of the potential curve of STE and then relax either radiatively or nonradiatively to the ground state either by slow tunneling to the ground state or by being trapped by defect states. Thus low fluorescence quantum efficiency of the 1D polymers is systematically explained by the proposed model shown by Figs. 3 and 4.

The author would like to express thanks to Drs. Yoshizawa and U. Stamm, and Mr. M. Taiji for the collaboration and to Prof. K. Yoshino for providing polythiophene samples and to Profs. Y. Toyozawa, E. Hanamura, S. Abe, and H. Sumi for valuable discussion.

References

[1] M. Yoshizawa, M. Taiji, and T. Kobayashi, IEEE J. Quantum Electron. **25**, 2532 (1989).
[2] U. Stamm, M. Taiji, M. Yoshizawa, T. Kobayashi, and K. Yoshino, Mol. Cryst. Liq. Cryst. **182A**, 147 (1990).
[3] T. Kobayashi, M. Yoshizawa, U. Stamm, M. Taiji,and M. Hasegawa, J. Opt. Soc. Am. **B7**, 1558 (1990).
[4] M. Yoshizawa, A. Yasuda, and T. Kobayashi, Appl. Phys., to be published.
[5] K. Nasu, J. Luminescence **38**, 90 (1987).
[6] H. Sumi, M. Georgiev, and A. Sumi, Rev. Solid State Sci., **4**, 209 (1990).
[7] S. Halle, M. Yoshizawa, H. Murata, T. Tsutsui, S. Saito, and T. Kobayashi, to be published.

Ultrafast Relaxation Dynamics of a Single-Crystal Polydiacetylene, PDA-DFMP

S.D. Halle[1], *M. Yoshizawa*[1], *H. Matsuda*[2], *S. Okada*[2], *H. Nakanishi*[2], *and T. Kobayashi*[1]

[1]**Department of Physics, Faculty of Science, University of Tokyo, 7-3-1 Hongo, Bunkyo-ku, Tokyo 113, Japan**
[2]**Research Institute for Polymers and Textiles, Tsukuba-shi 305, Japan**

Conjugated polymers, such as polydiacetylenes (PDAs), exhibit ultrafast photo-induced response times and large optical nonlinearities; consequently PDAs are attractive candidates for potential applications in nonlinear optical devices. Recently the ultrafast photoexcitation decay processes of several conjugated polymers have been investigated [1,2]. The ultrafast relaxation dynamics of these systems can be explained by the self–trapped exciton relaxation model [3]. By synthetically altering the sidegroups, R and R′, in the backbone structure $\nleftarrow$ RC–C≡C–CR′ $\nrightarrow_n$ the linear optical properties of PDAs such as the exciton transition energy can be modified. However, an understanding of how these ultrafast properties may vary with different chemical modifications of PDAs is still quite limited. In this work we report the first study of the ultrafast relaxation dynamics of a PDA that has sidegroups π–conjugated to the main chain. Femtosecond pump–probe time–resolved spectroscopy is employed to measure both the photo–induced spectral and temporal response of a single–crystal of PDA-DFMP, where R = R′ = 2,5-bis trifluoromethyl-phenyl.

The femtosecond apparatus consists of a colliding-pulse mode–locked dye laser amplified by a four stage dye amplifier using a 10 Hz Nd:YAG laser, yielding ≈100 fs pulses [2]. A pump–probe excitation geometry is employed using white–light continuum generated probe pulses. The pump photon energy at 1.98 eV is resonant with the ^1B_u exciton transition at 2.25 eV. The differential absorption spectra, measured at 297±2 K with linearly polarized laser beams parallel to the chain axis, are displayed in Fig. 1 for five pump–probe delay times.

At a delay time of t = 0 ps (as shown in Fig. 1), which corresponds to the temporal overlap of the pump and probe pulses, two peak minima are observed at 1.80 and 1.72 eV. These gain signals occur at Stokes shifts from the pump at ≈1500 and ≈2150 cm^{-1}, corresponding to the observed ground state C=C (ω_2) and C≡C (ω_1) stretching mode frequencies, respectively. These signals are attributed to (stimulated) Raman gain. The intensity of the Raman gain signals for ω_1 is larger than that for ω_2, in contrast to Raman gain intensities found in other PDAs [1]. This suggests the exciton–phonon coupling strength for the ω_2 mode is weaker than in PDAs such as PDA-3BCMU, due to a reduction of the C=C bond order of PDA-DFMP by the electron–withdrawing sidegroup.

The spectral peak with a photon energy of ≈1.3 eV at 0 ps is due to the transition from the free ^1B_u exciton state to either ^1A_g exciton states and/or the continuum. At longer delay times the spectral peak shifts to higher probe photon energies. This observation is consistent with both previous experiments in PDAs [1] and the model in which photoexcited free excitons for quasi one–dimensional systems can spontaneously relax to the self–trapped exciton [4].

From 0.6 ps to ≥ 100 ps delay times, a spectral peak is found at 1.48 eV. Previous studies have shown that the long–lived (≈100 μs) state at 1.35 eV in polydiacetlyene, PDA-PTS is due to a triplet state [5]. Therefore, the long–lived component of the spectral feature at 1.48 eV is attributed to be the triplet exciton state. When the linearly polarized

Springer Proceedings in Physics, Vol. 68 Time Resolved Vibrational Spectroscopy V
Ed.: H. Takahashi

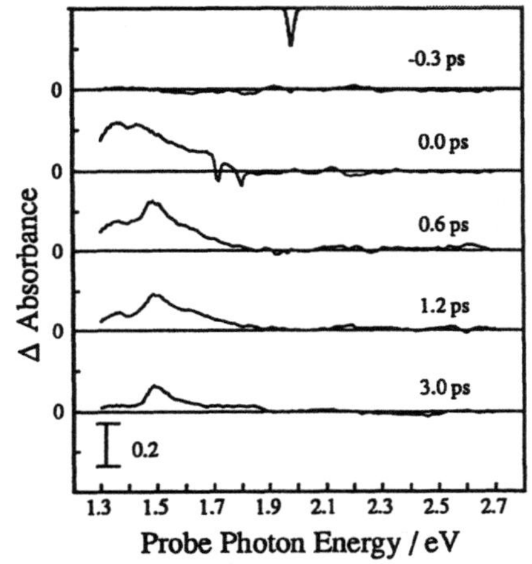

Figure 1. Transient absorbance change spectra of PDA-DFMP. The pump spectrum centered at 1.98 eV is shown.

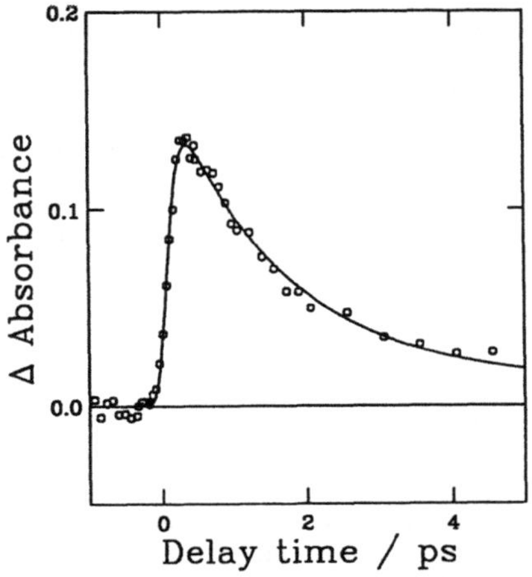

Figure 2. Time dependence of the absorbance change of PDA-DFMP recorded at a photon energy of 1.82 eV at 297 K.

pump beam is oriented perpendicular to the chain axis, no differential absorption signal can be detected. This suggests that the major contribution to the observed absorbance change signal is *not* an inter-chain photoexcitation process.

The photo-induced decay curve measured at 1.82 eV is fit to a single exponential decay plus a constant long-time component, yielding a relaxation time for the ^1B_u self-trapped exciton of 1.62±0.08 ps (see Fig. 2). The formation time of the self-trapped state is determined to be 100±30 fs. For this nonfluorescent polymer, the relaxation kinetics are similar to previous observations of "blue phase" PDAs, in which the self-trapped excition is believed to relax to the ground state mainly via tunneling [1,3].

References

[1] T. Kobayashi, M. Yoshizawa, U. Stamm, M. Taiji, M. Hasegawa, J. Opt. Soc. Am. B **7**, 1558 (1990).

[2] M. Yoshizawa, M. Taiji, T. Kobayashi, IEEE J. Quant. Elect. **25**, 2532 (1989).

[3] T. Kobayashi, in *Time-Resolved Vibrational Spectroscopy V*, (Springer, Berlin, 1991), these proceedings.

[4] Y. Toyozawa, in *Highlights of Condensed-Matter Theory*, ed. F. Bassanu (North-Holland, Amsterdam, 1985), pp.798.

[5] L. Robins, J. Orenstein, R. Superfine, Phys. Rev. Lett. **56**, 1850 (1986).

Femtosecond Time-Resolved Observation of Soft Mode Dynamics in Ferroelectric Crystals

G.P. Wiederrecht, T.P. Dougherty, M.H. Garrett, and K.A. Nelson*

Department of Chemistry (*Center for Materials Science and Engineering), Massachusetts Institute of Technology, Cambridge, MA 02139, USA

1. Introduction

Despite extensive study of the barium titanate (perovskite) family of ferroelectrics, the dynamics and microscopic mechanism for the phase transitions between the paraelectric cubic phase and the tetragonal, orthorhombic, and rhombohedral ferroelectric phases remain in doubt. From a dynamical point of view, the degree to which collective motion of the order parameter, i.e. the polarization, is vibrational (phonon-like) in character versus relaxational (hopping dynamics) is in dispute due to the conflicting results of recent low-frequency Raman spectroscopy studies [1,2]. In terms of microsopic mechanism, relaxational dynamics are indicatative of an order-disorder transition involving collective hopping of ions between different crystalline arrangements. Vibrational dynamics indicate a displacive transition involving motions of ions within a single free energy region whose minimum position shifts with temperature. The vibrational component of the order parameter is often very heavily damped (sometimes overdamped) and so Raman spectral responses are central peaks which merge with relaxational responses. It is often difficult to discern and analyze quantitatively the vibrational and relaxational dynamics.

Here we report results of impulsive stimulated Raman scattering (ISRS) experiments on several perovskites. Impulsive stimulated scattering is an entirely time-domain light scattering technique which is particularly well suited for study of low-frequency and/or heavily damped excitations [3]. It has been used for study of acoustic phonons in crystals near structural phase transitions, but not for direct observation of order parameter dynamics.

An "eight-site" model has been proposed for the series of phase transitions in the barium titanate family, which includes $KNbO_3$ and $BaTiO_3$. According to this model, the central ion (Nb or Ti), which is coordinated by six O atoms at the unit cell faces, occupies one of eight local potential energy minima displaced from the unit cell center along the <111> (cubic phase) axes (toward the K ions at the unit cell corners). In the cubic phase, all eight sites are of equal free energy. In the tetragonal phase, four sites in a plane are of lower energy than the other four. In the orthorhombic phase, two sites along a unit cell edge are allowed; the four sites along adjacent edges are of higher free energy, and the farthest two sites are of highest energy. In the rhombohedral phase, only one site is allowed. The eight site model can also account for both vibrational and relaxational order parameter dynamics, if it is assumed that the potential energy barriers between allowed sites are lower than the zero-point vibrational energy (resulting in oscillatory motion between allowed sites) and the barriers between allowed and higher-energy sites are higher (resulting in hopping dynamics between allowed and higher-energy sites). This implies different symmetries for the vibrational and relaxational modes, and so could be tested by Raman scattering with different polarization combinations. However, due to the difficulties described above, Raman spectroscopy has yielded conflicting results [1,2].

A further test of any model is provided by the Lyddane-Sachs-Teller (LST) relation which expresses the low-frequency dielectric constant ε_0 in terms of the frequencies of polar phonons (or for polar relaxational modes, the decay rates). None of

Springer Proceedings in Physics, Vol. 68 Time Resolved Vibrational Spectroscopy V
Ed.: H. Takahashi

the Raman results reported for $KNbO_3$ yielded both the scattering symmetries predicted by the eight-site model and the ε_0 value determined through capacitance measurements. In $BaTiO_3$, different Raman results yielded a wide range of values for ε_0, none agreed with that determined independently.

ISRS experiments were conducted to resolve unambiguously the symmetries of the vibrational and relaxational order disorder components and to carry out an improved comparison with capacitance measurements of ε_0.

2. Results and Discussion

The experiments were conducted with a transient grating geometry using amplified femtosecond pulses (620nm, 70 fsec, 6 μJ pulses at a 350Hz repetition rate)[4]. Single crystals of $LiTaO_3$ and $KNbO_3$ were obtained from Inrad and Virgo Optics respectively. A $BaTiO_3$ crystal was fabricated by the top-seeded solution growth technique and electrically poled to obtain single domain crystals in the tetragonal phase.

Data from lithium tantalate, a relatively simple crystal with a single paraelectric - ferroelectric phase transition and in which the phonons are lightly damped, is shown in Fig. 1. In this case the ISRS data yield frequencies and decay rates which can also be determined readily from the corresponding Raman spectrum. (Note that since the polar lattice modes have polariton character and are therefore dispersive, the frequency varies with Raman or ISRS scattering angle)[5] It may be possible to go beyond the information content of the Raman spectrum by driving coherent lattice vibrations whose amplitudes are large, and efforts to do this with one or many excitation pulses are under way[6].

Typical data from $KNbO_3$ are shown in Fig. 2 along with a fit to a damped oscillator response. In B_2-symmetry data of this type recorded at sample temperatures throughout the orthorhombic phase and at several scattering wave vectors, no additional contribution to the response due to any relaxational mode was found. The extent to which such contributions can be ruled out has been emphasized in some detail [7]. ISRS data (shown elsewhere) from modes of A_1 symmetry showed two relaxational features, corresponding to hopping into the two types of higher-energy sites [7].

The dispersive phonon-polariton frequencies and damping rates measured at different wave vectors permitted accurate determination of the uncoupled phonon parameters at various temperatures. As discussed in detail elsewhere, the temperature-dependent values of ε_0 calculated with these parameters through the LST relation agree well with ε_0 values determined through dielectric measurements [7]. Thus the qualitative (symmetry) and quantitative ISRS results from $KNbO_3$ all support the eight-site model.

Similar results were found for the tetragonal phase of $BaTiO_3$. Typical data is shown in Fig. 3. In this case the polariton mode is underdamped at small wavevector. As

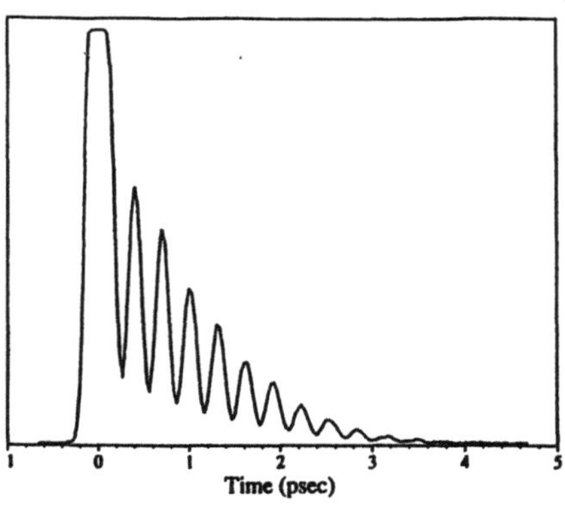

Figure 1 - ISS signal from the lightly damped A_1 polariton in $LiTaO_3$.

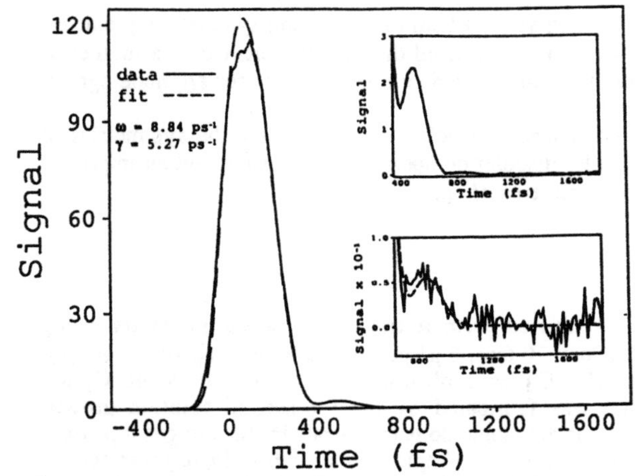

Figure 2- The B_2 polariton in $KNbO_3$ at 442.4K. The large dynamic range, indicated by the signal values in the insets, permits accurate determination of the vibrational dynamics.

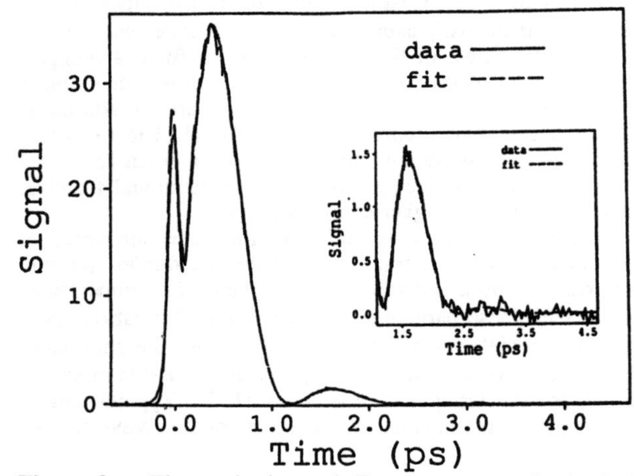

Figure 3 - The underdamped E-symmetry mode in $BaTiO_3$ at 392K.

in $KNbO_3$, no relaxational mode of the same symmetry was found. ISRS data permitted determination of uncoupled phonon parameters which we believe to be considerably more reliable than those determined from Raman spectra [8]. However, as in earlier studies the LST relation yielded disagreement with capacitance measurements of ε_0 in $BaTiO_3$. Additional dielectric measurements may be called for.

In summary, our results provide strong support for the eight-site model of phase transitions in the perovskite ferroelectrics. ISRS should prove useful for study of cooperative motion involved in many phase transitions.

This work was supported in part by NSF Grant No. DMR-9002279, by contributions from Dupont and Perkin-Elmer, and NASA contract NAG 1-996.

References

1. M.D. Fontana, A. Ridah, G.E. Kugel, and C. Carabatos-Nedelec, J. Phys. C **21**, 5853 (1988).
2. J.P. Sokoloff, L.L. Chase, and D. Rytz, Phys. Rev. B **38**, 597 (1988).
3. Y.-X. Yan, L.-T. Cheng and K. A. Nelson in *Advances in Infrared and Raman Specroscopy: v. 16,* ed. by R. J. H. Clark and R. E. Hester (Wiley,1987), p. 299.
4. S. Ruhman, A.G. Joly, and K.A. Nelson, IEEE J. of Quant. Elec. **24**, 460 (1988).
5. A.F. Penna, A. Chaves, P. da R Andrade, and S.P.S. Porto, Phys. Rev. B **13**, 4907 (1976).
6. A.M. Weiner, D.E. Leaird, G.P Wiederrecht, and K.A. Nelson, Science **247**, 1327 (1990).
7. T.P. Dougherty, G.P. Wiederrecht, and K.A. Nelson, to be published
8. T.P. Dougherty, G.P. Wiederrecht, M.H. Garrett, and K.A. Nelson, to be published.

The Tunneling Kinetics of NH_3D^+ Salts

H.L. Strauss, W. Burrows, H.-G. Cho, and J.E. Weier

Department of Chemistry, University of California,
Berkeley, CA 94720, USA

1. Introduction

We have found it possible to hole-burn the N-D stretching bands of a large number of crystalline ammonium salts containing small amounts of the NH_3D^+ ion [1,2,3]. The crystals that burn well are those with intermediate barriers to rotation, that is, barriers of the order of 1000 cm^{-1}. Such crystals often have the ammonium ions in asymmetric (C_1) sites, and these sites tend to have rather long and therefore rather weak hydrogen bonds. [Very symmetrical sites can have short and therefore strong hydrogen bonds, but low barriers to rotation due to the high symmetry.] The NH_3D^+ ion has four distinguishable positions and so gives rise to four N-D stretching bands -- one for each different orientation of the N-D bond in the site (Fig. 1). The holes can be burned and relax on the time scale of hours and so the process can be followed conveniently with an FTIR [4]. We use a diode laser and irradiate the sample while it is in place in the FTIR (Fig. 2). The FTIR detector is used to detect the laser radiation and thus to line up the

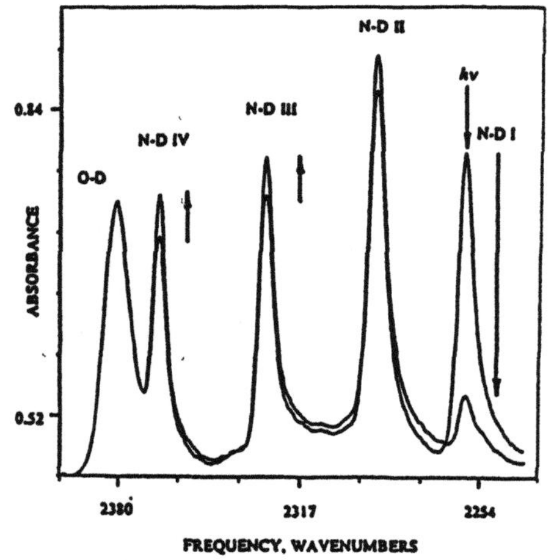

Figure 1. IR spectrum of the N-D stretching region of the NH_3D^+ ion in ammonium cobalt sulfate hydrate at 4 K. Two spectra are superimposed. One spectrum is taken before irradiation and the second thereafter. The second spectrum shows the decrease of the burned band (N-D I) and the concomitant increase in the other N-D bands. Note that the O-D does not change in intensity.

Springer Proceedings in Physics, Vol. 68 Time Resolved Vibrational Spectroscopy V
Ed.: H. Takahashi

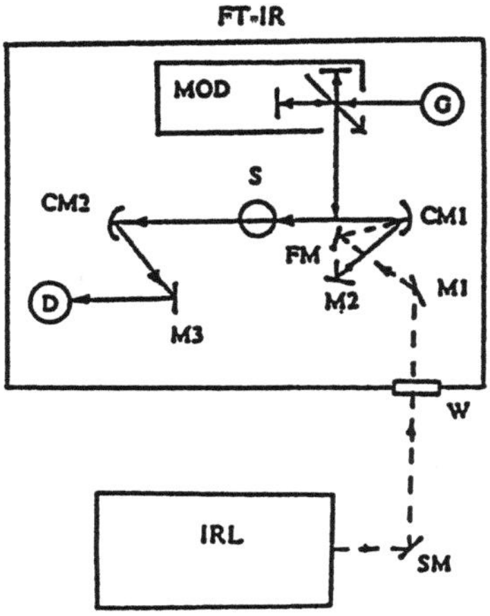

Figure 2. Schematic diagram for hole-burning. IRL is an infrared diode laser and monochromator, FM is a flat mirror that can be flipped up or down, S is the sample, D is the detector, and the other parts are standard for an FTIR.

laser beam. We use the sample optics of the FTIR to assure that the laser radiation is parallel to the FTIR beam. Irradiation of any one of the N-D bands results in the burning of that band and an increase in the other three bands of that ion (Fig. 1).

2. Tunneling Kinetics

Once the sample is "burned," relaxation back to equilibrium can be studied. The orientational distribution of the ions relaxes by a series of coupled first-order processes between orientation sites. The overall process consists of twelve coupled reactions, which can be analyzed by standard methods. As in most experimental kinetics, the problem is simplified by taking some of the many rates as equal. We have successfully analyzed a number of different regimes. By systematically burning each of the bands in turn and then analyzing the rate data, we have shown, for NH_4NO_3, that the rate model is consistent. Then by studying the temperature variation of the rates at low temperature, we find that the rates vary linearly with temperature (in $(NH_4)_2Ni(H_2O)_6(SO_4)_2)$) [2] (Fig. 3). This is the temperature dependence to be expected for phonon-assisted tunneling.

At low enough temperatures, only the lowest level in each well is occupied. Transitions between the wells occur strictly by tunneling, since the barrier is much too high to allow passage over the barrier. Since the sites and thus the wells are asymmetric, the lowest levels are not at equal energy. As a result, a lattice phonon must participate in the tunneling and the temperature factors for a one-phonon process result in the linear dependence. The slope of the variation depends on the value ΔE between the lowest levels, and this can be checked with the temperature dependence of the equilibrium populations. All this works out as expected [2].

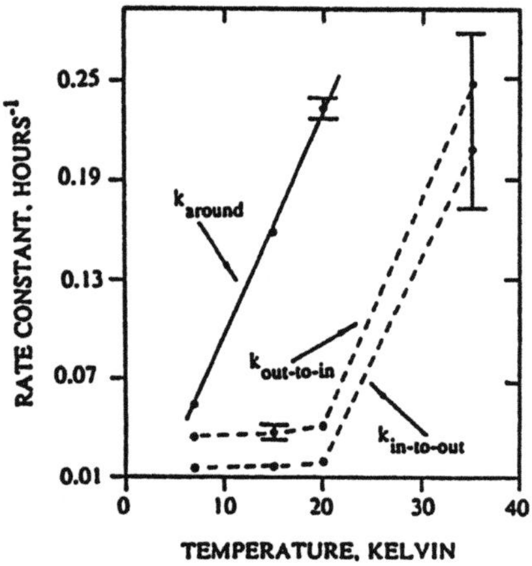

Figure 3. The temperature dependence of three of the rate constants for reequilibration of the orientation of NH_3D^+ in ammonium cobalt sulfate hydrate. The diagram shows the linear dependence on temperature expected at low temperature for tunneling. The curves are labelled with names that come from the kinetic model.

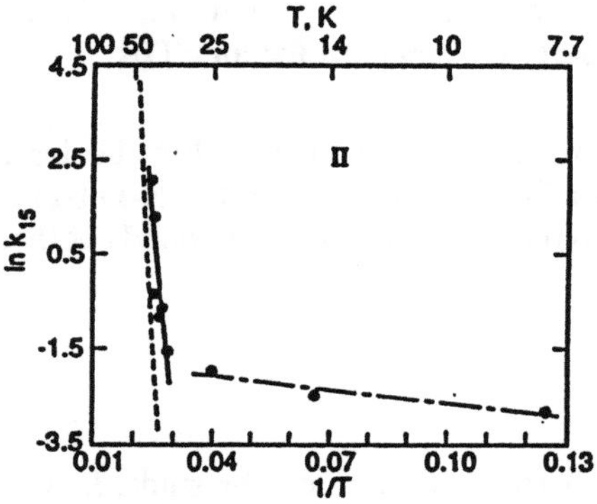

Figure 4. The temperature dependence of one of the rates of reorientation of NH_3D^+ in ammonium sulfate. Both the linear (low temperature) and exponential (intermediate temperature) regions are shown. The expected over-the-barrier rate is given by the dashed line.

As the temperature is raised, the slope of the rate vs. temperature abruptly increases [5] (Fig. 4). There are two possible explanations for this behavior. Theories that consider only the lowest two levels predict that the one-phonon process will be superseded by multiphonon processes with progressively higher slopes. However, before the temperature at which this happens is reached, the higher levels in each well become occupied. Tunneling between these higher levels is very much faster than that between the lower levels and thus the slope increases rapidly. We have fit the data for $(NH_4)_2SO_4$ to "simple quantum-mechanical transition-state theory," that is, to a sum over rates from a Boltzmann distribution of levels in the reactant well. The results are excellent (Fig. 4) and fit the *known* librational barrier for the $(NH_4)^+$ rotation.

3. The Faster Kinetics

Since the laser has a spectral width much less than the width of the spectral band, it is clear that the excitation process -- and likely the relaxation process -- is more

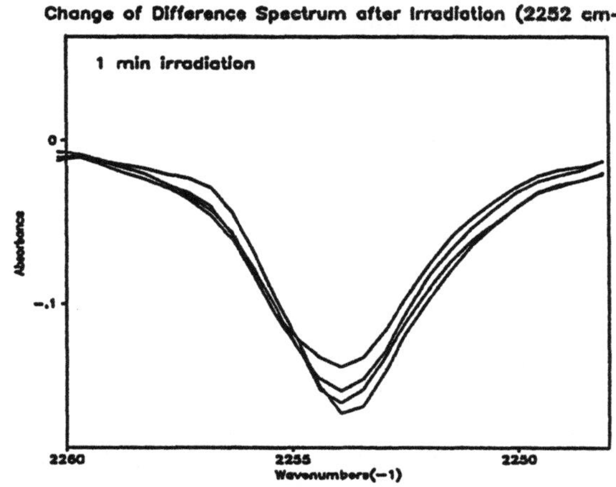

Figure 5. The time dependence of the narrow hole in ammonium sulfate. Note that the center of the band shifts with time (5-minute intervals).

complex than that implied by the discussion above. We have examined the faster processes in both excitation and recovery. To determine the excitation we burn at a fixed frequency and follow the transmission of the burning laser through the sample. The time constant for burning for $(NH_4)_2SO_4$ is a few minutes. The band can be burned off-center and the resulting hole is then off-center as expected (Fig. 5) and relaxes -- also with a few-minute time constant. In a very preliminary model, we guess that the excitation process puts the ions into the excited state of the N-D stretch, and then this relaxes to the upper orientational levels of the ground stretch state and sometimes crosses over to the next well. The hole fills in toward the center of the band, and so the spectral position of a given N-D is somehow lost in the process of formation and reequilibration.

Supported by the National Science Foundation

[1] A. P. Trapani and H.L. Strauss, J. Chem. Phys. 87, 1899 (1987).

[2] A. P. Trapani and H. L. Strauss, J. Amer. Chem. Soc. 111, 910 (1989).

[3] A. P. Trapani, S. W. Gensler, and H. L. Strauss, J. Chem. Phys. 87, 4456 (1987).

[4] J. E. Weier, A. P. Trapani, and H. L. Strauss, Spectroscopy Intl. 1, 49 (1989) [also Spectroscopy 3, 47 (1988)].

[5] W. Burrows and H. L. Strauss, J. Chem. Phys. 93, 7510 (1990).

Quantum Beats of Biphonons in Disordered NH_4Cl

F. Vallée

Laboratoire d'Optique Quantique du C.N.R.S., Ecole Polytechnique,
F-91128 Palaiseau Cédex, France

The observation of quantum beats of nearby vibrational excitations in time resolved coherent experiments has provided very precise information on the first-order spectra of liquids and molecular solids [1,2]. We have taken advantadge of the high sensitivity of these coherent methods to investigate very weak features in the two-phonon spectra of a molecular crystal: ammonium chloride (NH_4Cl).

This crystal is of particular interest since spontaneous Raman investigations indicate that introduction of ^{15}N impurities modifies the spectral properties of the $2\nu_4$ biphonon which is split off from the free two-phonon continuum by intramolecular anharmonicity in the perfect crystal [3] (ν_4 is an iontemal mode of the NH_4^+ ions). It was suggested [4] that these modifications result from the interaction of the nonlocal biphonon of the pure crystal with the corresponding local overtone of the isolated impurities. As these states are quasi-degenerated, they strongly interact through the harmonic intermolecular potential leading to the formation of two quasi-local biphonons with close frequencies and comparable Raman intensities, even for very low ^{15}N concentrations (natural concentration is 0.37%). Spontaneous Raman data are however insufficient to clearly establish the existence of this doublet and to access its spectral features.

Using a time resolved CARS technique, we have observed quantum beats in the $2\nu_4$ spectral region of the natural ammonium chloride crystal which yield evidence for the existence of a biphonon doublet (Fig.1 lower curve). Here the two modes are simultaneously coherently excited by coherent Raman scattering of two time coincident picosecond pulses with a frequency difference matching the biphonon frequencies. Coherent scattering of a time-delayed probe pulse gives rise to two anti-Stokes signals whose interferences lead to the characteristic oscillations of the measured signal. The beating period precisely measures the frequency splitting $\delta\nu$ of the two lines and for the natural crystal we determine: $\delta\nu = 0.7 \pm .04$ cm^{-1}. The dephasing times and the relative Raman intensity of the components of the doublet can also be extracted by fitting the experimental results (full line in Fig.1) [5].

The previous interpretation of the biphonon splitting can be tested by performing similar measurements in crystals containing various concentrations of impurities and in particular in a purified one. The results obtained in a crystal containing less than 0.01% ^{15}N impurities is shown in Fig.1 (upper curve). An exponential decay of the signal is observed over 5 orders of magnitude indicating that an unique homogeneously broadened biphonon is formed in the absence of impurities. This clearly demonstrates that the biphonon splitting observed in natural NH_4Cl originates from the strong interaction of the local and nonlocal overtones.

Springer Proceedings in Physics, Vol. 68 Time Resolved Vibrational Spectroscopy V
Ed.: H. Takahashi

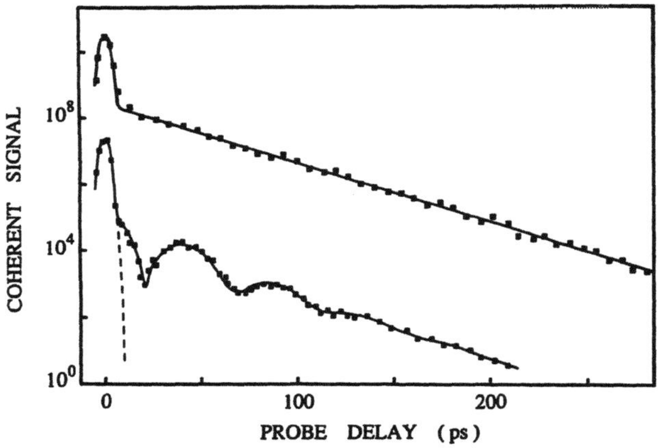

Figure 1: CARS signal from the $2\nu_4$ biphonons in a natural NH_4Cl crystal (lower curve) and in a purified crystals (upper curve, which is up-shifted for clarity). The temperature of the samples is 78K.

Similar measurements were also performed in crystals with impurity concentrations of: 1%, 2% and 5%. As expected, quantum beats were observed in these samples with a beating period decreasing as impurity concentration increases. This behavior is a consequence of the strengthening of the local/nonlocal mode interactions and can be correctly reproduced by a theoretical model based on a generalization of the coherent potential approximation (CPA) method to the two-phonon states[4].

References

[1] A.Laubereau and W.Kaiser, Rev. Mod. Phys. 50, 607 (1978)
[2] S.Velsko, J.Trout and R.M.Hochstrasser, J. Chem. Phys. 79, 2114 (1983)
[3] M.V. Belousov and D.E.Pogarev, J.E.T.P. Lett.11, 644 (1978)
[4] V.M.Agranovich and O.A.Dubovsky, in: Optical Properties of Mixed Crystals, eds. R.J.Elliott and I.P.Ipatova (Elsevier Sciences) p.297 (1988)
[5] F.Vallée, Chem. Phys. Lett. 175, 382 (1990)

LO-Phonon Dephasing in Semi-Insulating and n-Type GaAs

F. Vallée, F. Bogani, and C. Flytzanis*

Laboratoire d'Optique Quantique du C.N.R.S., Ecole Polytechnique, F-91128 Palaiseau Cédex, France
*Permanent address: Dipartimento di Fisica, Università di Firenze and Unità del Gruppo Nazionale di Struttura della Materia, Largo E. Fermi 2, I-50125 Firenze, Italy

Time-resolved coherent nonlinear techniques were extensively applied to the investigation of vibrational dephasing in liquids and molecular solids but were scarcely used for the analysis of optical phonon dynamics in semiconductors. In very commonly used III-V compounds such as GaAs or InP, the problem of LO-phonon dephasing has not yet been addressed in the time domain in spite of a poor understanding of the processes that governs their relaxation. The dynamics of these states in pure sample or in the presence of an electron plasma is however of both fundamental and technological interest since LO-phonons strongly influence most of the electronic properties of these polar semiconductors[1].

We have performed a direct time-domain determination of the LO-phonon dephasing time (T_2) in semi-insulating and n-type GaAs using an infrared time-resolved Coherent Anti-Stokes Raman Scattering (CARS) technique. Here a coherent phonon population is created in the crystal by coherent Raman scattering of two picosecond pulses of frequencies ω_L and ω_S, with $\omega_L - \omega_S = \omega_{LO}$. The loss of coherence of this excitation is then followed by coherent anti-Stokes Raman scattering at $\omega_{AS} = \omega_P + \omega_{LO}$ of a time-delayed probe pulse of frequency ω_P.

In such a bulk experiment the three frequencies must lie in the transparency region of the sample, i.e., in the near infrared for GaAs ($E_{gap} \sim 1.5$eV). These infrared pulses were produced by a modified version of the passively mode-locked Nd:glass laser system already described and utilized for visible CARS experiments [2]. The initial beam, delivering 5 ps pulses of 2 mJ at 1.054 µm is split into three parts. The first one (~ 10 µJ), after passing through a variable delay, is used as the probe beam (ω_P). The second one (~ 30 µJ) is utilized as one of the excitation beams (ω_L), while the third one is frequency shifted by stimulated Raman scattering in $CHBr_2Cl$ to generate the ω_S pulse (~ 3 µJ) at ~ 1.084 µm (ω_{LO} = 295 cm^{-1} in GaAs). After polarization adjustment, the three beams are focussed into a 350µm thick crystal with a (100)-orientation. The generated anti-Stokes signal is detected by a S1 photomultiplier after spatial spectral and polarization filtering.

The observed CARS signal in a semi-insulating sample is plotted on a logarithmic scale as a function of probe delay in Fig. 1 (curve-a) for a low focalization of the laser pulses. After an initial fast rise and fall limited by the system response function, we observe an exponential decrease of the coherent signal indicating a Lorentzian (homogeneous) broadening of the line. The slope of the decay

Springer Proceedings in Physics, Vol. 68 Time Resolved Vibrational Spectroscopy V
Ed.: H. Takahashi

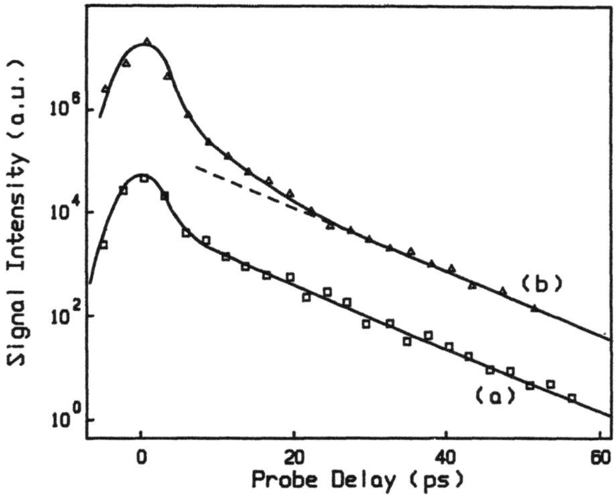

Figure 1 - Coherent anti-Stokes signal from the LO-phonon in semi-insulating GaAs at 78K obtained for - (a) a weak focalization of the laser beams -(b) a strong focalization (the diameter of the focal spot is decrease by a factor of about 2). The (b)-curve is up-shifted for clarity.

directly measures the intrinsic dephasing time $T_{2L}/2 = 6.4 \pm .4$ ps at a temperature of 78K. This value is in good agreement with the energy relaxation time (T_1=7±1ps) measured in time resolved Raman experiments [3,4] indicating that LO-phonon dephasing is dominated by population relaxation in semi-insulating GaAs.

Similar measurements were realized for various crystal temperature ranging form 10 to 320K. The observed temperature dependence of the dephasing rate allows to ascribe relaxation to the intraband LO-phonon decay into a transverse acoustic (TA) phonon ($\nu_{TA} \sim 60$ cm^{-1}) and a LO phonon ($\nu_{LO} \sim 235$ cm^{-1}) of opposite wave vector at the L critical point of the Brillouin zone in agreement with our preliminary results[5].

In this first experiment, the focussing of the beams was limited in order to avoid any influence of electrons photoexcited by two-photon absorption and by residual one-photon absorption from impurities. Interaction of photoexcited carriers with the LO phonons strongly modifies their dynamics [6], as examplified in Fig.1 (curve-b) obtained for an higher focussing of the laser beams. Here the photoexcited electron-plasma reachs a sufficient concentration to significantly accelerate the LO-phonon dephasing leading to an important modification of the measured decay of the CARS signal which is exponential only for long probe delay (note that the long-time signal decays with the previously measured intrinsic dephasing time $T_{2L}/2$). This behavior results from the spatial inhomogeneity of the created plasma leading to a variation of the effective LO-phonon dephasing rate with plasma density inside the excited zone. For sufficiently long time delay only phonons with the longer decay rate (i.e. $2/T_{2L}$) are observed.

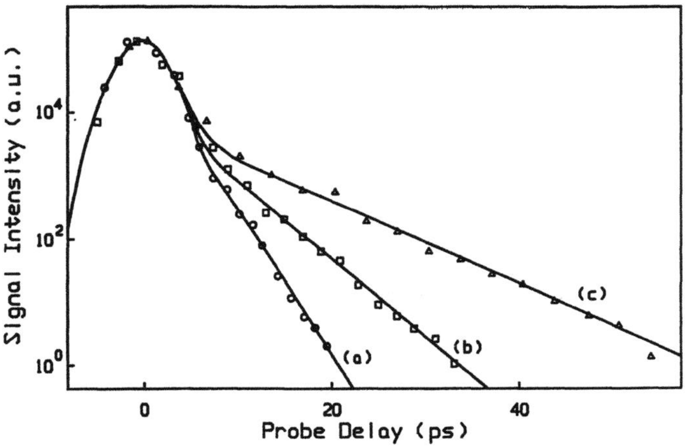

Figure 2 - CARS signal from the phonon-like plasmon-phonon hybrid mode in n-type GaAs at 78K. The electronic plasma density are (a)$10^{17}cm^{-3}$, (b) 5 $10^{16}cm^{-3}$ and (c)$\leq 10^{15}cm^{-3}$. The measured dephasing rate are respectively (a) 1.8ps, (b) 3.2ps and (c) 6.4ps.

The influence of electrons on the LO-phonon coherence can be investigated more quantitatively in n-type GaAs. In these crystals the longitudinal electric field of the LO-phonons interacts with the collective plasmon mode of the equilibrium electron-plasma resulting in the formation of mixed plasmon-phonon excitations [7]. Using the CARS technique described above, we have coherently excited and probed the phonon-like mixed mode. The results are shown in Fig.2 for three crystal doping (Si dopant): n_e= 1.1 $10^{17}cm^{-3}$, 4.7 $10^{16}cm^{-3}$ and $\leq 10^{15}cm^{-3}$. For the lower doping, the results are identical to those obtain in the semi-insulating crystal while in the more heavily doped samples the signal still decay exponentially but with a faster rate. This reflects the increase of the plasmon part of the phonon-like coupled mode from which information on plasmon dynamics can thus be extracted. In the long wavelenth region accessed in our experiment, the spectral response function of the hybrid mode can be calculated from the total dielectric function [7]:

$$\varepsilon = \varepsilon_\infty \left(1 + \frac{\omega_{LO}^2 - \omega_{TO}^2}{\omega_{TO}^2 - \omega^2 - 2i\omega/T_{2L}} - \frac{\Omega_P^2}{\omega(\omega + i/\tau)} \right) \qquad (1)$$

Here τ is the collision relaxation time of the electrons and $\Omega_P^2 = 4\pi n_e e^2/\varepsilon_\infty m^*$ is the plasma frequency. For the plasma density of the investigated samples, the LO-phonon mode is only weakly perturbed so that the frequency of the phonon-like hybrid mode is nearly identical with ω_{LO}. Due to the large plasmon damping, its lineshape is however deeply modified . Using (1) one can show that a Lorentzian broadening of the phonon-like mode is expected corresponding to an exponential decay of the CARS signal in the time domain, in agreement with our observation. The hybrid mode dephasing time T_2 is given by:

$$1/T_2 = 1/T_{2L} + (\omega_{LO} - \omega_{TO})\,\Omega_P^2/\omega_{LO}^3\,\tau \qquad (2)$$

Using the value of T_{2L} determined in the semi-insulating sample, the relaxation time τ can be evaluated from the measured hybrid-mode dehasing time. We find: τ~60fs for n_e=4.7 $10^{16}cm^{-3}$ and τ~50fs for n_e=1.1 $10^{17}cm^{-3}$ at 78K. Because of its small plasmon content, the phonon-like mixed mode thus constitutes a sensitive probe of the plasmon properties, opening-up new possibilities for the investigation of very fast electronic processes.

Our measured value of T_2 are smaller than the population relaxation times T_1 measured in semi-insulating GaAs in the presence of a photoexcited plasma of comparable density [4]. This suggests that scattering by ionized impurities plays an important role in the mixed mode dephasing of our doped sample. Quantitative T_2 measurements after photoexcitation of a two-component plasma in semi-insulating GaAs and its comparison with the T_1 data should be particularly interesting for the understanding of both plasmon and LO-phonon relaxation processes.

We wish to thank Professor G.Guillot, I.N.S.A. Lyon, France and Doctor R.Magnanini, M.A.S.P.E.C. Parma, Italy, for providing the samples. We are also indebted to C. Gaonach, L.C.R.-Thomson, France for measuring the carrier concentrations.

References

[1] S.A. Lyon, J. of Lum. 35, 121 (1986)
[2] F.Vallée, G.M.Gale and C.Flytzanis, Phys. Rev. Lett. 61, 2102 (1988)
[3] D.von der Linde, J.Kuhl and H.Klingenberg, Phys. Rev. Lett. 44, 1505 (1980)
[4] J.A. Kash and J.C. Tsang, Solid State Electr. 31, 419 (1988)
[5] F.Vallée and F.Bogani, to be published in Phys. Rev. B
[6] C.L. Collins and P.Y. Yu, Solid State Comm. 51, 123 (1984)
[7] G.Abstreiter,M.Cardona and A.Pinczuk, in " Ligth Scattering in Solids IV ", eds M.Cardona and G.Guntherodt, p.5, Springer Verlag (1984)

Dephasing Decay of the Vibrons in $Ar_x(N_2)_{1-x}$ Mixed Crystals

J. De Kinder[1], E. Goovaerts[2], A. Bouwen, and D. Schoemaker

Physics Department, University of Antwerp (U.I.A.),
B-2610 Wilrijk (Antwerp), Belgium
[1]Research Assistent and [2]Research Associate of the National Fund for Scientific Research Belgium (NFWO)

In the low temperature α phase of solid N_2, vibrational excitations on individual $^{14}N_2$ molecules couple to form delocalized states with an excitonic character, the so-called vibron states. This coupling wipes away small differences in local site energy, thereby reducing the inhomogeneous line broadening. Linewidth measurements of the $^{14}N_2$ vibrons and the isolated $^{15}N_2$ stretching vibrations in α-$(^{15}N_2)_x(^{14}N_2)_{1-x}$ mixed crystals [1] showed that also the homogeneous linewidth is reduced for the vibron states. For increasing temperatures, the vibrational coupling among the individual molecules is destroyed [1] by scattering on librational motions which become more important towards the α-β phase transition.

We have applied time-resolved coherent anti-Stokes Raman scattering to measure the linewidth of the vibrons. In this four-photon light scattering process, we measure the decay of the coherence of the vibrational population created in the sample. The time resolution of our setup is 6 ps. Homogeneous and inhomogeneous contributions to the decay were determined by fitting [1,2] the product of an exponential and a Gaussian decay function to the observed dephasing signals.

We have studied the dephasing of the $\mathbf{k} \simeq 0$ vibrons in $Ar_x(N_2)_{1-x}$ mixed crystals as a function of temperature (7 - 40 K) and Ar concentration ($0 < x < 0.15$). For small Ar concentrations only an increase of the inhomogeneous linewidth (see Fig. 1) could be resolved. Above $x = 0.02$, both the homogeneous and the inhomogeneous broadening increase with Ar concentration, while the latter becomes temperature dependent, as can be seen in Fig. 1.

The increase of the inhomogeneous linewidth at low temperature is due to the different lattice parameter for solid argon, which is 6 % smaller than for nitrogen. The free rotation of the N_2 molecules in the α phase is prohibited by their anisotropy and by the electrostatic quadrupole-quadrupole coupling among them, so that librational states are formed. The substitution of the N_2 molecules by Ar atoms reduces the corresponding potential wells. Consequently, the librational amplitude increases — thereby reducing the coupling between the individual molecules — and suppresses the effect of line narrowing. This is already reflected in the homogeneous and the inhomogeneous broadening at low temperature. With increasing temperature, the librational amplitudes increase, which in turn cause a temperature dependence of the inhomogeneous linewidth.

Springer Proceedings in Physics, Vol. 68 Time Resolved Vibrational Spectroscopy V
Ed.: H. Takahashi

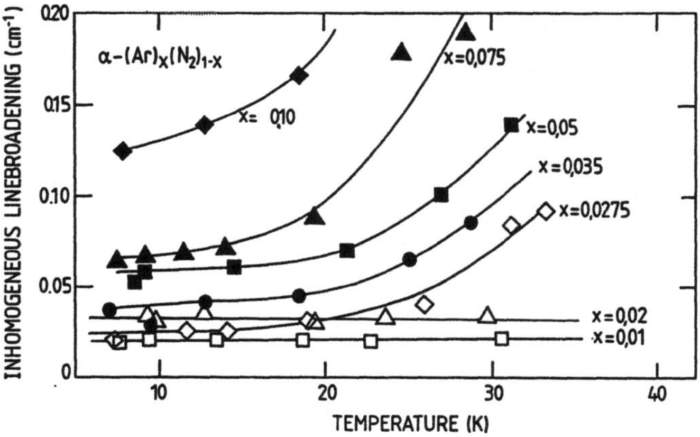

Fig. 1 Inhomogeneous line broadening as a function of the temperature and the Ar concentration x. The temperature independent value for pure α-N_2, i.e., 0.010 cm^{-1} was not indicated in the figure.

In the α phase, the stretching vibration is split into an A_g and a T_g mode. This leads to a beating pattern superimposed on the dephasing signal, from which we could determine the crystal-field splitting. We found a similar temperature dependence as for the pure α-N_2 crystals, with a splitting shifted towards higher values for higher Ar concentrations. The increase of the low temperature value was almost linearly as a function of the Ar concentration and was about 17 % for an Ar concentration $x = 0.05$.

References

[1] J. De Kinder, E. Goovaerts, A. Bouwen, and D. Schoemaker, Phys. Rev. **B 42**, 5953 (1990).

[2] J. De Kinder, E. Goovaerts, A. Bouwen, and D. Schoemaker, J. Lumin. **45**, 423 (1990).

Time-Resolved Infrared Study of Electric-Field-Induced Reorientation of Nematic Liquid Crystal Molecules

T. Urano[1,2], *M. Sakamoto*[2], *and H. Hamaguchi*[1]

[1]Kanagawa Academy of Science and Technology (KAST), KSP East 301, 100-1 Sakato, Takatsuku, Kawasaki 213, Japan

[2]Toshiba Research and Development Center, 1 Saiwaiku, Kawasaki 210, Japan

1. Introduction

It is well-known that liquid crystal (LC) molecules reorientate under the applied electric field due to their electric anisotropy. Nevertheless, the detailed mechanism of reorientation at molecular levels is not known. For further study of this reorientation mechanism, it is desirable to obtain information on the time course of the reorientation and the subsequent recovery, which is most suitably studied by time-resolved spectroscopy. In the present paper, μs-time-resolved infrared spectroscopy is applied to the study of the time response of 4-n-pentyl-4'-cyanobiphenyl (5CB) to the applied electric field.

2. Experimental

The sample of 5CB was held in a 5μm-thick cell constructed from two Si electrodes. The cell surface was treated with rubbed polyimide membrane to give a homogeneous alignment of 5CB molecules. The alignment of the sample was checked by the polarized infrared measurement.

Time-resolved polarized infrared spectra were obtained with a dispersive μs-time-resolved infrared spectrometer described previously [2]. The axis of the polarizer was set parallel to the rubbed direction. A burst electric field (FIG. 1) was applied to the sample cell by using a multifunction synthesizer. A boxcar integrator was used for the time-resolved spectral measurement with a gate width of 78 μs. The time response of each band was measured with a digital sampling oscilloscope with a time resolution of 2 μs.

3. Results and discussion

FIG. 2 shows the time-resolved infrared spectrum taken 1.9 ms after the application of the electric field. Six positive peaks in the spetrum represent the decrease of the absorption. These bands are assigned to the modes whose transition moments are parallel to the molecular axis. Three negative peaks, on the other hand, represent the increase of the absorption. These bands were assigned to the modes whose transition moments are perpendicular to both the molecular axis and the cell surface.

The time response to the burst electric field was measured for six bands as shown in FIG. 1. It is found that the methylene CH anti-symmetric stretching band (2916 cm^{-1}) and methylene CH deformation band (1400 cm^{-1}) show markedly different

Springer Proceedings in Physics, Vol. 68 Time Resolved Vibrational Spectroscopy V
Ed.: H. Takahashi

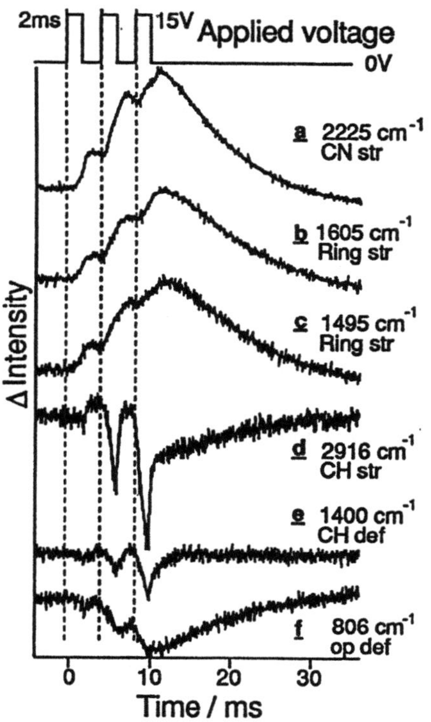

FIG. 1 Time response of TRIR bands to the burst electric field

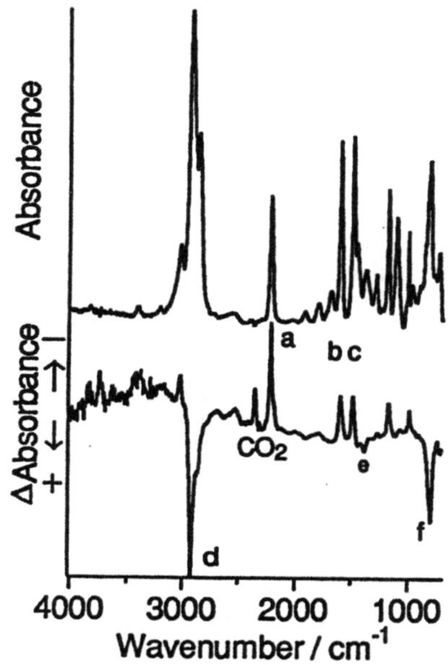

FIG. 2 TRIR spectrum of 5CB in the burst electric field

time response (relaxation with a double exponential with τ ~ 10ms, and 0.5ms) from the other vibrational bands (a single exponential with τ ~ 10ms). This observation suggests that the motion of the methylene group is markedly different from those of the functional groups in the biphenyl and the CN moiety. The fast decay component may be attributable to the internal rotation around C-C bond in the alkyl moiety.

References

[1]G. Labrunie and J. Robert, J. Appl. Phys. 44, 4869 (1973).
[2]K. Iwata and H. Hamaguchi, Appl. Spectrosc. 44, 1431 (1990).

1D and 2D Time-Resolved FT-IR Studies of Liquid Crystal Reorientation Dynamics

H. Toriumi

Department of Chemistry, College of Arts and Sciences,
University of Tokyo, Komaba, Meguro, Tokyo 153, Japan

1. Introduction

FT-IR time-resolved spectroscopy (FT-IR TRS) with simultaneous temporal and spectral resolution can provide detailed information concerning both structural and orientational changes of molecules in transient systems. This paper reports the application of 1D and 2D FT-IR TRS to the study of electric field-induced reorientaion dynamics of liquid crystals.

2. Results and Discussion

2.1 Rapid-Scan 1D FT-IR TRS of Nematic Liquid Crystals

Fundamentals of 1D FT-IR TRS experiment using a rapid-scan spectrometer and the design of a variable temperature liquid-crystal cell have been detailed in our recent papers [1,2]. Nematic liquid crystals undergo a homogeneous-to-homeotropic transition when a sufficiently high electric field is applied. The average orientation of bond segments in homogeneously oriented nematic liquid crystal samples (5CB and 5PCH) is determined from static dichroism measurements. The observed temperature dependence of the molecular orientational order parameter agrees quantitatively with the prediction of the Maier-Saupe theory. Analysis of time-resolved dichroic spectra measured by using 1D FT-IR TRS reveals that the homogeneous-to-homeotropic transition requires an induction period of several millisecond and completes in the 10-20 ms region. It is also found that the rigid core segment reorients as a unit in response to the applied electric field and the reorientaion rate increases at elevated temperatures as the rotational viscosity decreases.

2.2 Step-Scan 2D FT-IR TRS of Nematic Liquid Crystals

To enhance the dynamic spectral analysis of the liquid crystal reorientation kinetics, we used a step-scan 2D FT-IR frequency correlation spectroscopy [3]. The synchronous 2D FT-IR correlation map confirms the above observation in 1D TRS experiment that the rigid core of 5CB reorients as a unit. The asynchronous correlation map, however, exhibits cross peaks between the rigid-core bands and the pentyl-chain C-H deformation band, indicating that different time-dependent responses exist between the rigid cyanobiphenyl core and the flexible pentyl chain. The fact that these cross peaks are negative implies that the pentyl chain reorients faster than the rigid core (in the decay process from the electrically oriented homeotropic state to the homogeneous state).

2.3 FT-IR TRS of Ferroelectric Smectic Liquid Crystals

We used three different TRS techniques, which cover a time-resolution range from millisecond down to microsecond, to analyze the elementary molecular process of the ferroelectric transition and the electroclinic transition exhibited by smectic liquid crystals [4]. For a smectic sample with a terphenyl core (MPBIC), the ferroelectric transition in the smectic C* phase occurs with a response time of about 40 μs while the electroclinic transition in the smectic A phase has a much shorter response time of 10 μs. The dynamic dichroic spectra show that the ferroelectric transition involves at least two different elementary processes. The transmitted light intensity

Springer Proceedings in Physics, Vol. 68 Time Resolved Vibrational Spectroscopy V
Ed.: H. Takahashi

measured in the layer normal direction shows significant changes during the ferroelectric transition, indicating that the molecular reorientation process in the smectic C* cannot simply be described by a currently accepted model (which assumes uniform precession of molecules on the cone surface) but requires a new description of the ferroelectric transition mechanism. No corresponding changes has been observed for the electroclinic transition in the smectic A phase, i.e., the molecules undergo uniform precession on the cone surface.

3. Conclusion

The results obtained in these studies demonstrate that FT-IR TRS, with the capability of obtaining simultaneous temporal and spectral resolution, can provide valuable new information concerning the elementary molecular process underlying the electric field-induced reorientation of liquid crystals. However, it should be noted that the quality of dynamic IR spectra of anisotropically oriented liquid crystals sensitively depends on the sample preparation. If the samples were poorly prepared, various unpredictable artifacts would appear. This is particularly so for smectic liquid crystals [4]. The key to success is to prepare uniform, highly oriented surface-stabilized ferroelectric liquid crystal (SSFLC) samples. For such carefully prepared samples the time-resolved vibrational spectroscopy is quite powerful and profitable. The time-resolved vibrational spectroscopy would probably be the only technique that can obtain real-time dynamic spectral information concerning the reorientation kinetics of liquid crystals at the segmental level.

References

1. H. Toriumi, H. Sugisawa, and H. Watanabe, Jpn. J. Appl. Phys., 27, L935 (1988).
2. V. G. Gregoriou, J. L. Chao, H. Toriumi, and R. A. Palmer, Chem. Phys. Lett., in press (1991).
3. H. Sugisawa, H. Toriumi, and H. Watanabe, J. Chem. Phys., submitted.
4. H. Toriumi, H. Sugisawa, H. Watanabe, K. Iwata, H. Hamaguchi, I. Nishiyama, Ferroelectrics, submitted (1991).

Critical Speeding-up of the Fluorescence Lifetime near the Structural Phase Transitions

T. Shigenari, K. Abe, Y. Ino, K. Morita, and E. Kojima

Department of Applied Physics and Chemistry, The University of Electro-Communications, Chofu, Tokyo 182, Japan

The influence of the structural phase transitions on the impurity-induced fluorescence from the dielectric crystals were investigated by the time correlated single photon counting method. A mode-locked picosecond dye-laser(100mW, 10ps) and its second harmonics were used as the excitation source.

In the α-β phase transition of resorcinol ($C_6H_4(OH)_2$, m-dioxybenzene) at $T_c(\approx 380K)$, being a strongly first-order transition, it was observed that the lifetimes of the fluorescence show the critical speeding-up as $T \rightarrow T_c$. In particular, the lifetime of the faster component τ_f, varies from τ=0.9ns at room

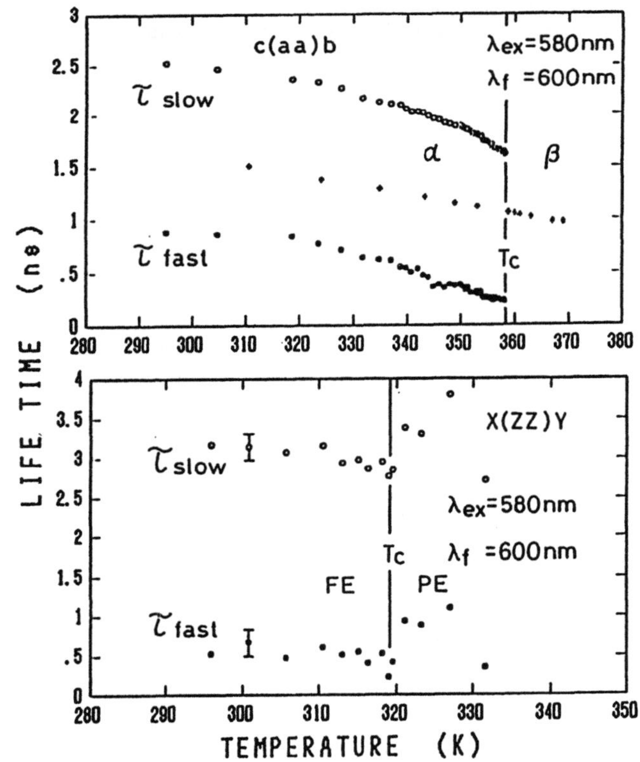

Fig.1 Temperature dependence of the fluorescence lifetime in the first-order phase transition in resorcinol

Fig.2 Temperature dependence of the florescence lifetime in the second-order phase transition in TGS

Springer Proceedings in Physics, Vol. 68 Time Resolved Vibrational Spectroscopy V
Ed.: H. Takahashi

temperature to 0.2ns at close to T_c (Fig.1) and its intensity increases rapidly towards T_c. (ref.1)

In case of TGS(triglycine sulfate)crystals(Fig.2), which undergoes a typical second-order transition, the decrease in the lifetimes near $T_c(322K)$ was not as significant as in the resorcinol case, in spite of the fact that the dielectric response of TGS shows the critical slowing-down from well below T_c.

The results, especially that for resorcinol, are surprising since in a strongly first-order transition, any change associated with the transition are expected to occur suddenly at T_c. One cannot foresee a first-order transition since it is not preceded by a precursor like soft mode or critical slowing-down. Actually, the frequencies of the vibrational modes measured by Raman scattering did not vary with temperature, that is, no soft mode exists in resorcinol.

The critical speeding-up might be interpreted as follows: The first-order phase transitions do not accompany the divergence of the macroscopic susceptibility. However, it would be still probable that, as $T \to T_C$, the atomic displacements (or molecular vibrations) are highly excited in a short time period in a microscopic region. Then, the large and local displacements of atoms would disturb the crystal field and make the lifetime of the excited electronic state short. Since the structural difference caused by the transition is very large in resorcinol and small in TGS, the effect of the momentary atomic displacement must be much more remarkable in resorcinol than TGS.

When the crystal is excited by a UV light (λ=290nm), the speeding-up was observed only in the faster($\tau <$ 1ns) component and not in the slower decay component($\tau >$ a few ns). This indicates that, among many decay routes, only the fast decay process in which the decay-rate is not governed by the thermal relaxations among the excited states are sensitive to the phase transitions. The present method would be useful for the detection of precursors in other first-order phase transitions.

1. T.Shigenari et al. : Phys.Rev.Lett. 66, 2112 (1991).

Photophysics of Fullerenes: C60 and C70

T.W. Ebbesen[1], *K. Tanigaki*[1], *S. Kuroshima*[1], *H. Hiura*[2], *and H. Takahashi*[2]

[1]NEC Fundamental Research Laboratories, Tsukuba 305, Japan
[2]Department of Chemistry, School of Science and Engineering, Waseda University, Tokyo 169, Japan

1. INTRODUCTION

The successful preparation of C60 and C70 in large quantities [1-3] has resulted in surge of interest in their basic properties [4-11]. A number of excited triplet state properties of C60 and C70 have been measured by different groups [8-11]. In particular, their triplet lifetimes, phosphorescence and EPR spectra were well established by Wasielewski et al. [9]. Recently, we measured the correct C60 and C70 triplet and singlet excited state absorption spectra, their molar absorption coefficients and their lifetimes in toluene [10,11]. These results will be summarized here together with Raman spectra of the ground state C60 and C70. Raman spectra of C60 and C70 were first reported by Bethune et al [4]. Although our results are in good agreement with their work, we report here additional new features of interest.

2. EXPERIMENTAL

The fullerenes were prepared according to the Kratschmer - Huffman technique [1-3]. Soot containing the fullerenes was generated by discharge between two graphite rods under 100 torr of He. The fullerenes were then extracted in a soxhlet extractor with benzene. C60 and C70 were purified from the extract using medium pressure chromatography (first silica gel with n-hexane then neutral alumina with hexane -toluene mixtures). The high purity (> 99%) of the samples was confirmed by absorption, IR and mass spectrometries.

The excited state properties were measured with picosecond and nanosecond laser flash photolysis equipment [12]. For these experiments the 354 nm and 532 nm excitation wavelengths of the YAG lasers were used giving the same results except that with the 532 nm multi-photonic degradation appears at much higher powers.The samples dissolved in Uvasol grade Merck toluene were degased with ultra-pure argon. The Raman equipment has been described elsewhere [13].

3. Excited State Properties of C60 and C70 [10,11]

After starting these experiments, it was found that both C60 and C70 are easily photodegraded when using 353nm laser excitation. The intensity dependence of the photodegradation shows that this is a biphotonic process, most likely

Springer Proceedings in Physics, Vol. 68 Time Resolved Vibrational Spectroscopy V
Ed.: H. Takahashi

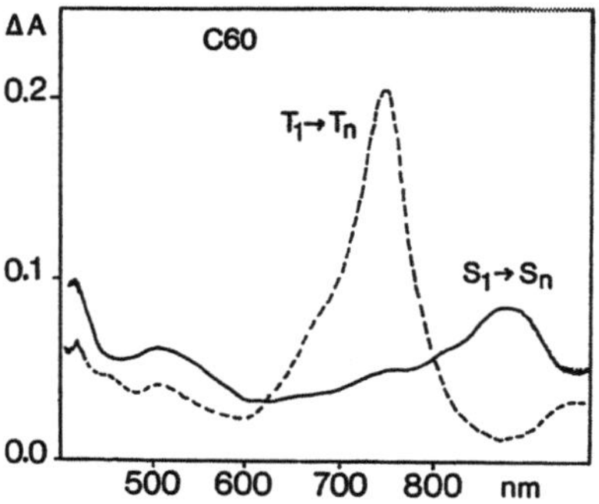

Fig.1

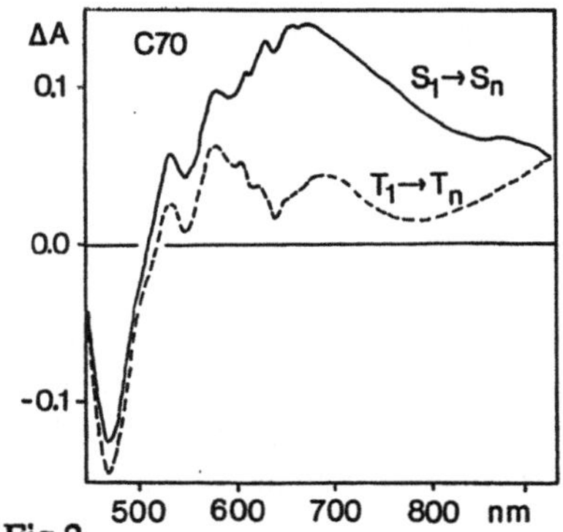

Fig.2

photoionization. In order to avoid this problem the 354 nm laser intensities must be kept below $4 mJ/cm^2$ for C60 and half that for C70 when using picosecond pulses. The photodegradation not only distorts the transient spectra but also gives rise to shorter apparent lifetimes.

The excited singlet and triplet state differential spectra are shown in Fig.1 (C60) and Fig.2 (C70). The singlet state spectra were recorded immediately after excitation and the triplet state at 10 ns. The decay kinetics of both C60 S_1 and C70 S_1 were single exponentials with lifetimes of 1.2 ns and 670 ps respectively (Table 1). The triplet decay kinetics showed both second order and first order components due to the presence of triplet - triplet annihilation. From the intensity dependence study of the triplet decay kinetics, the triplet -triplet annihilation rate constants were found to be diffusion limited (Table 1). Those bimolecular rate constants are the lower limits assuming a triplet quantum yield of unity. The lifetimes of C60 T_1 and C70 T_1 were extrapolated at zero intensity. The true lifetimes are probably much longer since impurity quenching will easily reduce the lifetimes of such long lived states. The triplet lifetimes measured in solid matrices in the absence of impurity diffusion must be the more accurate values [9]. These are given in the Table 1 for the purpose of comparison.

The differential extinction coefficients of the excited singlet states and the product of the differential extinction coefficients and the triplet quantum yields were also determined by using as reference either the benzophenone triplet state or the zinc

Table 1:

	$\Delta\varepsilon_S$ or $\phi_T\Delta\varepsilon_T$ $M^{-1}cm^{-1}$	τ	k_{TT} $M^{-1}s^{-1}$
C60 S_1	6,300 (885nm)	1.20 ns	
C70 S_1	3,800 (675nm)	0.67 ns	
C60 T_1	15,000 (747nm)	>0.28 ms (.41ms)	4.8×10^9
C70 T_1	1,800 (579nm)	>0.20 ms (51ms)	8×10^9

tetraphenyl porphyrin triplet state [10,11]. The values were determined from the initial slopes of the intensity dependence of the transient absorbances measured down to ca. 60 uJ/ pulse for accuracy. The resulting values are given in Table 1.

4. Raman Studies of C60 and C70

In view of the unusual superconductive properties of C60 [5-7] it is very important to understand both the vibrational properties of the individual fullerene molecules and that of their crystals. So efforts were made to measure the Raman spectra of C60 and C70 both in solution and as thin films. Figure 3 shows the Raman spectra of C60 on KBr substrate taken in two different experimental conditions using a 532 nm probing beam. The first (a) was taken at low intensity and corresponds well to the published spectrum [4]. There should be only 10 Raman active peaks and 8 of those can clearly be seen. Choosing the two remaining is difficult since there are a number of bumps close to the detection limit. After long exposure to relatively high laser probing intensity, the spectrum changed as shown in Fig.3 (b). The spectrum looks very similar to the first one (a) however nearly all the peaks have shifted by 10 cm^{-1} or less. This change was reversible when the sample was left for a couple of hours without irradiation. One possible explanation for this reversible

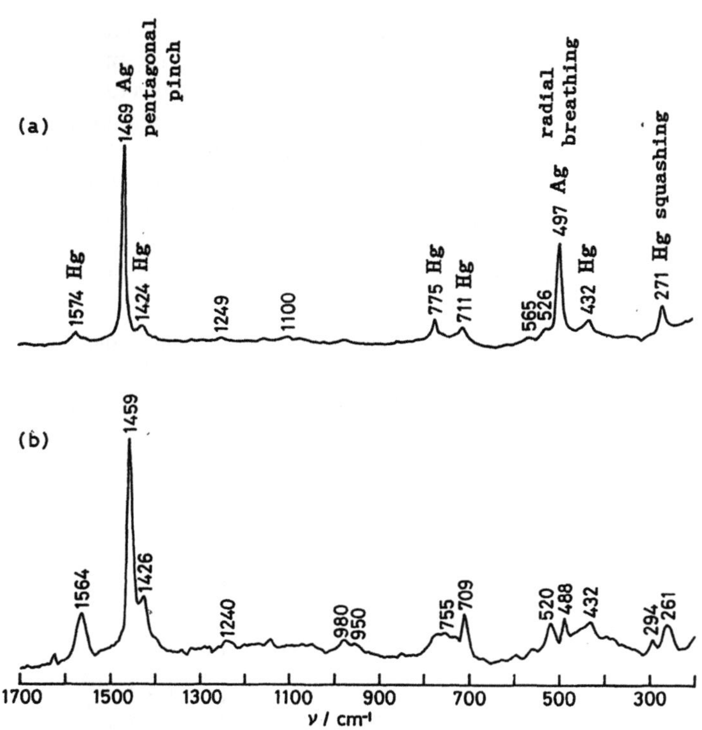

Fig.3 Raman of C60 thin film

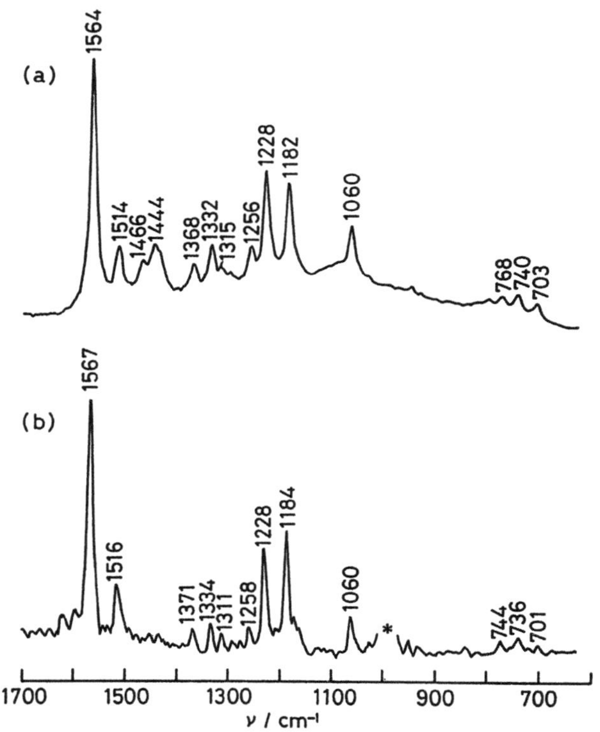

Fig.4 Raman of C70 in benzene (a), solid (b).

conversion might be a crystalline phase change. C60 is known to have two possible crystal structures HCP and FCC. TEM analysis of the thin crystalline film revealed that the sample was mainly HCP. This appears to be thermodynamically the most stable structure. It is therefore possible that upon irradiation the sample heats up sufficiently to upset the HCP structure resulting in the observed changes. Further studies are underway to check whether this hypothesis is correct or not.

While we have so far been unable to measure the Raman spectrum of C60 in solution, that of C70 could be measured with the 532 nm probe beam both in solution and solid phase as shown in Fig 4 (a) and (b) respectively. The two spectra ressemble each other except for the additional peaks at 1466 and 1444 cm^{-1} in the solid phase. Besides being sensitive to the state of the sample, solid or in solution, the intensities of these two Raman peaks exhibit a markedly different wavelength dependence on the probing wavelength as compared with other peaks. In solution, these two peaks seem to be resonance-enhanced not by the absorption band at 472 nm, but by the bands at wavelengths shorter than 400 nm. As a result, in contrast to the other peaks, these two lose intensity rapidly as the probing wavelength increases and are no longer detectable at wavelengths beyond 520 nm.

Assigning the Raman peaks of C70 will require further calculations currently underway. The assignement is much more difficult than C60 because of the lower symmetry of C70 (D5h) results in many more peaks.

Presently efforts are underway to measure the crystal lattice vibrations which are obviously very important for understanding phonon-electron interactions in the superconducting C60 materials.

REFERENCES:

1) W. Kratschmer, L.D. Lamb, K. Fostiropoulos, and D.R. Huffman, Nature (London) 347, 354 (1990). 2) R. Taylor, J.P. Hare, A.K. Abdul-Sada, and H.W. Kroto, J. Chem. Soc., Chem. Commun. 94, 1423 (1990). 3) R.E. Haufler, J. Conceicao, L.P.F. Chibante, Y. Chai, N.E. Byrne, S. Flanagan, M.M. Haley, S.C. O'Brien, C. Pan, Z. Xiao, W.E. Billups, M.A. Ciufolini, R.H. Hauge, J.L. Margrave, L.J. Wilson, R.F. Curl, and R.E. Smalley, J. Phys. Chem. 94, 8634 (1990).4) D.S. Bethune, G. Meijer, W.C. Tang, H.J. Rosen, Chem. Phys. Lett. 174 (1990) 219. 5) A.F. Hebard, M.J. Rosseinsky, R.C. Haddon, D.W. Murphy, S.H. Glarum, T.T.M. Palstra, A.P. Ramirez, A.R. Kortan, Nature 350, 600 (1991). 6) K. Holczer, 0. Klein, S.-M. Huang, R.B. Kaner, K.-J. Fu, R.L. Whetten, and F. Diederich, Science 252, 1154 (1991). 7) K. Tanigaki, T.W. Ebbesen, S. Saito, J. Mizuki, J.S. Tsai, Y. Kubo, and S. Kuroshima, Nature, in press. 8) J.W. Arbogast, A.P. Darmanyan, C.S. Foote, Y. Rubin, F. Diederich, M.M. Alvarez, S.J. Anz, and R.L. Whetten, J. Phys. Chem. 95, 11 (1991). 9) M.R. Wasielewski, M.P. O'Neil, K.R. Lykke, M.J. Pellin, D.M. Gruen, J. Am. Chem. Soc. 113, 2774 (1991).10) T.W. Ebbesen, K. Tanigaki, S. Kuroshima, Chem. Phys. Lett. in press. 11) K, Tanigaki, T.W. Ebbesen, and S. Kuroshima, Chem. Phys. Lett., submitted. 12) T.W. Ebbesen, Rev. Sci. Instrum. 59 1307 (1988). 13) S. Hasimoto, A. Shimojima, T. Yuzawa, H. Hiura, J. Abe and H. Takahashi, J. Mol. Struct. 242, 1 (1991).

Part VIII

Theoretical Aspects

Interferometric Coherent Raman Spectroscopies with Incoherent Light: Time and Frequency Resolution via Radiation-Matter Oscillations

M.A. Dugan and A.C. Albrecht

Department of Chemistry, Cornell University, Ithaca, NY 14853-1301, USA

1. Introduction

The study of dynamics through light/matter interaction has been pushed to ultrashort time resolution with the advent of femtosecond lasers. But in the early 1980's [1] it was realized that apparently ultrashort timing could also be realized using light whose phase noise is sufficient to exhibit femtosecond scale autocorrelation times, even with nanosecond pulses. We have recently compared ultrashort timing with transform limited femtosecond light pulses (amplitude modulated light) and timing with incoherent light (phase modulated light), with abundant reference to the literature [2]. The object of this report is to briefly summarize our findings.

1.1 What is There to Time?

Since time dependent properties of an ensemble of molecules are fully determined by the time dependent density matrix (in an appropriate basis), the task of the dynamicist is to establish the time behaviour of its matrix elements, both the diagonal (populations) and the off-diagonal (coherences). Typically, a system under study is divided into entities of direct spectroscopic interest, conveniently called 'chromophores', and a surrounding 'bath' (including the black-body radiation field) to which they are coupled. This partitioning leads to the defining basis set for the evolving density matrix (the eigenstate basis of the chromophore). The bath serves as both a source for energy conservation in population changing events, and also as the locus of stochastic processes which elastically modulate the Bohr frequencies of the chromophore, thereby initiating pure dephasing of the coherences. A favorite description regards them as several independently acting Gaussian stochastic processes with zero mean, leading to time dependent pure-dephasing rate 'constants' as given by Kubo and others. Studies in coherence dynamics aim to elucidate these stochastic events.

1.2 How to Time?

Timing in spectroscopy can be achieved indirectly in the frequency domain by line shape analysis, or directly in the time domain by observing signal decay in real time using short-pulsed coherent light (SPCL). Pure dephasing rate constants that might be derived from these entirely different approaches, when in their extremes, may correspond to the opposite limits of stochastic modulation and therefore can be truly complementary. Timing with long-pulsed incoherent light (LPIL) having ultrashort autocorrelation time potentially represents an intermediate case. However it seems that for purely vibrational coherences at room temperature all three methods appear to probe the stochastic events responsible for pure dephasing in the same fast modulation limit. In this case the methods are complementary only with regard to technique.

2. Comments On Coherent Raman Spectroscopies

Coherent spectroscopies in their purest form require wave vector conservation (phase matching). This can only be realized when the signal photon is created from signal fields derived from at least two spatially separated chromophores. At the dipole approximation of light/matter interaction, signal photons generated from one chromophore contain no reference to the wave vectors of the driving fields.

The basic CRS experiment, whether CARS or CSRS, has three driving fields with unique wave vectors but only two frequencies ω_1 and ω_2 such that $|\omega_1 - \omega_2| \sim \omega_R$ where ω_R is the Bohr frequency of a given Raman transition (such as libration, rotation, vibration, electronic). In the time-resolved experiment, first one field at ω_1, another at ω_2, act together to set up a Raman polarization. Then at a controlled time, τ, later a delayed portion of one of these fields, assigned its own wave vector, adds one more step to the second-order Raman coherence, provided it has not decayed. The third-order polarization at an optical frequency $|\omega_1 - 2\omega_2|$, say, serves as a source for the new signal field which appears at $\omega_S \sim |\omega_1 - 2\omega_2|$ as the third order polarization collapses.

Whether working with transform limited femtosecond pulses (SPCL), or with incoherent light having a short coherence time (LPIL), spectral width is necessarily introduced into the polarization and several Raman Bohr frequencies may be embraced (particularly when these are closely grouped). Now the third-order polarization contains, among other things, sum and difference Raman beats and will confer upon the interchromophore derived signal oscillatory behaviour in time, τ, that contains both frequency and decay constant information. In fact, as reported in the literature, degenerate (single color) 4WM can carry low frequency Raman resonances whenever the driving fields carry sufficient spectral width.

3. New Features in CRS with Incoherent Light

Any nonlinear spectroscopy necessarily involves multiple point time correlators. When LPIL principally involves phase noise modulation, the (convenient and therefore standard) assumption of circular complex Gaussian noise statistics is introduced, and these many time point correlators factor into field pair time correlators. Now when in CRS experiments 'twin' 2, 2' LPIL fields, derived from the same noisy light source and relatively τ-shifted are used, (together with a monochromatic filed at ω_1), the fields 2 and 2' act in-phase (that is, with the same-signed Fourier component) on a given chromophore. Since fields 2 and 2' coexist in the medium for times many orders of magnitude longer than dephasing times, the polarization is in a steady-state condition and the role of fields 2 and 2' in the time ordered polarization of a chromophore is interchangeable. One field must appear during the second order Raman polarization, the other acting in the third step. Whatever their ordering, because fields 2 and 2' act in phase only high-frequency 2,2' time correlators are found in the polarization on a single chromophore. These vanish upon cycle averaging. It is the low frequency time pair correlators that survive cycle averaging and lead to τ dependence in the signal. They appear only at the level of the conjugate interchromophore cross-terms. Thus time (τ) dependence in the LPIL-based CRS is derived entirely from conjugate interchromophore time correlation. By significant contrast, τ-dependence in CRS using SPCL, which leads to amplitude modulation of the polarization, can also be derived from τ-dependence in the third order polarization on a single chromophore. Furthermore the time ordering of femtosecond pulsed fields 2 and 2' in the polarization of the chromophore is not interchangeable but is fixed by experiment as long as 2 and 2' do not overlap in time.

3.1 'Noise Nulling' and a CRS - τ, ω_S Raman Surface [2]

Because LPIL-based CRS involves principally phase-based modulation, τ dependence is derived from time correlation across conjugate processes on two different chromophores. As long as field 1 is noise-free (monochromatic) and the detected signal, S, is spectrally resolved at high resolution, the noise contained in the conjugate events is due only to the identical (but phase-shifted) phase noise contained in each of the twin fields 2 and 2'. With the aid of phase memory across the ensemble of chromophores, their conjugate combination permits exact 'noise-nulling' in the τ dependent signal for τ less than the dephasing time of the material. Surviving is a new kind of radiation/matter beat. It is a Rabi detuning oscillation in τ whose frequency, Δ, is just an algebraic sum of the detuning from the Raman resonance at second order, and detuning from Raman resonance between the third acting field 2 (or 2') and the (precise) monochromator setting, ω_S. Thus for any Raman-like resonance $\Delta = [2\omega_R \pm (\omega_S - \omega_1)]$, (CARS(+), CSRS(-). All reference to fields 2, 2' is absent. Thus in LPIL-based CRS any Raman-like resonance can be observed in an entirely new fashion - as a CRS vs τ, ω_S surface, the 'Raman Surface' involving two parameters, the Bohr frequency, ω_R, and the (fast modulation limit) dephasing rate constant, γ_R.

Whenever more than one Raman oscillator is resonantly polarized because of spectral width in fields 2 and 2', but the phase filtering remains sufficiently sharp, the resulting surface appears to be analyzable in terms of independent contributions from each Raman oscillator. Such mixed surfaces, always bichromophoric in origin, can be built from multiple Raman oscillators from a given chemical species, or from mixed species. Most significantly, if the CRS process is also electronically resonant, Raman oscillator(s) from the excited electronic state can also contribute to the surface.

As soon as noise (nonzero bandwidth) is introduced into field 1, and/or into field S (by opening the slit), and/or a large inhomogeneaous distribution of the Raman frequencies is present, the Raman Surface begins to collapse. In the limit, only pairwise Bohr frequency difference beats are seen that decay with pairwise mean dephasing rate constants. In SPCL based CRS a single Raman resonance produces a coherence spike followed by a much weaker, exponentially decaying signal in the τ. The Raman detuning surface is absent. When more than one Raman oscillator is polarized, only the decaying Bohr frequency beats are seen.

3.2 Sub-Raman linewidth Spectral Features Induced by Noise

As soon as a nonzero Lorentzian bandwidth is assigned to field 1, or the monochromator field S, and that bandwidth does not exceed the Raman dephasing rate constant, theory reveals a dramatic change in the Raman Surface. It now possess two ridges that are ultra narrow in the frequency variable, ω_S. One is located at the exact vanishing of the Rabi detuning oscillation: $\Delta = 0$, or when $2\omega_R = \pm(\omega_1 - \omega_S)$. The other sharp ridge appears at the exact vanishing of the alternative detuning frequency, that among the fields alone, when $2\omega_2 = \omega_1 + \omega_S$. These ridges possess a spectral width determined only by that of field 1 and the monochromator. The contribution from the natural Raman linewidth appears with opposite sign from each chromophores and thus is vanishes for a single component system.

Experimentally, only the ridge at the vanishing of the Rabi detuning frequency has been seen and its sub-Raman linewidth properties have been confirmed [2]. Theory shows that when Gaussian (instead of Lorentzian) spectral densities are used for the spectral functions of fields 1 and the monochromator slit, the purely light related ridge is absent. The experimental spectra of the noisy fields in fact are not Lorentzian, thus possibly accounting for the presence of only one ridge. Most importantly, this ridge locates a Raman frequency with sub-Raman linewidth precision. Furthermore theory predicts that two Raman oscillators, separated in frequency by an amount much less than the sum of their bandwidths, should be easily resolved unless their bandwidths are principally inhomogeneous in origin. If such features prove ubiquitous, one has not only encountered an interesting new spectroscopic feature (absent in purely SPCL based CRS), but possibly one of considerable analytic consequence.

References

1. N. Morita and T. Yajima, Phys. Rev. A 30, 2525(1984). R. Beach and S. R. Hartmann, Phys. Rev. Letters 53, 663(1984).

2. M.A. Dugan and A. C. Albrecht, Phys. Rev. A 43, 3877, 3922(1991).(Also M.A. Dugan, J. S. Melinger, and A. C. Albrecht, Chem. Phys. Letters, 147, 411(1988)).

Simulation of the Ultrafast Optical Response of Water

L.E. Fried and S. Mukamel

Department of Chemistry, University of Rochester,
Rochester, NY 14627, USA

There has been considerable experimental and theoretical interest in ultrafast optical measurements[1]. In this paper we present a simulation of the two-pulse photon echo of a vibrationless chromophore in water. We find the presence of *intermolecular* solvent vibrations in the calculated signal.

1 Theory

In this section we discuss the formulas used in the calculation of the photon echo signal. A full theoretical treatment of the photon echo experiment has been given by Yan and Mukamel [2]. We consider a single chromophore immersed in water. The chromophore is assumed to have two relevant electronic states, $|g\rangle$ (the ground state) and $|e\rangle$ (the excited state). The corresponding adiabatic Hamiltonians of the solvent plus solute system are H_g and H_e,respectively.

In the photon echo experiment, a laser with wavevector k_1 and frequency ω_1 sets up an optical coherence in the sample. After the pulse, the coherence decays during the delay period because of dephasing processes. The part of this decay due to inhomogeneous broadening, however, can be reversed by the application of a second pulse with wavevector k_2 at time $t = 0$. The echo signal is produced with wavevector $k_S \equiv 2k_2 - k_1$. We will consider here an ideal photon echo experiment with infinitely short pulses.

The experimental observable in the photon echo experiment is the total number of photons produced in the k_S direction as a function of the delay time τ between the two pulses. The signal $S(\tau)$ is proportional to the square of the third order macroscopic polarization with wave vector k_S: $S(\tau) = \int_0^\infty dt |P^{(3)}(k_S, t)|^2$.

Given the above assumption of ultrashort pulses, the photon echo experiment directly measures the following *nonlinear response function*: $P^{(3)}(k_S, t) \propto R(t,\tau) \equiv \mathrm{Tr}[G_{eg}(t-\tau)G_{ge}(\tau)\rho_g]$ Here, ρ_g is the thermal density matrix in the ground electronic state, whereas the propagator G_{eg} is defined through its action on an operator A: $G_{eg}(t)A \equiv \exp(-iH_e t)A\exp(+iH_g t)$.

We consider here a classical approximation for the nonlinear response function[3,4]

$$R(t,\tau) = \left\langle \exp(i \int_\tau^{t+\tau} U(Q[t']dt') \exp(-i \int_0^\tau U(Q[t'])dt')\rho_g(Q,P) \right\rangle . \quad (1)$$

Here, $U \equiv V_g - V_e$ is the difference between the ground and excited state potential energy surfaces. $Q[t]$ is the initial set of all nuclear coordinates propagated forward a time t by classical mechanics. $\rho_g(Q,P)$ is the classical Boltzmann distribution of the ground state adiabatic surface.

Yan and Mukamel[2] have proposed a semiclassical formula for $R(t,\tau)$ based on a cumulant expansion of the response function in powers of U. They find

$$R(t,\tau) = \exp[-g(t) - 2g^*(\tau) - g^*(t) + g^*(t+\tau)], \tag{2}$$

where $g(t) = i\langle U\rangle t + \int_0^t dt' \int_0^{t'} dt'' \langle U(t'')U\rangle - \langle U^2\rangle$. Eq. (2) is expected to be valid when U is composed of many independent contributions, so that the central limit theorem holds.

2 Computational method

We have performed a molecular dynamics (MD) simulation of the photon echo signal of a spherical chromophore immersed in water. The ground and excited state potentials were of the following form: $V_g = \sum_{i<j} V_{SPC/E}(R_i, R_j) + \sum_i V_{LJ}(r_c, R_i^O)$; $V_e = V_g + \sum_{i,\alpha} q_C q_\alpha / |r_c - R_i^\alpha|$ Here, R_i^α ($\alpha = O,\ H_1\ H_2$) is the position of atom α on molecule i. r_c is the position of the chromophore. q_C is a charge placed on the center of the chromophore in its excited state. q_α is the charge on site α of a water molecule. $V_{SPC/E}$ represents the enhanced simple point charge model of water[6], whereas V_{LJ} denotes the Lennard-Jones interaction. The Lennard-Jones σ of the chromophore was varied from 2 to 6 Å, while the ϵ was fixed at 70 cm^{-1}. The chromophore mass was taken to be 50 amu, while its excited state charge was taken to be $+0.44|e|$.

The simulation was carried out for a system of 100 water molecules and 1 chromophore. Periodic boundary conditions were employed using a cubic simulation box. The box size L was adjusted so that the density of the water excluding the volume of the chromophore was equal to 1 g / cm^3. Chromophore diameters of 2, 4, and 6 $\AA$ were used. The potential was cut off at $R_c = 1/2L$. A switching function centered on the O atom of the water and the center of the chromophore was used between $R_c - 0.5\AA$ and R_c. Long range corrections to the Coulomb force were incorporated via the reaction field method.

Initial conditions were chosen randomly, then equilibrated through a 20,000 cycle Monte Carlo calculation done at 298K. The final configuration of this Monte Carlo simulation provided the initial positions for a subsequent MD simulation. Initial velocities were chosen from the Boltzmann distribution. A constant energy MD simulation was done for 64 ps using a timestep of 1.8 fs. The velocities were rescaled to prevent energy drift. During the simulation, $U(t)$ was calculated at each time step. The $U(t)$ values were then used to calculate the lineshape, through either Eq. (1) or Eq. (2).

3 Results

In Figure 1 we show the photon echo signal as a function of the delay time τ between the pulses for chromophore diameters d of 2, 4, and 6 Å calculated on the basis of the cumulant formula of Eq. (2). First, we note that the signal has a strong dependence on chromophore size. Dielectric continuum theory can be used to estimate the autocorrelation function of $U(t)$. The predictions of

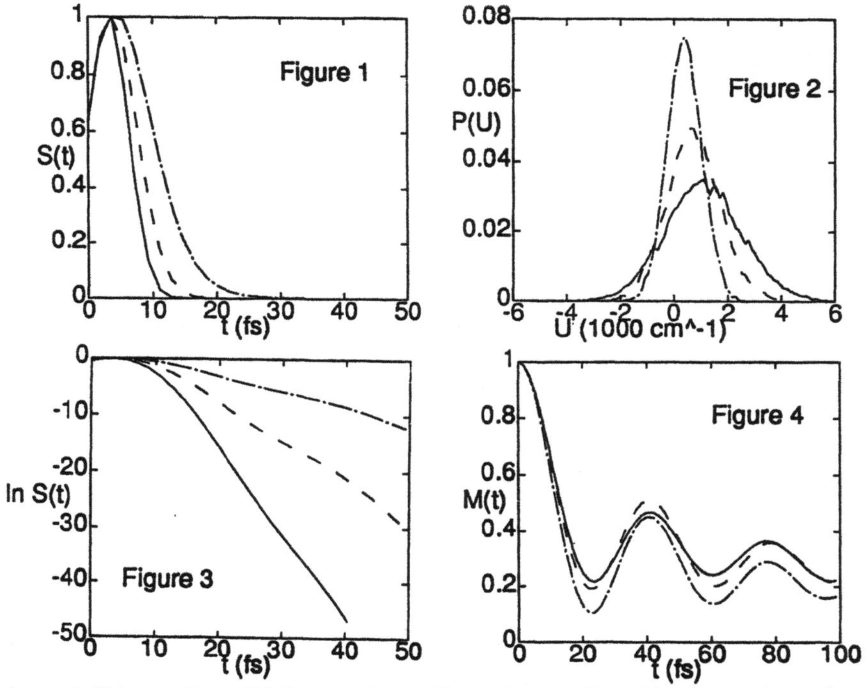

In each Figure, the solid line corresponds to chromophore diameter $d = 2$ Å, the dashed line has $d = 4$ Å, and the dash-dot line has $d = 6$ Å.

dielectric continuum theory, however, are independent of solute size. Thus, we can readily see that the photon echo signal shows pronounced deviation from dielectric continuum behavior. This is directly analogous to the strong non-continuum behavior seen in measurements of the time resolved Stokes shift. In fact, the time resolved Stokes shift signal can also be related to the autocorrelation of U. Another interesting feature of the results is that the decay is more rapid for a small chromophore than for a larger chromophore. This is reminiscent of the simulation of solvation in water by Maroncelli and Fleming[5], who found that water molecules in the neighborhood of the chromophore produced the fastest component of the solvation response.

The cumulant expansion assumes that $U(t)$ is a Gaussian (possibly non-Markovian) random process. To explore the validity of this approximation, we have calculated the probability of a given U value $P(U)$ along the trajectory. If $U(t)$ is a Gaussian random process, then $P(U)$ will be a Gaussian. Figure 2 shows $P(U)$ for the three simulations reported above. In each case the distributions are close to Gaussian. Preliminary calculations based on Eq. (1) also show that the cumulant approximation is highly accurate. This provides further corroborating evidence for the hypothesis that $U(t)$ is a Gaussian random process.

There are other important features of $P(U)$ apparent in Figure 2. P is broader for small chromophore sizes than for large sizes. This is because the smaller chromophores are more sensitive to fluctuations in the local environment. As the chromophore size increases, the importance of specific interactions decreases as the "first solvation shell" contains a rapidly increasing number of water molecules.

Next, we consider the extent to which $U(t)$ is non-Markovian. In the limit of a Markovian $U(t)$, the log of the signal should decay linearly. In this case, the slope of that decay is given by homogeneous decay time T_2: $\ln S(\tau) = -4\tau/T_2$. Figure 3 shows $\ln S(\tau)$ for our simulations. While the decay is approximately linear, there is a small but noticeable modulation in $\ln S(\tau)$, reminiscent of vibrational motion. These *intermolecular* vibrations are made much more apparent if we look directly at the autocorrelation function of U itself. In Figure 4 we show $M(t) \equiv (\langle U(t)U(0)\rangle - \langle U^2\rangle)/\langle U^2\rangle$. The existence of a simple damped vibration is highly apparent in Figure 4. $M(t)$ shows this vibrational structure more clearly than the signal itself because the signal is related to the double integral of $M(t)$, and hence masks damped oscilliatory behavior.

We have shown here that there can be subtle oscillations in the photon echo due to intermolecular motions. These oscillations can be traced back to the nature of $M(t)$, which shows a damped oscillatory behavior. This suggests that analysis of $M(t)$ in terms of a Brownian oscillator model[2] may be quite illuminating.

The support of the National Science Foundation, the Air Force Office of Scientific Research, and the Petroleum Research Fund, administered by the American Chemical Society, is gratefully acknowledged. L.E.F. acknowledges support through the NSF Postdoctoral Fellowship in Chemistry.

References

[1] J.Y. Bigot, M.T. Portella, R.W. Schoenlein, C.J. Bardeen, A. Migus, and C.V. Shank, Phys. Rev. Lett. **66**, 1138(1991).

[2] Y.J. Yan and S. Mukamel, J. Chem. Phys. **94**, 179(1991).

[3] L. E. Fried and S. Mukamel, J. Chem. Phys. **93**, 3063(1990).

[4] A. M. Walsh and R. F. Loring, to appear in J. Chem. Phys. (1991).

[5] M. Maroncelli and G. R. Fleming, J. Chem. Phys., **89**, 5044(1988).

[6] H.J.C. Berendsen, J.R. Grigera, and T.P. Straatsma, J. Chem. Phys. **91**, 6269(1987).

Vibrational Wave Packet Dynamics of Diatomic Molecules Under Intense Ultra-Short Laser Pulses

M. Sugawara and Y. Fujimura

Department of Chemistry, Faculty of Science,
Tohoku University, Sendai 980, Japan

1. Introduction

Recently, investigation of above-threshold dissociation (ATD) of diatomic molecular ions under intense laser field has been paid a considerable attention. The ATD refers to photodissociation processes above dissociation threshold after the bound-free and free-free photon absorptions take place.[1] The purpose of this paper is to present a vibrational wave packet treatment of the ATD of H_2^+ ion under femtosecond intense laser field condition. Use of such time-dependent treatment with the dressed state method makes it possible to describe the stimulated free-free emission during the ATD process as the crossing problem between the dressed adiabatic potential surfaces, which gives us the direct information about the interplay of such optical transitions in ATD. The wave packet dynamics in momentum space is obtained by applying the Fourier Transform to the nuclear vibrational wave packet in coordinate space.

2. Theory

Total system Hamiltonian $\hat{H}(t)$ is given as the summation of the Hamiltonian of the molecular ion $\hat{H}_M$, the radiation field $\hat{H}_R$, and the interaction between the molecular ion and time-dependent radiation field $\hat{H}_{MR}(t)$, as

$$\hat{H}(t) = \hat{H}_M + \hat{H}_R + \hat{H}_{MR}(t)\,. \tag{1}$$

The time evolution of nuclear wave packets on l-th dressed potential surfaces, $\phi_l(t)$ defined as $\phi_l(R_i, t) = \langle l|\Psi(t)\rangle$, is given by utilizing the sliced propagator technique as

$$|\Psi(t+\varepsilon)\rangle = \exp[-\frac{i}{\hbar}\hat{H}(t)\varepsilon]|\Psi(t)\rangle\,, \tag{3}$$

where ε denotes an infinitesimal time step.

3. Wave packet dynamics in ATD

Fig.1 shows the wave packet dynamics in coordinate space in the case in which the laser pulse has the duration of 10fs as FWHM and 1.0×10^{14} cm^{-1}/W as peak intensity. The lower and upper dressed potentials correspond to the ATD channels through 2-photon(channel 2) and 3-photon(channel 3) absorption, respectively. When the wave packet excited on the channel 3 comes to the curve crossing point, it is mostly transferred onto the channel 2 surface due to the stimulated one-photon emission. In Fig.2 the time evolution of the wave packets in momentum space is shown. In Fig.2c each peak in the momentum distribution corresponds to the fragment H^+ dissociated via each channel after passing

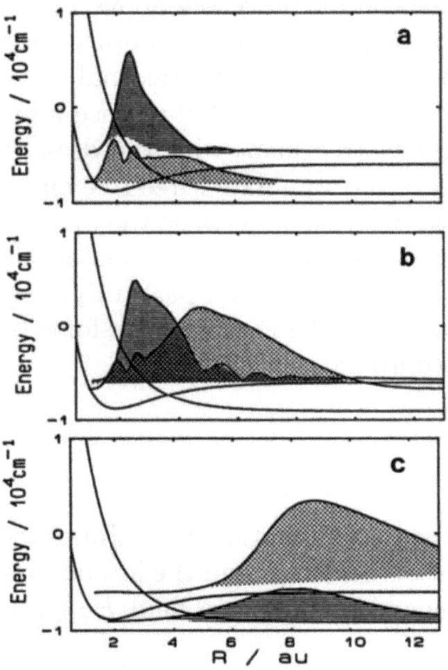

Fig.1. The evolution of wave packets in coordinate space. The wave packets in Figs.1a, 1b and 1c are evaluated at t=-1.5, 3.5 and 11fs, respectively, while the origin of the time is taken to be at the maximum peak of the laser pulse envelope. The duration of laser pulse is 10fs as FWHM and the peak intensity is $1.0\times10^{14}cm^{-1}/W$. The areas dotted and hatched by straight lines represent the wave packets associated with channels 2 and 3, respectively.

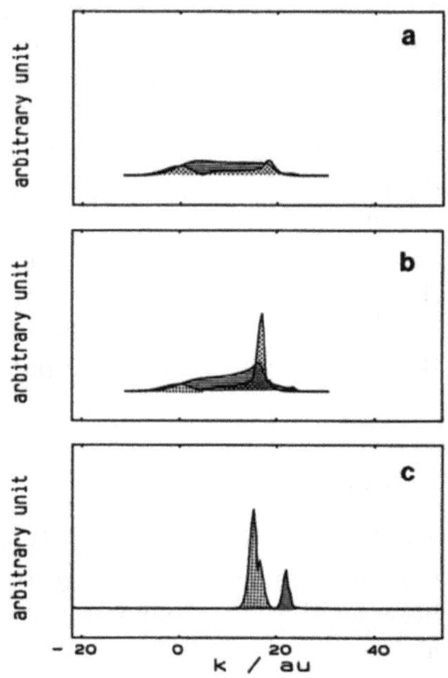

Fig.2. The evolution of wave packets in momentum space. The wave packets in Figs.2a, 2b and 2c are evaluated at t=-1.5, 3.5, and 23 fs, respectively. The laser pulse used is the same as that in Fig.1.

the crossing point, and it is seen that channel 2 is the dominant channel in this case.

References

[1] A.Giusti-Suzur, X.He, O.Atabek, and F.H Mies, Phys.Rev.Lett.**64** (1990) 515.

Theoretical Treatment of Low-Frequency Motions in Excited State Molecules

J.H. Frederick and J.E. Hadder

Department of Chemistry/216, University of Nevada, Reno, NV 89557, USA

In this report, we describe new techniques developed for the study of the low frequency motions of large molecules in excited states and apply them to a detailed study of the symmetric phenyl torsion vibration in excited singlet benzophenone.

I. Theoretical Modeling

Quantitative *ab initio* surfaces for excited states of large molecules are unavailable, necessitating the construction of empirical potentials. To model the potential energy surface for benzophenone and similar molecules along the low frequency coordinates, we write the potential as the sum of several contributions: $V = V_{conj} + V_{steric} + V_{other}$, where V_{other} treats the localized bonding interactions. For the conjugation potential, we use a Fourier series in an appropriate torsional angle, ϕ,

$$V_{conj} \approx {}^1\!/_2\, V_2 (1 - \cos 2\phi) + {}^1\!/_2\, V_4 (1 - \cos 4\phi) = V_2 \sin^2\phi + V_4 \sin^2 2\phi \qquad (1)$$

where V_2 and V_4 are parameters to be determined by a spectral fitting procedure. The steric potential, V_{steric}, is represented by using a standard non-bonded H-H or C-C model potential or by treating it as a simple exponential repulsion.

Once the coordinates associated with low frequency motions have been identified, we derive appropriate kinetic energy expressions for the reduced model system. The total kinetic energy of an arbitrary vibrating, rotating molecule is given by

$$T = {}^1\!/_2\, \omega \cdot \mathbf{I} \cdot \omega + \omega \cdot \ell + T_{vib} = {}^1\!/_2\, \mathbf{L} \cdot \mathbf{I}^{-1} \cdot \mathbf{L} - {}^1\!/_2\, \ell \cdot \mathbf{I}^{-1} \cdot \ell + T_{vib} \qquad (2)$$

where $\mathbf{I}$ is the moment of inertia tensor, ω is the rotational angular velocity of the molecular axis system, and T_{vib} is the kinetic energy in the molecular axi frame.[1] For $\mathbf{L} = \mathbf{0}$, the total kinetic energy is given by the last two terms in Eq.(2). One should note that the relative magnitude of these two terms depends on the choice of the molecular axis system, but their sum is *completely independent of this choice.* Thus, one is free to choose an axis frame that is computationally convenient. Our procedure is to express the Cartesian positions and velocities of all the atoms in terms of the desired internal coordinates, **q**, using a suitable choice of internal axis frame, and use Eq.(2) to obtain the kinetic energy in terms of **q** and $\dot{\mathbf{q}}$.

Springer Proceedings in Physics, Vol. 68 Time Resolved Vibrational Spectroscopy V
Ed.: H. Takahashi

II. Application to Benzophenone

We have applied our analysis to the symmetric phenyl torsion, ϕ, of benzophenone, for which the (n,π^*) transition has recently been measured at high resolution.[2,3,4] The kinetic energy expression for the symmetric torsion is found to be

$$T(\phi) = {}^1/_2\, \mu_\phi \left[\, 1 - \mu_\phi\, (1 + \cos\alpha)/_{2I_{zz}} \,\right] \dot{\phi}^2 \qquad (3)$$

where the molecule is held rigid except for ϕ. In Eq.(3), μ_ϕ and I_{zz} are the moments of inertia for the phenyl ring and the whole molecule about their symmetry axes, respectively, and α is the central C-C-C bond angle.

Two qualitatively different possibilities for the excited state surface emerged from the potential parameter fitting procedure: one having a small barrier at ϕ=0 and large barrier at ϕ=90°, the other having the opposite behavior. The latter was eliminated by computing the relative intensities and comparing to experiment. Results for the best fit potential are shown in Fig. 1 below. The barriers for this potential are 863 cm^{-1} and 6080 cm^{-1} at the planar and perpendicular geometries, respectively.

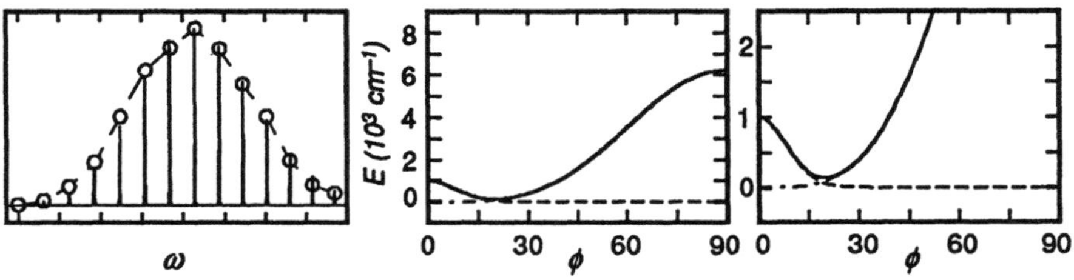

Figure 1. Results for the best empirical potential surface. (a) Comparison of the computed and observed spectra. Observed line positions and intensities are denoted by the open circles; tick marks are spaced by 100 cm^{-1} and start at the band origin on the left. (b), (c) The empirical surface for different energy ranges. The two dashed lines indicate the contributions from V_{steric} and V_{conj}.

Acknowledgment is made to the Donors of the Petroleum Research Fund, administered by the American Chemical Society, for the support of this work.

References

[1] J.H. Frederick, G.M. McClelland, and P. Brumer, *J. Chem. Phys.* **83**, 190 (1985).
[2] K.W. Holtzclaw and D.W. Pratt, *J. Chem. Phys.* **84**, 4713 (1986).
[3] S. Kamei, T. Sato, N. Mikami, and M. Ito, *J. Phys. Chem.* **90**, 5615 (1986).
[4] J.H. Frederick, E.J. Heller, J.L. Ozment, and D.W. Pratt, *J. Chem. Phys.* **88**, 2169 (1988).

Ultrafast Hole Burning in a Strongly-Coupled Localized-Electron-Phonon System

M. Hama[1] *and M. Aihara*[2]

[1]Department of Applied Physics, Faculty of Engineering, Osaka City University, Osaka 558, Japan
[2]Department of Physics, Faculty of Liberal Arts, Yamaguchi University, Yamaguchi 753, Japan

With the development of ultrashort pulsed laser techniques, the extremely fast transient phenomena in several condensed materials have become a subject of much investigation.

In this paper, we theoretically predict a new type of optical hole burning in a strongly-coupled localized-electron-phonon system. The hole arises because the ensemlbes of the multiphonon structure play the similar role as the inhomogeneous line broadening in the extremely short time region; this is closely related to the photon echo phenomenon in a strongly-coupled sysytem studied by one of these authors (M. A.) [1, 2].

By the analysis based on the microscopic density operator formalism, we obtain the third-order nonlinear optical polarization corresponding to the pump-probe type optical processes caused by a two-pulse excitation.

We consider a linearly-coupled localized-electron-phonon system. Inhomogeneous line broadening in the conventional sense is assumed to be negligibly small. Thermal average is evaluated by the cumulant expansion method.

In the numerical calculations, S (the electron-phonon coupling strength) and γ_p (the damping rate of interaction phonon mode) are set at 50 and $0.4\omega_p$, respectively, where ω_p is the mean phonon frequency. The spectral width δ of pump pulse and probe one is set at ω_p. The center frequency ω_1 of pump pulse is selected at the Franck-Condon energy ε_0. For emphasizing the hole spectrum, the value of $|dE_1^0|$ (Rabi frequency for pump light) is set at ω_p, where d and E_1^0 denote the dipole moment and the amplitude of electric field of pump light, respectively.

We obtain the intensity of transmitted probe light by performing the triple numerical integration. Figure 1 shows the result of transient spectrum for pulse separation $\tau_s=-1/\omega_p$. Hole spectrum clearly appears at the center frequency of pump light. It should be noted that hole burning is not caused by the inhomogeneous broadening in the conventional sense, but it arises due to the distribution of the lattice

Springer Proceedings in Physics, Vol. 68 Time Resolved Vibrational Spectroscopy V
Ed.: H. Takahashi

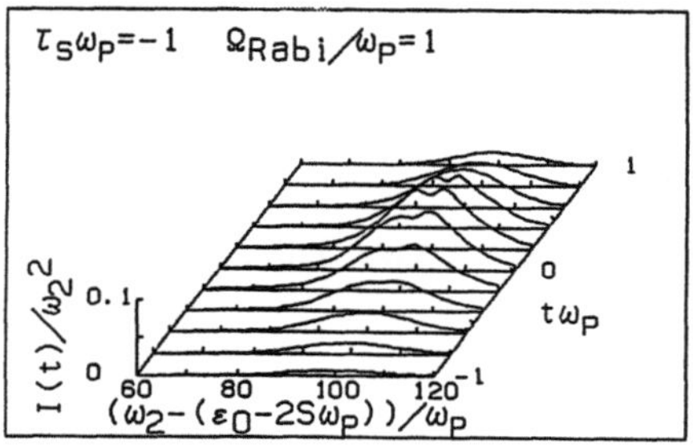

Fig. 1. transient spectrum for pulse separation $\tau_s=-1/\omega_p$.

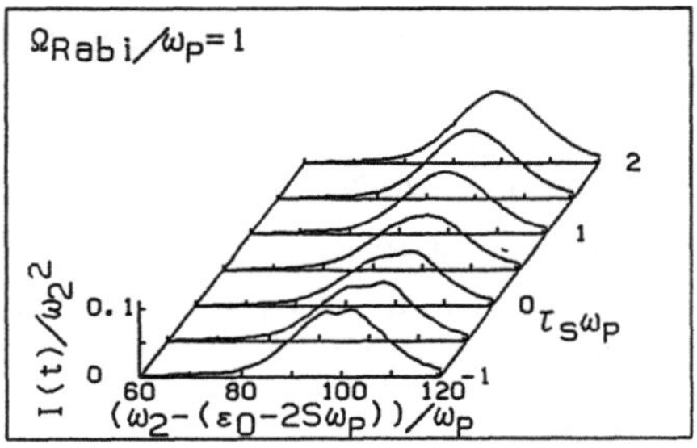

Fig. 2. the spectrum of time integrated intensity.

distortion in the electronic ground state. That is, the assembly of the nearly independent Franck-Condon transitions in the configuration coordinate space before the lattice relaxation, plays the role of inhomogeneous broadening. In Fig. 1, we find that hole burning exist in the time region shorter than the excitation pulse duration, because the hole is smeared out by the motion of phonon wave packet in the excited state.

The transient hole burning can also be found in the spectrum of time integrated intensity as a function of τ_s, as is shown in Figure 2. This has the distinctive advantage in that the dynamics of lattice relaxation is reflected in the τ_s dependence of the time-integrated intensity, because the experiment for Fig. 2 is much easier than the direct measurement of time dependence of radiation intensity (Fig. 1).

Transient hole burning provides us with a new tool of investigating a dynamics of strongly-coupled localized-electron-phonon system. Detailed studies will appear elsewhere.

One of these authors (M. H.) would like to thank Research Center for Computer Science at Hannan University for the aid of numerical calculations performed by the FACOM M770/6.

References

[1]M. Aihara, Phys. Rev. B 21, 2051 (1980).
[2]M. Aihara, Phys. Rev. B 25, 53 (1982).

Time Behavior of Raman Scattering Under Pulsed Light Excitation

T. Kushida[1] *and S. Kinoshita*[2]

[1]Department of Physics, Osaka University, Toyonaka, Osaka 560, Japan
[2]Research Institute of Applied Electricity, Hokkaido University, Sapporo 060, Japan

1. Introduction

In order to clarify the time characteristics of Raman scattering under short light-pulse excitation near the resonance region, we have calculated time-resolved spectra and temporal profiles of the frequency-integrated intensities of secondary radiation components using a stochastic model [1-3]. We discuss the results adopting the definitions that Raman scattering is a single coherent quantum process, while luminescence is a two-step process consisting of separate light absorption and emission [4,5].

2. Theory

Let us consider a secondary radiation process under sufficiently weak light excitation of a three-level system consisting of the initial (g), intermediate (m) and final (f) states of the transition. This system is coupled with a reservoir, which does not interact with the radiation field directly but perturbs the system. We simulate the effect of the adiabatic perturbation on the system by the random modulation of the energy of the state m, and that of the nonadiabatic one by the population decay of the state m with a constant rate. For simplicity, we neglect such effects on the states g and f. Further, the energy modulation is assumed to be expressed by a Gaussian-Markovian process.

The secondary radiation intensity is proportional to

$$\langle|\ddot{P}(t)|^2\rangle = |\langle\ddot{P}(t)\rangle|^2 + \langle|\ddot{P}(t) - \langle\ddot{P}(t)\rangle|^2\rangle, \tag{1}$$

where $\ddot{P}(t)$ is the second time derivative of the electric dipole moment induced by the exciting light, and the square brackets denote the average over the perturbation due to the system-reservoir interaction. We call the secondary radiation components due to the first and the second terms in the right-hand side of Eq. (1), which arise from the mean value of the second time derivative of the emission dipole and its deviation from the mean value, as the A and B components, respectively. It is known that the secondary radiation consists of a sharp Raman line, a

Springer Proceedings in Physics, Vol. 68 Time Resolved Vibrational Spectroscopy V
Ed.: H. Takahashi

broad Raman band and a luminescence band. These three are ascribable to the secondary radiation components emitted from the photoexited system before, during and after the (pure) dephasing relaxation, respectively [5]. For the stationary monochromatic-light excitation, the A component gives a sharp line at the energy expected for the ordinary Raman scattering, and it is evident that this component corresponds to the sharp Raman line. On the other hand, the B component consists of a luminescence band and a broad Raman band. Since the intensity of the latter band is usually negligible, we consider only the A component as the Raman scattering.

3. Results of Calculation and Discussion

We calculated the time characteristics of the frequency-integrated intensities of the A and B components using the above model. The exciting pulse was assumed to have a Gauusian time profile with a shorter duration than the lifetime of the state m. We did not consider the inhomogeneous broadening of the absorption band. It was found that the time behavior of the A component under far off-resonance excitation is essentially the same as that of the exciting pulse, except for a tail whose decay rate is roughly equal to the homogeneous width (FWHM) $\Delta\omega_h$ of the absorption band. The intensity of this tail increases as the detuning energy is decreased, and at exact resonance the A component consists of only this part. This indicates that Raman scattering occurs, under just-resonance excitation, within a duration around the reciprocal homogeneous absorption bandwidth. This is consistent with our definition of Raman scattering, because the inverse of the homogeneous width of the absorption band corresponds to the relaxation time of coherence between the g and m states.

Our calculation showed that the fraction of the A component in the time-integrated secondary radiation intensity under just-resonance excitation is almost equal to $1/\tau_m\Delta\omega_h$ irrespective of the exciting pulse duration, where τ_m is the lifetime of the state m. Therefore, we compared the experimental value of $1/\tau_m\Delta\omega_h$ with the fraction of the A component in the secondary radiation under just-resonance stationary-light excitation in several dye solutions [6]. Fairly good agreement was obtained between these two, which result supports our interpretation that Raman scattering occurs, under just-resonance excitation, within a duration around the reciprocal homogeneous absorption bandwidth.

We also calculated the time-resolved secondary radiation spectra under pulsed light excitation for various detuning energies of the exciting light from resonance to off-resonance using the above stochastic model. The temporal response function of the detection system was assumed to be the same Gaussian as the time profile of the exciting pulse. As shown in Fig. 1, under just-resonance excitation, only a single

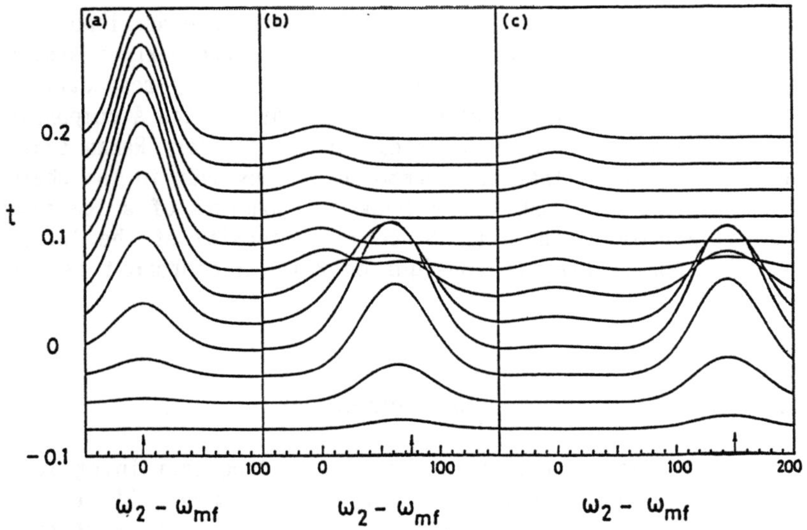

Fig. 1. Time-resolved secondary radiation spectra calculated for a three-level system under pulsed-light excitation. The arrows show the energy positions expected for the ordinary Raman scattering. The detuning energy of the exciting light $\omega_1 - \omega_{mg}$ is (a) 0, (b) 75, and (c) 150, and the parameter values of D = 40, τ_c = 1/4000 and τ_m = 0.5 were employed, where D and τ_c are the amplitude and the correlation time of the energy modulation of the state m due to the system reservoir interaction. The exciting pulse duration is 0.06.

band appears around $\omega_2 = \omega_1 - \omega_{fg} = \omega_{mf}$ and its spectral shape is insensitive to time. Here, ω_1 and ω_2 are the energies of the incident and secondarily emitted photons, and ω_{ij} is the difference between the mean energies of the states i and j. Under off-resonance excitation, a fast component appears around $\omega_2 = \omega_1 - \omega_{fg}$ in addition to a slow component located around $\omega_2 = \omega_{mf}$. For the near-resonance excitation, on the other hand, the secondary radiation spectrum in the early stage changes with time, and the peak energy does not agree with that expected for the ordinary Raman scattering.

In this near-resonance excitation case, a dip was found to appear in the time profile of the total secondary radiation intensity when the spectral width of the detection system is increased sufficiently. The calculation of the time characteristics of the frequency-integrated intensities of the A and B components showed that the dip is present only in the A component, and the B component shows an ordinary slow response. Further, the interference effect between the dominant term in the Raman scattering which has a peak around $\omega_2 = \omega_{mf}$ and that around $\omega_2 = \omega_1 - \omega_{fg}$ reproduced the time behavior of the A component fairly well. Therefore, the dip is interpreted in terms of the quantum interference between the resonant and nonresonant Raman processes. This interference appears because the Raman scattering is a coherent quantum process, and the presence of this dip indicates that Raman scattering takes place even under resonant short-pulse excitation, where the Raman component is not separable from the luminescence band.

It was also found that the total secondary radiation intensity under nearly resonant excitation cannot be expressed as the sum of the fast component having the same time profile as the exciting pulse and the slow one with the same time profile as the total secondary radiation under just-resonance excitation. One might consider that the time behavior of Raman intensity is always the same as that of the exciting light. However, the above result is difficult to explain if one adopts this interpretation.

It can be said that the time needed for the Raman scattering to occur is determined generally by the detuning energy and the homogeneous width of the absorption band. This is reasonable because they are related to the time for the system to stay in the excited state under off-resonance excitation and that for the coherence between the ground and excited states to be maintained, respectively. The former dominates the nonresonant contribution in the Raman scattering, while the latter the resonant contribution. These two contributions interfere with each other and produce the dip in the time profile of the secondary radiation intensity.

1. T. Kushida and S. Kinoshita, Ind. J. Phys. (Raman Centenary Special Issue) 72, 147 (1988).
2. T. Kushida and S. Kinoshita, Phys. Rev. A 41, 6042 (1990).
3. S. Kinoshita and T. Kushida, Phys. Rev. A 42, 2751 (1990).
4. T. Kushida, Solid State Commun. 32, 33 (1979).
5. T. Kushida and S. Kinoshita, Vib. Spectra Struct. 17B, 495 (1989).
6. T. Kushida and S. Kinoshita, J. Lumin. 45, 416 (1990).

Dynamical Effects on Ultrashort Time-Resolved Nonlinear Coherent Scattering from Molecules in Condensed Phases

Y. Fujimura

Department of Chemistry, Faculty of Science,
Tohoku University, Sendai 980, Japan

1. Introduction

One of the characteristics in the nonlinear coherent scattering spectroscopy is capability of creation of intermolecular coherence in molecular ensemble, that is, pairs of molecules at various sites are simultaneously excited, and the time evolution of the molecular pairs is detected through the coherent probing process. For example, the creation of intermolecular vibrational coherence is reflected in the appearance of quantum beats in the time-resolved nonlinear coherent scattering profiles of molecules in condensed phases [1]. The beat frequency is equal to the frequency difference between two Raman transitions taking place at various sites. In principle, dynamical information associated with intermolecular interactions between molecular pairs, such as intermolecular dephasing, intermolecular coherence transfer etc. can be obtained by analyzing the time profiles of the coherent spectroscopy.

The purpose of this paper is to investigate the effects of the intermolecular interactions on the time-resolved nonlinear coherent scattering profile from molecules in condensed phases from the microscopic point of view. For this purpose, we present the mechanisms of the intermolecular dephasing and the intermolecular coherence transfer in a binary molecular mixture. We call the intermolecular interaction mode the common heat bath mode.

2. Equation of motion for intermolecular vibrational coherence

We consider a molecular ensemble which consists of two kinds of molecules, *i.e.*, one is characterized by two Raman transitions $b \leftarrow a$ and $c \leftarrow a$, and the other a Raman transition $b' \leftarrow a'$ as is shown in Fig.1.

The equation of motion for the intermolecular coherence is expressed in terms of the density matrix elements of two-site $(\alpha\beta)$ density operator, $G^{\alpha\beta}_{ca:a'b'}(t) = << ca : a'b' \mid \rho_{\alpha\beta}(t) >>$, as

Springer Proceedings in Physics, Vol. 68 Time Resolved Vibrational Spectroscopy V
Ed.: H. Takahashi

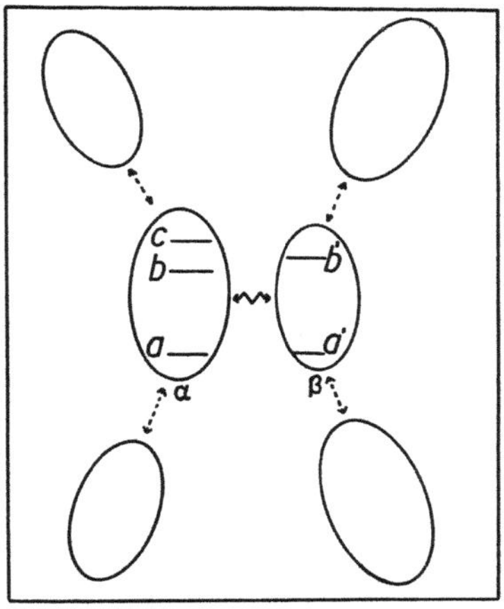

Fig.1 Model for intermolecular interactions between two kinds of molecules in condensed phase. Here, molecule at site α has two Raman transitions $b \leftarrow a$ and $c \leftarrow a$, and molecule at site β a Raman transition $b' \leftarrow a'$. In time-resolved nonlinear coherent scattering processes, both molecules are simultaneously excited by nonlinear coherent light pulses, and the time evolution of the intermolecular coherence is detected through the coherent probing process.

$$\frac{d}{dt}G^{\alpha\beta}_{ca:a'b'}(t) = -[i\omega^{\alpha\beta}_{ca:b'a'} + \Gamma^{\alpha\beta}_{ca,a'b':ca,a'b'}]G^{\alpha\beta}_{ca:a'b'}(t)$$

$$+\Gamma^{\alpha\beta}_{ca,a'b':ba,a'b'}G^{\alpha\beta}_{ba:a'b'}(t), \qquad (1)$$

in which $\omega^{\alpha\beta}_{ca:b'a'}$ and $\Gamma^{\alpha\beta}_{ca,a'b':ca,a'b'}$ denote frequency difference between two Raman transitions, $c \leftarrow a$ at site α and $b' \leftarrow a'$ at site β, and the intermolecular dephasing constant relevant to the two Raman transitions, respectively. Constant $\Gamma^{\alpha\beta}_{ca,a'b':ba,a'b'}$ represents the magunitude of transfer of the intermolecular vibrational coherence from $c \leftrightarrow a$ and $a' \leftrightarrow b'$ to $b \leftrightarrow a$ and $a' \leftrightarrow b'$, and called the intermolecular vibrational coherence transfer constant. Equation (1) has been derived within the Markoff approximation [2]. The intermolecular dephasing constant originates from fluctuation of the two Raman transitions induced by the intermolecular interaction. The relation between the intermolecular dephasing constant and the ordinary, intramolecular dephasing constant is discussed elsewhere [3].

3. Mechanism of intermolecular vibrational coherence transfer

The system Hamiltonian is assumed to be expressed as

$$\hat{H} = H_0 + H'_{MB}, \tag{2}$$

where the zeroth order Hamiltonian H_0 is given by the sum of the two types of molecules of interest and the common heat bath mode, and the interaction between the molecules and the heat bath mode H'_{MB} is expressed as

$$H'_{MB} = \hat{V}_\alpha \hat{R}_B + \hat{V}_\beta \hat{R}_B. \tag{3}$$

In Eq.(3), $\hat{V}_\alpha$ ($\hat{V}_\beta$) and $\hat{R}_B$ denote the operators of the molecule α (β) and the heat-bath mode, respectively. In the lowest order approximation, the intermolecular vibrational coherence transfer constant is expressed as

$$\Gamma^{\alpha\beta}_{ca,a'b':ba,a'b'} = \frac{\pi}{\hbar^2} \sum_i \sum_f p_i \left| < i \mid \hat{R}_B \mid f > \right|^2 < c \mid \hat{V}_\alpha \mid b >$$

$$\times \{ < b' \mid \hat{V}_\beta \mid b' > - < a' \mid \hat{V}_\beta \mid a' >$$

$$- (< b \mid \hat{V}_\alpha \mid b > - < a \mid \hat{V}_\alpha \mid a >) \} \delta(\omega_{B_{if}}), \tag{4}$$

where i and f denote the initial and final states of the common heat bath mode, $\omega_{B_{if}}$ the frequency difference, and p_i the initial distribution. Equation (4) indicates that the coherence transfer is induced by two types of the molecule-heat bath interactions, *i.e.*, exchange and elastic interactions. We note that the elastic interaction does not make any contribution when the molecular system consists of the same kind of molecules. In this case, the magnitude of the coherence transfer constant vanishes in the lowest approximation, and the coherence transfer is induced by the exchange interactions at both sites α and β. This is a higher order mechanism than that given by Eq.(4). In other words, the coherence transfer between different kinds of molecules is important compared with that between the same kind of molecules. This qualitative view is consistent with experimental results by Leonhardt *et al.* [1]. The coherence transfer taking place at local sites α and β brings about a frequency shift of quantum beats. A slow decrease in the amplitude of the quantum beats appears after tracing out the site-dependence of the coherence transfer constant [4].

In summary, the mechanism of the intermolecular vibrational coherence transfer has been discussed, and the effects of the intermolecular coherence transfer on the time-resolved nonlinear coherent scattering profiles have been clarified.

This work was supported by Grant-in-Aid for Scientific Research on Priority Area of "Molecular Approach to Non-equilibrium Processes in Solutions" from the Ministry of Education, Science and Culture, Japan (03231202).

References

1. R.Leonhardt, W.Holzapfdel, W.Zinth, and W.Kaiser, Chem.Phys.Lett. 133,373(1987) ; N.Kohles and A.Laubereau, Chem.Phys.Lett.138, 365(1987) ; S.Ruhman, A.G.Joly, and K.A.Nelson, J.Chem.Phys.86, 6563(1987) ; H.Okamoto and K.Yoshihara, J.Opt.Soc.Am.B7,1702(1990).
2. M.Hayashi, M.Sugawara, and Y.Fujimura, Phys.Rev.A43, 2416(1991).
3. M.Hayashi, M.Sugawara, and Y.Fujimura, elsewhere in this volume of "Time-Resolved Vibrational Spectroscopy V".
4. M.Hayashi and Y.Fujimura, J.Opt.Soc.America,B.7, 1653(1990).

Effects of Intermolecular Vibrational Dephasing on Time-Resolved Coherent Anti-Stokes Raman Scattering from Molecules in Liquids

M. Hayashi, M. Sugawara, and Y. Fujimura

Department of Chemistry, Faculty of Science,
Tohoku University, Sendai 980, Japan

1. Introduction

With the development of femtosecond time-resolved coherent-scattering spectorscopy, transient behaviors of intermolecular dynamics of molecules in liquids have been studied both experimentally and theoretically with great interest. This is mainly due to the fact that the time development of two molecules interacting with each other at various positions can be directly observed by using such a coherent spectroscopy. For example, quantum beats originated from creation of intermolecular coherence between Raman active vibrational transitions of molecules at various sites appear in time-resolved CARS (coherent anti-Stokes Raman scattering)[1]. Effects of dephasing of the intermolecular vibrational coherence on time-resolved CARS profile were not clarified so far.

The purpose of this paper is to present an expression for the intermolecular dephasing constant taking into account the molecules heat-bath interaction, and to clarify how the intermolecular dephaing effects appear in the time-resolved CARS profiles.

2. Effects of Intermolecular Vibrational Dephasing

An expression for the time-resolved CARS profile $I(\tau)$ as a function of the delay time τ between pumping and probing processes is derived by using the perturbative density matrix method [2]. We confine ourselves to the CARS from an ensemble consisting of the same kind of molecules with two Raman active modes. Within the Markov approximation, the intermolecular dephasing constant consisits of the sum of population decay and intermolecular pure dephasing constants. The latter constant originates from an elastic interaction between the relevant molecules and a common heat bath mode.

In the case in which two molecules are correlated through a common heat bath mode, intermolecular dephasing constant $\Gamma^{ll'}_{ca,ca:ca,ca}$ for the vibrational transition $c \leftrightarrow a$ at site l and $c \leftrightarrow a$ at site l' is given by

$$\Gamma^{ll'}_{ca,ca:ca,ca} = (\Gamma^{l}_{aa:aa} + \Gamma^{l}_{cc:cc} + \Gamma^{l'}_{aa:aa} + \Gamma^{l'}_{cc:cc})/2 \tag{1}$$

Springer Proceedings in Physics, Vol. 68 Time Resolved Vibrational Spectroscopy V
Ed.: H. Takahashi

and that of the transition $c \leftrightarrow a$ at site l and $d \leftrightarrow a$ at site l' is expressed as

$$\Gamma^{ll'}_{ca,da:ca,da} = (\Gamma^{l}_{aa:aa} + \Gamma^{l}_{cc:cc} + \Gamma^{l'}_{aa:aa} + \Gamma^{l'}_{cc:cc})/2 + \Gamma^{(d)ll'}_{ca,da:ca,da} \tag{2}$$

where $\Gamma^{l}_{ii:ii}$ and $\Gamma^{(d)ll'}_{ca,da:ca,da}$ denote the population decay constant for the state i of the molecule at site l and the intermolecular pure dephasing constant for the intermolecular coherence between $a \leftrightarrow c$ and $a \leftrightarrow d$, respectively. Equations (1) and (2) are associated with the diagonal part of the time-evolution matrix element of the intermolecular coherence and the off-diagonal one, respectively. Equations (1) and (2) indicate that the amplitude of the quantum beats in the time-resolved CARS decrease faster than that of the average decay in the presence of the intermolecular dephasind process. Quantum beats appeared in the time-resolved profile of CARS from neat pyridine, which have been reported by Leonhardt *et.al.* [1], show fast decrease in the amplitude of the beats compared with the magnitude of average decay of the CARS profile. This feature can be explained as a results of the participation of the intermolecular dephasing constant in Eq.(2). That feature cannot be explained by using the conventional CARS theory because a sum of the intramolecular dephaing constant relevant to each Raman transition is simply adopted as the intermolecular dephasing constant. In the case in which two molecules are uncorrelated, *i.e.*, each molecule interacts with the heat bath independently, the intermolecular dephasing constant for the transition $a \leftrightarrow c$ at site l and $a \leftrightarrow d$ at site l' reduces to

$$\Gamma^{ll'}_{ca,da:ca,da} = \Gamma^{l}_{ca:ca} + \Gamma^{l'}_{da:da} \tag{3}$$

where $\Gamma^{l}_{ca:ca}$ represents the intramolecular dephasing constant for the transition $a \leftrightarrow c$ of the molecule at site l. In this case the amplitude of the quantum beats of time-resolved CARS shows the parallel decay with that of the average decay.

References

1.R.Leonhardt, W.Holzapfdel, W.Zinth, and W.Kaiser, Chem.Phys.Lett. 133,373(1987).; N.Kohles and A.Laubereau, Chem.Phys.Lett.138, 365(1987).; H.Okamoto and K.Yoshihara, J.Opt.Soc.Am.B7,1702(1990).
2.M.Hayashi, M.Sugawara, and Y.Fujimura, Phys.Rev.A43,2416(1991).

Theory of Inhomogeneous Gaussian Broadening of Resonance Raman and Coherent Anti-Stokes Raman Scattering Excitation Profiles for Polyatomic Molecules in Solution

M.G. Zakaraya

Institute of Inorganic Chemistry and Electrochemistry of the Georgian Academy of Sciences, Jikiya 7, 380086 Tbilisi, Georgia

Reliable quantitative determination of parameters characterizing the time evolution of secondary radiation processes with polyatomic molecules in solutions, such as energy relaxation, elastic solute-bath collisions (dephasing), etc, from experimental data for the absorption spectra and the excitation profiles of the resonance Raman scattering (RRS) and coherent anti-Stokes Raman scattering (CARS) essentially depends on the correct theoretical account of the effects of the medium's influence on the characteristics for isolated molecules. The medium exerts double influence on the characteristics of both the single-photon spectra of absorption (fluorescence) by molecules in solution and the RRS and CARS excitation profiles (REPs and CEPs): dynamic, of homogeneous nature, i.e. having effect even on a single molecule via the electron-phonon coupling, and static, of essentially inhomogeneous origin, evoked by variations of the local surroundings of each molecule of the ensemble under study.

With respect to solute-solvent interactions, Raman excitation profiles are conveniently referred to the same broad classification as for single-photon transitions. In particular, we have recently shown /1/ that in the limit of strong solute-solvent coupling the excited-state decay properties are unimportant and the REP is solely determined by the solvent and local mode coupling. In the weak-coupling limit the individual vibronic components are Lorentzians and dominated by the excited-state lifetime. In this limit, the overall lineshape is easily distorted by other broadening mechanisms, of which in homogeneous broadening is particularly important.

In Refs /2-3/ we have investigated the REP when two broadening mechanisms operate at the same time in the weak-coupling limit of solute-solvent interaction. One of these is Lorentzian and representative either of finite excited-state lifetimes or of weak solute-solvent coupling. The other is Gaussian and most adequately representative of inhomogeneous environmental broadening /4/. The REP is calculated analytically, and by means of recent optical bandshape theory for Voigt profiles /5-6/ it can be recast into rapidly converging and easily tractable series expansions well suited for numerical data fitting. The profile expressions also apply to arbitrary values of the broadening parameters and in these respect improve on several previous approaches to inhomogeneous broadening. Notable outcomes are that the REP takes the form of series of Franck-Condon modulated Voight profiles when the broadening parameters are small compared with the vibrational frequencies of the Raman-active modes. In this limit only diagonal terms in the sum over vibrational quantum numbers contribute. When the broadening parameters are not small, non-diagonal terms are important and lead to additional broadening, profile maximum shifts and bandshape distortion.

Springer Proceedings in Physics, Vol. 68 Time Resolved Vibrational Spectroscopy V
Ed.: H. Takahashi

For the case of CARS the problem is even more complicated. The CEPs are described by the function

$$P_{n_g}(\omega_1) \simeq |\chi^{(3)}_{n_g}(\omega_1) \equiv \sum_{k_e}\sum_{m_e} \chi^{k_e,m_e}_{n_g}(\omega_1)|^2 \tag{1}$$

with a fixed value of the vibrational quantum number n_g of the ground (g) electronic state. $\chi^{(3)}$ is the third order susceptibility tensor, and the sum is taken over all vibrational states of the excited (e) electronic state. Correct averaging over the Gaussian distribution function of the purely electronic transition energy ω ($\hbar=1$) with respect to the average value $\bar{\omega}$ and the dispersion σ^2 has been conducted in /7/. Here we present the expression for the CEP of the fundamental tone $n_g=1$ in the simplest case of only one active intramolecular degree of freedom on the assumption that $\Gamma_k = \Gamma_m \equiv \Gamma$. (Note that CEPs show increased sensitivity to the decay constants). In the case at hand, the main contribution into the double sum (1) is made by terms with correspond to the so-called double resonance /7/, i.e. the terms with m = k +1 (k =0,1,...). The average $\langle\langle \chi^{(3)}_{1g}(\omega_1)\rangle\rangle$ does not have singularities and can be truthfully approximated by the simple sum:

$$\sum_{k=0}^{\infty} \langle 0 | k+1 \rangle \langle k+1 | 1 \rangle \langle 1 | k \rangle \langle k | 0 \rangle \, [\delta\psi(v_k;\xi)/\delta v_k + i\delta\varphi(v_k;\xi)/\delta v_k], \tag{2}$$

where $\psi(v;\xi)$ and $\varphi(v;\xi)$ are the real and imaginary parts, respectively, of the complex function

$$W(v;\xi) = \exp[(\xi - iv)^2]\,[1 - \operatorname{erf}(\xi - iv)] \tag{3}$$

with arbitrary values of $\xi=\Gamma/\sqrt{2}\,\sigma$ and dimensionless detuning $v=(\omega - \bar{\omega} - \varepsilon_k)/\sqrt{2}\,\sigma$. Analytical expressions for $\psi(v;\xi)$ and $\varphi(v;\xi)$ are obtained in /5-6/. They may help to write the real and imaginary parts of $\langle\langle \chi^{(3)}_{1g}(\omega_1)\rangle\rangle$, i.e. the derivation in (2):

$$\delta\psi(v;\xi)/\delta v = 2\xi\varphi(v;\xi) - 2v\psi(v;\xi); \tag{4}$$

$$\delta\varphi(v;\xi)/\delta v = 2/\sqrt{\pi} - 2\xi\psi(v;\xi) - 2v\varphi(v;\xi). \tag{5}$$

Finally, here is one more expression relating ψ and φ:

$$\varphi(v;\xi) = (2\xi)^{-1}\,\delta\psi(v;\xi)/\delta v + (v/\xi)\psi(v;\xi). \tag{6}$$

This expression relates the real and imaginary parts of $\langle\langle \chi^{(3)}_{1g}(\omega_1)\rangle\rangle$ as well as REPs and CEPs, thus aiding the reliable determination of the desired main parameters of the studied molecules and the solvent by comparing the above theory with the experiment.

1.M.G.Zakaraya and J.Ulstrup, Chem.Phys. 135, 49 (1989).
2.M.G.Zakaraya et al. J.Raman Spectrosc. 20, 359 (1989).
3.M.G.Zakaraya, J.Raman Spectrosc. 21, 401 (1990).
4.A.V.Lukashin and M.D.Frank-Kamenetskii, ibid 12, 234 (1982).
5.M.G.Zakaraya, Opt.Spectrosc. (USSR) 59, 282 (1985).
6.M.G.Zakaraya and J.Ulstrup, Opt.Commun. 62, 107 (1988).
7.M.G.Zakaraya and J.Ulstrup, Opt.Spectrosk. 71, 134 (1991).

Femtosecond Time-Resolved Resonance CARS Spectroscopy: Theory

V.F. Kamalov[1] *and Y.P. Svirko*[2]

[1]Photochemistry Department, Institute of Chemical Physics, Russian Academy of Sciences, Kosygin str. 4, 117977 Moscow, Russia
[2]Institute of General Physics, Russian Academy of Sciences, Vavilov str. 38, 117942 Moscow, Russia

Considerable progress has been made in recent years in studying the femtosecond intramolecular dynamics in the excited electronic states of large organic molecule. High intensity reaches by femtosecond lasers gives high opportunities for applications of nonlinear spectroscopy, especially four wave mixing technique for study ultrafast relaxational processes. Separation of femtosecond pulses in time used for CARS gives us new opportunities as compared with RR spectroscopy for study of intramolecular dynamics. In this work the resonance CARS response is calculated within the temporal approach. Two femtosecond CARS schemes for studying the relaxational processes in excited electronic state are discussed.

1. THEORY

On the basis of the density matrix equation induced molecular dipole moment at frequency $\omega_a = \omega_1+\omega_3-\omega_2$ was calculated within the time-dependent theory [1]:

$$p_i(\omega_a) = (-i/\hbar^3) \sum_\gamma \kappa^{ik}_{g\gamma}(\omega_a) e_{3k} \int_{-\infty}^{+\infty} dt E_2^*(t) \int_0^\infty d\tau' E_3(t+\tau') \int_0^\infty d\tau'' E_1(t-\tau'')$$
$$<(\vec{\phi}_\gamma \vec{e}_2)|(\vec{\phi}_g(\tau'')\vec{e}_1)> \exp\{i\omega_1\tau'' - \Gamma\tau''\} \exp\{i(\omega_1-\omega_2-\omega^{\gamma g}_{00})\tau' - \Gamma^{\gamma g}_{00}\tau'\}$$

where

$$\kappa^{ij}_{g\gamma}(\omega_a) = \int_0^\infty d\tau <\phi^i_g|\phi^j_\gamma(\tau)> \exp\{i\omega_a\tau - \Gamma\tau\}$$

is the Raman scattering tensor. The suming is carried out with respect to all vibrational levels of the ground electronic state, Γ being the homogeneous width of electronic transition. Dipole moment and Raman tensor are represented as overlap of wave packets [1] $<\vec{\phi}_{g,\gamma}| = <\Psi_{0g,\gamma}\vec{M}|$, $|\vec{\phi}_{g,\gamma}(t)> = |\exp\{-i\hat{H}t/\hbar\}\vec{M}\Psi_{0g,\gamma}>$ where $\Psi_{0\alpha}$ are wave functions in ground electronic state, index "g" labels the lowest vibrational level, $\vec{M}$ is the operator of dipole moment, H is the molecular Hamiltonian.

2. FEMTOSECOND CARS EXPERIMENTAL SCHEMES

Two CARS schemes have been considered. In the first scheme [2] two CARS-probing pulses have femtosecond duration, $\tau_p<1/\Gamma$, and the third pulse duration lies in the picosecond range (Fig.1a). In the second scheme three interacting pulses are femtosecond (Fig.1b). In the last case the dependence of CARS signal on Δt_2

Springer Proceedings in Physics, Vol. 68 Time Resolved Vibrational Spectroscopy V
Ed.: H. Takahashi

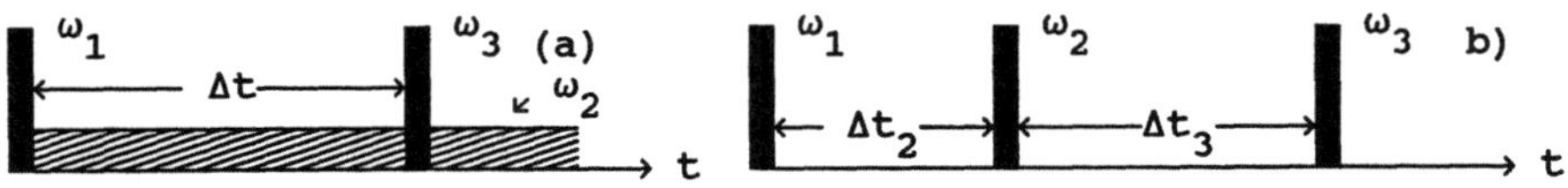

Fig.1. Two (a) ind three (b) femtosecond pulses CARS schemes.

is directly determined by the temporal evolution of molecule wave packet .

CARS depolarization ratio $\rho^{CARS} = \chi_{xyyx}/\chi_{xxxx}$ was calculated by orientational averaging of hyperpolarizibility for an isotropic medium. Temporal evolution of wave packet depends on the type of adiabatic potential of exited electronic state. When Γ is determined by electron-vibrational interaction and the shift between equilibrium positions of ground and excited adiabatic potentials is not high, the temporal evolution of wave packet leads to small changes of depolarization ratio in RR [3] and CARS. In other cases when others fast relaxational processes (such as femtosecond isomerization in exited electronic states) determine Γ essentially the temporal evolution of wave packet can lead to changes of ρ^{RR} [4] and ρ^{CARS}.

CARS depolarization ratio can be calculated using the simple model of wave packet evolution:

$$|\vec{\phi}_{g,\gamma}(t)> \quad = \{d_{g,\gamma}(t), d_{g,\gamma}(t), 0\}\exp\{-i\omega_e t\}$$

where ω_e is frequency of electronic transition. In this case $3\rho^{CARS}(\Delta t) = [1+f(\Delta t)]/[1+f(\Delta t)/6]$. For the first experimental scheme (Fig.1a)

$$f(\Delta t)=\sum \kappa^{xy}_{g\gamma}(\omega_a)\kappa^{xy}_{\gamma g}(\omega_1,\Delta t)/\sum \kappa^{xx}_{g\gamma}(\omega_a)\kappa^{xx}_{\gamma g}(\omega_1,\Delta t), \text{ and}$$

for the second scheme (Fig.1b)

$$f(\Delta t_2)=\sum \kappa^{xy}_{g\gamma}(\omega_a)d^x_\gamma d^y_g(\Delta t_2)/\sum \kappa^{xx}_{g\gamma}(\omega_a)d^x_\gamma d^x_g(\Delta t_2).$$

Steady state experiments can be used to check this model of wave packet evolution. In this case the depolarization ratios RR and resonance CARS prove to be connected:

$$(3\rho^{RR}-1)/(3\rho^{CARS}-1) \cong 2.$$

REFERENCES

1. Heller E.J., Sundberg R.L., J. Phys.Chem. 86,1822 (1982).
2. Okamoto H., Yoshihara K., Chem.Phys.Lett. 172, 323 (1990).
3. Mayers A., J.Phys.Chem. 95, 1536 (1991).
4. Hamaguchi H., J.Phys.Chem. 93, 7 (1989).

Part IX

Innovative Methods/Techniques

Consideration About Optimal Conditions for Backward Scattering Signals in Four-Wave-Mixing Techniques

A. Lau, M. Pfeiffer, and W. Werncke

Central Institute of Optics and Spectroscopy,
Rudower Chaussee 6, O-1199 Berlin, Fed. Rep. of Germany

Introduction

The investigation of opaque samples is very important for many fields of research in physics (semiconductors), biology, chemistry and medicine in finding correlations between structures and other optical and electrical qualities. For these samples the use of backscattering arrangements are possible only. In vibrational spectroscopy such schemes for IR and Raman techniques are well known. But the introduction of coherent vibrational methods are still very difficult and limited to only a few special cases. But the supplementary informations which are inherent in coherent methods demand for their corresponding application. Some results are already obtained by the method of Sum Frequency Generation [1], but with restrictions to special frequency regions, due to the lack of proper IR sources. Experiments obtained by Degenerate Four Wave Mixing Techniques show also very interesting results, but they are limited to investigations of electronic states [2]. For Nondegenerated Four Wave Mixing techniques up to now only one result is published concerning ps-CARS at silicon [3].
The aim of this investigation is to find conditions for which coherent Raman arrangements can be applied successfully.

CARS backward arrangements

Our considerations start with the Fresnel equation at boundaries, as was done for Degenerated Four Wave Mixing. For CARS it follows, that the condition for the tangential components of the wave vectors

$$k_{a//} = k_{p1//} + k_{p2//} - k_{s//}$$

has to be fullfilled.
In Fig. 1 one geometrical arrangement with the corresponding wavevector diagram is given from which the phasemismatch value "d" and the direction of the CARS signals can be calculated. The CARS intensity calculation must take into account the phase mismatch for the backward signals by using the corresponding equation.

$$I_{cars} \sim |X^{(3)}|^2 I_p^2 I_s \frac{\sin^2(d*x/2)}{(d/2)^2}$$

If phasematching in forward direction is fullfilled, it exists a unique connection between the angles of all beams.
From Fig.2 can be deduced, that the CARS backward signal strongly depends on its angle "ba"(function of all other angles)in which this signal is generated. For small interaction-

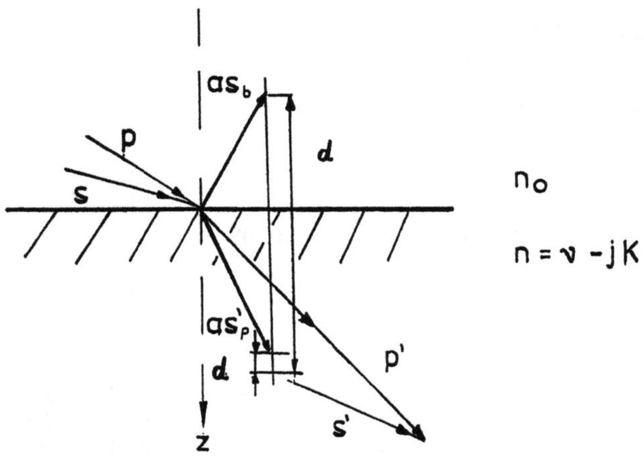

Fig. 1 Scheme of the CARS backward scattering arrangement with the corresponding k-vector diagram

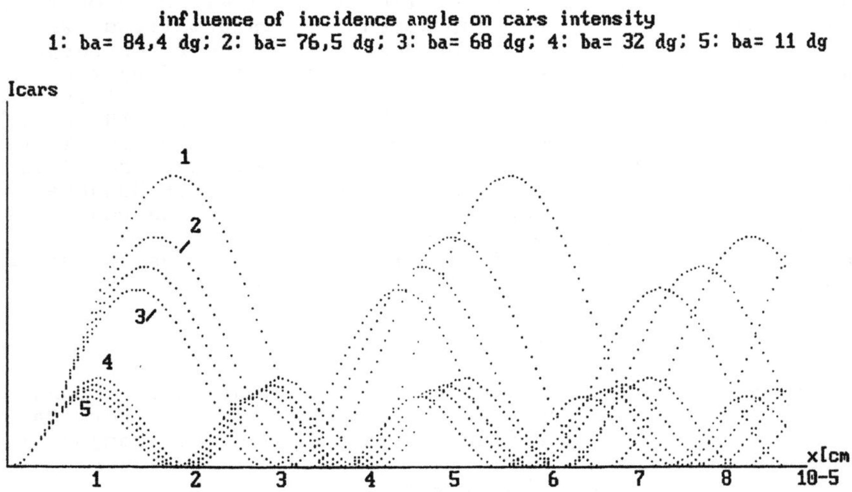

Fig. 2 Influence of incidence angle on CARS intensity for different CARS angles ba
1: ba= 84,4 dg; 2: ba= 76,5 dg; 3: ba= 68 dg; 4: ba= 32 dg; 5: ba= 11 dg

lenghts the CARS intensity reaches maximum for highest possible CARS angles.
In Fig. 3 the dependence of the CARS backward signal on the interactionlenght is given for a sample with a very high refractive index, high dispersion and for different absorption coefficients "a" (Ge-crystal). The high mismatch value leads to a fast oscillation of the intensity with a first max. at about 50 nm. The absorption of the exciting beams washes out the oscillations of the signal´s intensity, due to the fact, that the strongly damped signal, generated after a certain in-

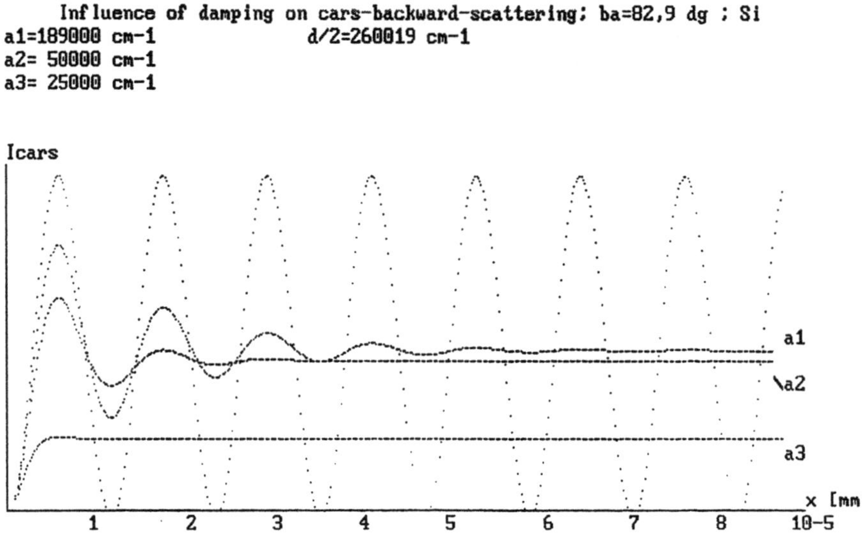

Fig. 3 Influence of damping on CARS-backward-scattering; ba= 82,9 dg; Ge
for d= 260018,5 cm^{-1} and a3= 189000 cm^{-1}; a2= 50000 cm^{-1}; a1= 25000 cm^{-1}

teractionlenght, can give neither constructive nor destructive interferences with the signals generated in the first part of the sample.

To carry out a successful experiment it is very important to know the absolut value of the expected signal´s intensity. For a corresponding impression we show in Fig. 4 the CARS intensities in forward and backward direction (i_f and i_b respectively). At about 200 nm interactionlenght the undamped CARS intensity in forward direction is about 3. times larger than the backward intensity. Though both intensities are comparable the absolute value in comparison with usual samplelengths (>10µm) are many orders of magnitude smaller (10^3 to 10^5). Therefore for transparent samples it will be nearly impossible to observe the backward signal. The aim was but to find schemes for opaque samples. For these samples a comparison of curve 2 and 4 in Fig. 4 shows, that the intensities in forward and backward directions are approximately equal. The absolut value depends strongly on the enhancement factor which have to be expected due to the Resonance Raman effect. Resonance Raman spectra (spontaneous as well as coherent Raman effects) are correlated to the absorption spectra. Generally high absorbances are related to high enhancement factors, taking into account modifications due to the actual linewidth of the corresponding electronic transitions. In backward scattering schemes it is not only possible but even necessary to use as high concentrations (absorbances) as possible or even pure substances (crystals) which will at least partly compensate for the smaller interactionlengths in comparison with forward CARS schemes. For the latter much lower concentrations (lower enhancement factors) has to be used to allow the exciting and signal beams to penetrate through the sample for samplelenghts of a-

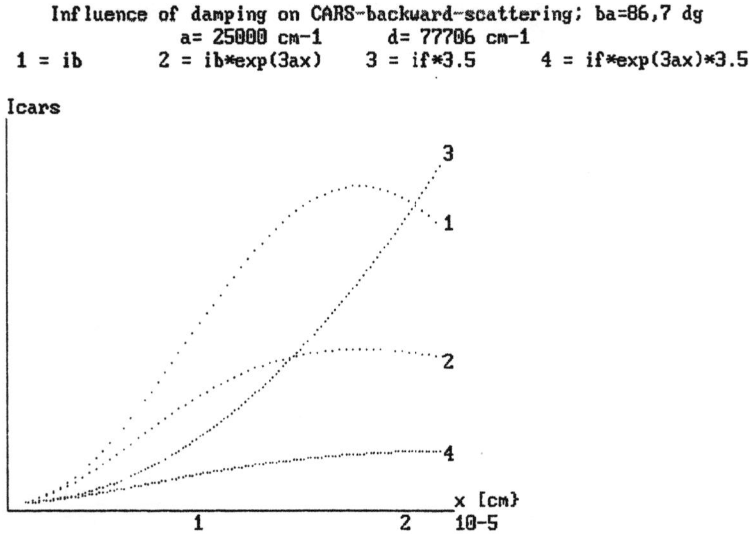

Fig. 4 Comparison of CARS intensity for a CARS angle of ba= 86,7 dg and an absorption coefficient a= 25000 cm^{-1} for the backward direction i_b and forward direction i_f; 1= i_b; 2= i_b* exp(3ax); 3= i_f*3,5 ; 4= i_f*exp(3ax)*3,5

bout 100 µm. As a result the intensity of the backward CARS signal can be expected in the same order of magnitude as the forward signal.
Nevertheless it is difficult to obtain CARS-signals in backward direction due to the generation of plasma by irradiating samples of high absorbances and other problems as will be discussed on the basis of first preliminary experiments.

We thank Mrs. R. Goleschny for her engaged technical assistance.

References

[1] A.L. Harris, N.J. Levinos; SPIE vol. 1056, 66 (1989)
[2] J. Kuhl, A. Honold, L. Schultheis, C.W. Tu Festkörperprobleme 29, 157 (1989)
[3] S.V. Govorkov, V.N. Zadkov, N.J. Koroteev, I.L. Shumay JETP Lett. 44, 98 (1986)

Molecular Ion Vibrational Spectroscopy by a Time-Resolved REMPI Threshold Photoelectron Technique

M. Takahashi and K. Kimura

Institute for Molecular Science,
Okazaki National Research Institutes, Okazaki 444, Japan

1. Introduction

During the last decade, laser photoelectron spectroscopy using REMPI (resonantly enhanced multiphoton ionization) has been developed as an important tool to study excited states and ionic states [1]. Figure 1 is a schematic diagram of REMPI. In photoelectron spectroscopy, it is desirable to observe rotational as well as clearly defined vibrational structure.

Collection of 'threshold photoelectrons' as a function of the laser wavelength is an ideal way to improve the spectral resolution. Schlag et al. [2] have measured 'nearly zero energy electrons' (called ZEKE) in two-color REMPI experiments with very high resolution. In this laboratory, we have recently designed and developed two kinds of analyzers with which photoelectrons less than a few cm^{-1} can be collected. Hereafter we abbreviate 'threshold photoelectron spectroscopy' as TES.

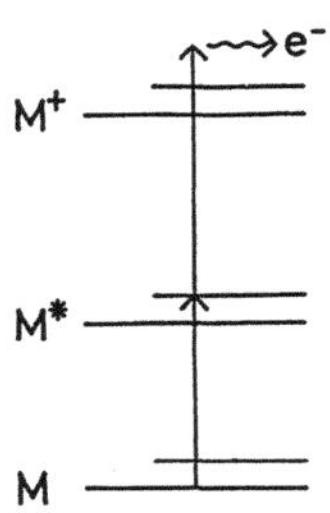

Fig. 1. Two-color REMPI process ejecting a photoelectron

2. Compact TES Analyzer

Our capillary-type [3] and deflection-type threshold photoelectron analyzers are schematically shown in Fig. 2. The capillary type consists of a pair of electrodes (P_1, P_2), a capillary plate (CP), a ground electrode (P_3), and an electron multiplier (EM Ceratron). A capillary plate of l/d = 20 has been used, positioned 10 mm from the ionization point (Q), where l and d are the length and the diameter of the capillary hole, respectively [3]. The deflection type has a pair of deflection plates (D).

In the capillary-type analyzer, threshold electrons (lower than a few cm^{-1}) are separated from other energetic electrons by 'time-resolved discrimination' as well as by 'angular discrimination'. Most energetic electrons are removed by the capillary plate. In the deflection-type analyzer, a deflector is used to remove energetic electrons. Threshold electrons less less than a few cm^{-1} should remain in the volume (10 mm^2 x 30 mm) surrounding Q for as long as 500 ns after each laser shot.

Therefore, we can collect the threshold electrons by applying a pulsed electric field (a few V/cm) across P_1 and P_2 at typically 500 ns. An Ar-

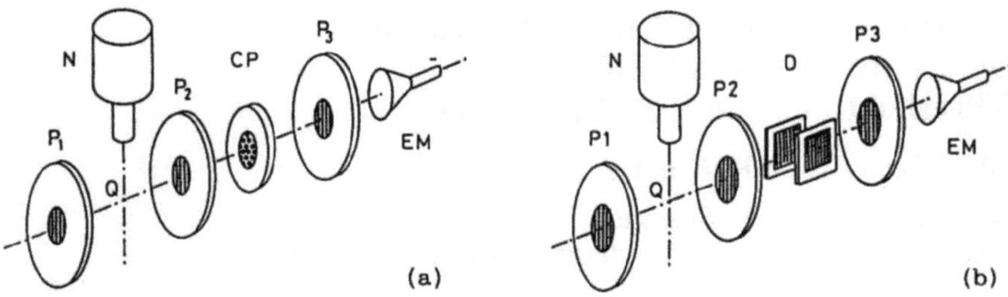

Fig. 2. (a) Capillary-type analyzer and (b) deflection-type analyzer

seeded molecular jet is irradiated by two Nd-YAG pumped dye lasers (ω_1 and ω_2). At each laser shot, photoelectrons are produced at zero electric field. The deflection type is much better than the capillary type in brightness, but the latter is much easier to handle.

3. NO^+ Cation

A two-color (1+1') TES spectrum obtained for NO via the $A^2\Sigma^+$ state (v'=0, N'=7) by using the deflection-type analyzer is shown in Fig. 3. Well-resolved rotational structure of NO^+ due to $\Delta N = N^+ - N' = 0, \pm1, \pm2$, and $\pm$ 3 transitions is observed. Although the energy resolution (fwhm) depends on the delay time of the electric field applied across P_1 and P_2 as well as on the field strength [4], we have obtained an energy resolution of 2 cm^{-1} (fwhm) for the threshold photoelectron spectrum shown in Fig. 3.

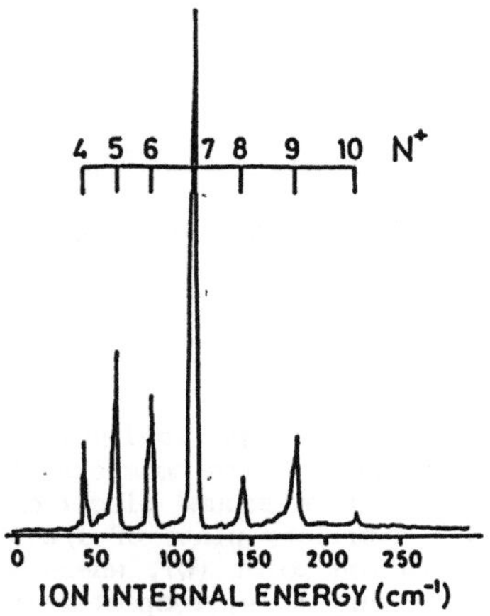

Fig. 3. REMPI-TES spectrum showing a rotational structure

4. *Trans* and *Gauche* *n*-Propylbenzene$^+$ Cation

Jet-cooled *n*-propylbenzene exists in the *trans* and *gauche* forms. The $S_1 \leftarrow S_0$ transition energy of *trans* is known to be 49 cm^{-1} larger than *gauche* [5]. By tuning ω_1 in two-color (1+1') experiments, it is possible to completely separate the two isomers from each other in TES spectra. The TES spectra

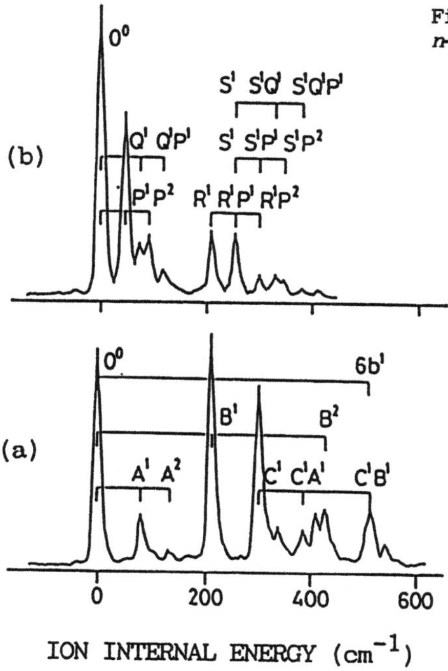

Fig. 4. REMPI-TES spectra of *n*-propylbenzene. (a) *Trans* (b) *Gauche*

thus obtained via the $S_1{}^0$ state, showing vibrational structures of the *trans* and *gauche* cations below 600 cm^{-1} are shown in Fig. 4.

The first bands shown in Fig. 4 correspond to the adiabatic ionization potentials 70265 ± 8 cm^{-1} (*trans*) and 70407 ± 8 cm^{-1} (*gauche*). The *trans* vibrational structure may be explained in terms of combination bands of three low-frequency vibrational modes labeled A, B, and C whose frequencies are 82, 212, and 300 cm^{-1}, respectively. On the other hand, the vibrational structure of the *gauche* cation may be explained in terms of four low-frequency vibrational modes labeled P, Q, R, and S whose frequencies are 46, 73, 207, and 252 cm^{-1}, respectively. These vibrational modes are possibly associated mainly with the *n*-propyl group.

In order to study these vibrational modes, we have carried out *ab initio* calculations using STO-2G basis sets (GAUSSIAN 82 program). After the

Table 1. Observed and calculated vibrational frequencies (cm^{-1}) of the *trans* *n*-propylbenzne cation for low-frequency modes

Calc.	Exptl.
32 A"	
79 A"	
79 A'	82 (A)
214 A'	212 (B)
233 A"	
304 A'	300 (C)
335 A"	
345 A"	
434 A'	
503 A"	504 (6b)
542 A'	

geometry optimization, normal mode calculations were carried out. The calculated values (multiplied by a factor of 0.9) are listed in Table 1, in good agreement with the experimental ones. From Table 1 it is seen that all the Franck-Condon active vibrational modes (A, B, and C) in the *trans* spectrum (Fig. 4a) belong to A' symmetry. From this fact, the rotational isomer excited here should be assigned to the *trans* form with C_s symmetry.

5. (Phenol-Water Complex)$^+$ Cation

A two-color (1+1') REMPI-TES spectrum of the 1:1 phenol-water complex via the S_1 0^0 level is shown in Fig. 5. (The first laser ω_1 was tuned to the complex S_1 origin.) A photoionization efficiency curve in the two-color (1+1') process of this complex has recently been observed by Lipert and Colson [6], who reported an adiabatic ionization potential of 63980 cm^{-1}. From the first band shown in Fig. 5, we have obtained an adiabatic ionization potential of 64015 ± 4 cm^{-1} (at the zero electric field). The reported value is slightly lower than our value probably due to the effect of field ionization.

The main vibrational progression shown in Fig. 5 is due to the intermolecular stretching modes whose frequency is about 240 cm^{-1}. From the anharmonicity of this vibrational mode, we have obtained a value for the dissociation energy (hydrogen bond energy) of the phenol-water cation of $D_e(D_0)$ = 6021 cm^{-1}. Furthermore, the dissociation energies of the neutral complex in the S_1 and S_0 states have been determined as D_e = 1723 cm^{-1} and 1367 cm^{-1}, respectively, using the ionization potentials of the phenol-water complex and phenol [7] and the intermolecular stretching frequencies in S_1 and S_0 [8,9].

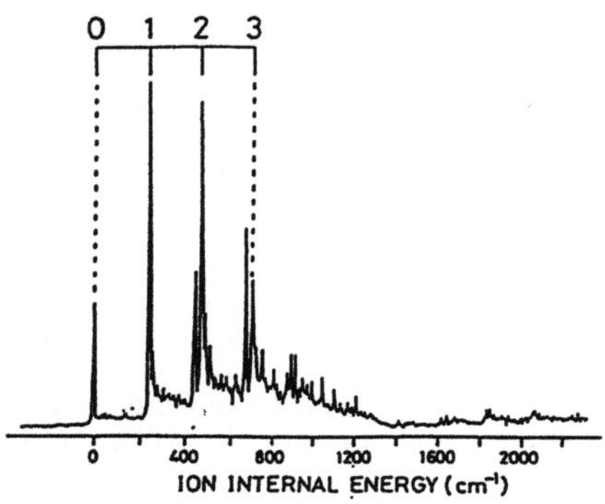

Fig. 5. REMPI-TES spectrum of the 1:1 phenol-water complex

References

[1] For example, K. Kimura, *Adv. Chem. Phys.*, **60**, 161 (1985).
[2] L. A. Chewter, M. Sander, K. Müller-Dethlefs, and E. W. Schlag, *J. Chem. Phys.*, **86**, 4737 (1987).
[3] M. Takahashi, K. Okuyama, and K. Kimura, *J. Mol. Struct.*, in press.
[4] M. Takahashi, H. Ozeki, and K. Kimura, *Chem. Phys. Lett.*, in press.
[5] J. B. Hopkins, D. E. Powers, and R. E. Smally, *J. Chem. Phys.*, **72** (1980) 5039.
[6] R. J. Lipert, and S. D. Colson, *J. Chem. Phys.*, **89**, 4579 (1988).
[7] M. Takahashi and K. Kimura, to be published.
[8] A. Oikawa, H. Abe, N. Mikami, and M. Ito, *J. Phys. Chem.*, **89**, 3642 (1985).
[9] T. Ebata, M. Furukawa, T. Suzuki, and M. Ito, *J. Opt. Soc. Am. B.*, **7**, 1890 (1990).

Two-Dimensional Infrared (2D IR) Spectroscopy Based on a Time-Resolved IR Measurement

I. Noda, A.E. Dowrey, and C. Marcott

The Procter and Gamble Company, Miami Valley Laboratories, P.O. Box 398707, Cincinnati, OH 45239-8707, USA

1. Introduction

Two-dimensional infrared (2D IR) spectroscopy [1-5] is a novel analytical technique based on a time-resolved IR measurement. The basic concept of 2D IR is somewhat analogous to the 2D correlation technique used extensively in NMR [6]; however, the experimental approach used in 2D IR is quite different. In 2D IR, a spectrum defined by two independent wavenumbers is generated by applying a correlation analysis to the dynamic fluctuation of infrared signals induced by an external macroscopic perturbation. A perturbation, such as small-amplitude oscillatory strain with an acoustic-range frequency, creates time-dependent reorientation of molecular constituents which can be monitored with a polarization-modulated IR probe produced by a photoelastic modulator [7]. Because of the specificity of absorption to individual chemical groups, fluctuations of IR signals contain detailed information about highly localized submolecular-level reorientation and relaxation processes.

2D IR spectroscopy is especially suited for the study of multiphase or multicomponent systems, such as polymer blends, where interactions among functional groups play an important role [3,4]. Distinct peaks appear on a 2D spectral plane when the time-dependent fluctuations of IR signals are strongly correlated with each other. The existence of such correlated fluctuations in turn provides information on connectivity and chemical interactions among functional groups associated with the IR bands. Spectral resolution is often enhanced in 2D IR, since peaks are spread over the second spectral dimension. Thus, complex IR spectra consisting of many overlapped peaks are simplified, and unambiguous assignments can be given by analyzing bands which are selectively coupled by various interaction mechanisms.

2. Instrumentation

In order to measure the perturbation-induced dynamic fluctuations of infrared signals needed for the generation of 2D IR spectra, we constructed a time-resolved spectrometer coupled with a dynamic mechanical perturbation cell. The measurement of dynamic IR signals is based on a simple triple-modulation experiment (Fig. 1). Since a detailed description of the instrument and its operation is reported elsewhere [7], only a brief description of the spectrometer is given here.

The IR beam from the source is labeled first by a chopper (∿1kHz) and polarized after going through an interferometer [5] or a monochromator [7]. The plane of polarization is then rapidly altered (∿100kHz) between directions parallel and

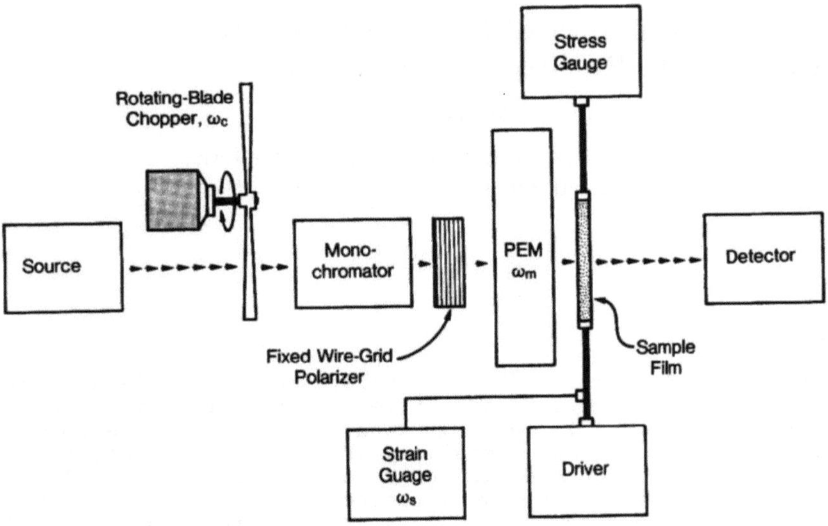

Fig. 1. Schematic diagram of the triple-modulation time-resolved dynamic infrared linear dichroism spectrometer [7].

perpendicular to the sample axis with a dynamic birefringence cell known as a photoelastic modulator (PEM). The polarized light is also modulated by the dynamic fluctuation of directional absorbances of the sample induced by the applied strain (~20Hz). Thus, the optical signal arriving at the detector is triply modulated by the chopper, PEM, and sample. The signal is then demodulated by a series of lock-in amplifiers.

3. Background

A directional external perturbation applied to a system will selectively induce time-dependent reorientations of electric dipole-transition moments associated with the individual normal modes of molecular vibration in the system. Such reorientations will result in the generation of dynamic infrared linear dichroism (DIRLD) signals [7]. If a small-amplitude sinusoidal strain is applied to the system, the time-dependent IR dichroism response also becomes sinusoidal. Interestingly, however, the phase angle between the dynamic dichroism and applied strain varies strongly as a function of the wavenumber of the IR probe. This result indicates that reorientations of individual dipole-transition moments representing different molecular vibrations are not synchronized and may proceed at different rates. Such a phenomenon can be visualized most effectively by applying a simple correlation analysis to the time-dependent variation of IR dichroism signals [4].

A pair-wise correlation of IR signals measured at two different wavenumbers yields an orthogonal set of correlation intensity spectra known as the <u>synchronous</u> and <u>asynchronous</u> 2D IR spectra. These two-dimensional spectra are used to accentuate the similarity and difference of the reorientation rates among dipole-transition moments. The synchronous spectrum characterizes the degree of coherence between the dynamic fluctuations of IR signals measured at two different wavenumbers; the correlation intensity becomes significant only if the time-dependent variations of the two IR signals (i.e., reorientation rates of dipole-transition moments) are similar to each other. Strong chemical interactions or connectivities are known to synchronize the reorientations of functional groups [4]. The asynchronous spectrum, on the other hand, characterizes the independent or uncoordinated fluctuations of dichroism signals. Unlike the synchronous correlation, the correlation intensity becomes nonvanishing only if the time-dependent variations are different. Asynchronous spectra are often used to differentiate highly overlapped IR bands.

4. Examples

Fig. 2 shows the synchronous 2D IR correlation spectrum of atactic poly(methyl methacrylate). Peaks observed on the diagonal of the spectrum near 2952 and 2997 cm^{-1} represent the perturbation-induced local reorientation of α- and ester-methyl groups of PMMA. The peaks near 2842 and 3039 cm^{-1}, on the other hand, arise exclusively from the contribution of ester-methyl groups [8]. Cross peaks appearing at the off-diagonal positions indicate the strong correlations between the various methyl bands. The existence of correlation with the 2842 and 3039 cm^{-1} ester-methyl bands implies that the fluctuation of IR signals for the 2952 and 2997 cm^{-1} bands also arises mostly from the motions of ester-methyl groups.

Fig. 3 shows an asynchronous 2D IR spectrum of human-hair keratin. Because the asynchronous spectrum accentuates the differences among the reorientation rates of individual functional groups or local conformations experiencing different molecular environments, it provides an exceptionally high

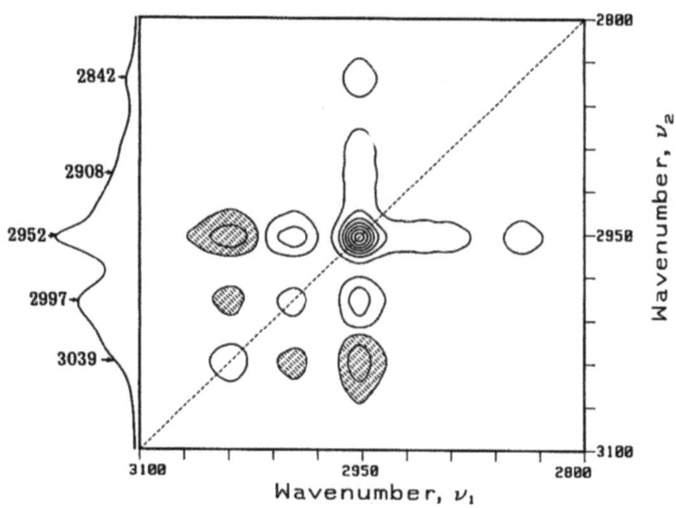

Fig. 2. Synchronous 2D IR spectrum of atactic PMMA. Normal IR absorbance spectrum is also shown for reference.

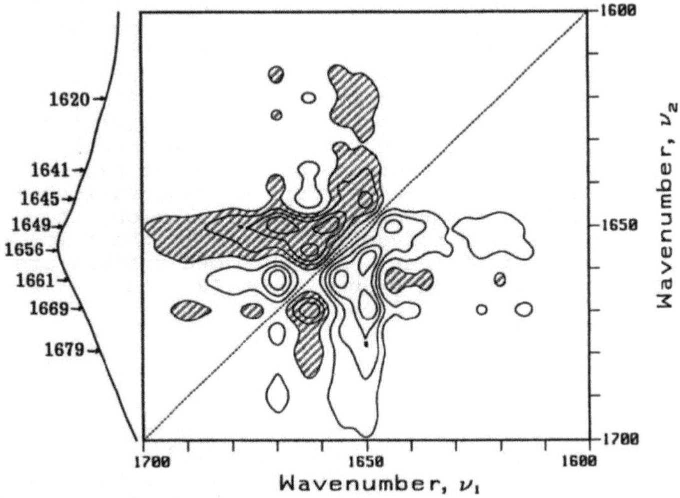

Fig. 3. Asynchronous 2D IR spectrum of human-hair keratin.

spectral resolution. We can observe the differentiation of IR bands at 1661 and 1649 cm^{-1} for the α-helix conformation, β-like turns and extended chains at 1679, 1669, 1645, 1641 and 1620 cm^{-1}, and disordered structures at 1656 cm^{-1}.

References

1. I. Noda, Bull. Am. Phys. Soc. 31, 520 (1986).
2. I. Noda, A. E. Dowrey, and C. Marcott, Microchim. Acta [Wien] 1, 101 (1988).
3. I. Noda, J. Am. Chem. Soc. 111, 8116 (1989).
4. I. Noda, Appl. Spectrosc. 44, 203 (1990).
5. R. A. Palmer, C. J. Manning, J. L. Chao, I. Noda, A. E. Dowrey, and C. Marcott, Appl. Spectrosc. 45, 12 (1991).
6. H. Kessler, M. Gehrke, and C. Griesinger, Angew. Chem. Int. Ed. Engl. 27, 490 (1988).
7. I. Noda, A. E. Dowrey, and C. Marcott, Appl. Spectrosc. 48, 203 (1988).
8. S. K. Dirlikov and J. L. Koenig, Appl. Spectrosc. 33, 555, (1979).

Novel Asynchronous Time-Resolved Fourier Transform IR Spectroscopy

K. Masutani[1], *A. Yokota*[2], *Y. Furukawa*[2], *and M. Tasumi*[2]

[1]JEOL Ltd., 3-1-2 Musashino, Akishima, Tokyo 196, Japan
[2]Department of Chemistry, Faculty of Science,
The University of Tokyo, 7-3-1 Hongo, Bunkyo-ku, Tokyo 113, Japan

We have developed a novel ASYNCHRONOUS time-resolved FT-IR spectrophotometer based on a conventional continuous-scan interferometer. In contrast to the existing method, this new method does not require the synchronization between the signal for time-resolving and the sampling timing of the A/D converter. The naming of "asynchronous time-resolved" comes from this feature.

The setup used for this method is schematically shown in Fig.1, where a signal processing assembly required for time-resolved measurements is given in the middle. This assembly consists of a pulse generator, a pulse delay circuit, a gate circuit, and a lowpass filter. This method features that the lowpass filter is placed between the gate circuit and the A/D converter for the purpose of cutting high-frequency components and converting discrete time-resolved signals from the gate circuit to analog signals. The analog signals can then be processed by the A/D converter at its own sampling

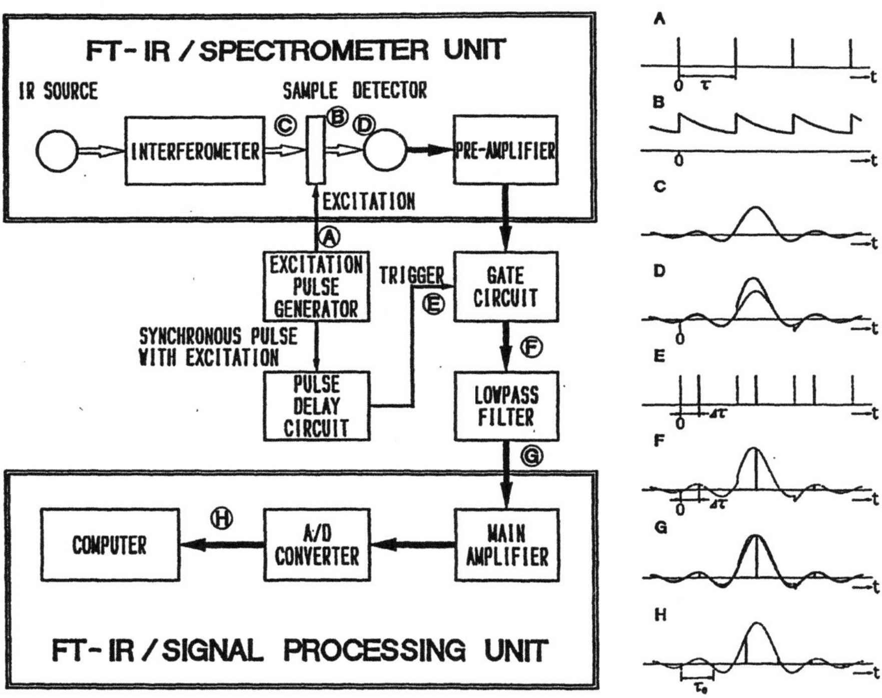

Fig.1 The setup used for this method and the output signal from each unit.

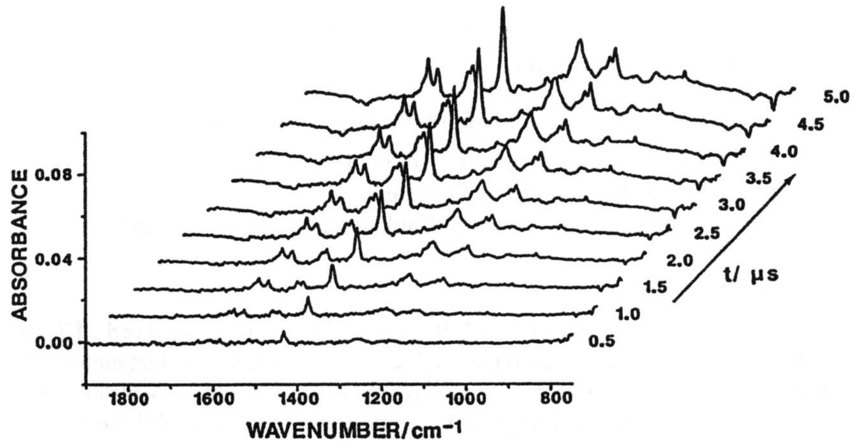

Fig.2 Time-Resolved IR spectra of a ferroelectric liquid crystal.

timing which has no relation with the gate timing for time-resolving. The signals in the setup are also shown in Fig.1.

The advantage of this method is that the signal processing assembly can in principle be attached to any commercial FT-IR spectrophotometers, and this simple addition enables time-resolved measurements without any further modifications in either hardware or software of the existing instrument. Another advantage is that it is possible to make the real time measurement of the time-resolved interferogram.

Our present system is best suited for observing transient phenomena with repetition rates of higher than 6kHz and lifetimes of 10^{-6}-10^{-4}s. However, these limits are variable.

For an application of this method, a time-resolved FT-IR measurement of reorientational dynamics in a ferroelectric liquid crystal induced by the change of external electric field was carried out. In order to observe the reorientation behavior of each functional group of the liquid crystal molecule, a polarized measurement was done. The obtained polarized infrared time-resolved spectra are shown in Fig.2. The polarization axis was fixed at 45° to the orientation direction of the liquid crystal. Applied rectangular pulses of electric field had ±10V peak voltage and 30μs repetition time. The duration of the gate signal was 0.5μs and the wavenumber resolution was 4 cm^{-1} . The vertical axis represents the absorption difference to the case of zero time delay from the excitation. We can conclude that these time-resolved spectra have a time resolution of better than 0.5μs.

Index of Contributors

Springer Proceedings in Physics

Managing Editor: H. K. V. Lotsch

1 *Fluctuations and Sensitivity in Nonequilibrium Systems*
Editors: W. Horsthemke and D. K. Kondepudi

2 *EXAFS and Near Edge Structure III*
Editors: K. O. Hodgson, B. Hedman, and J. E. Penner-Hahn

3 *Nonlinear Phenomena in Physics* Editor: F. Claro

4 *Time-Resolved Vibrational Spectroscopy*
Editors: A. Laubereau and M. Stockburger

5 *Physics of Finely Divided Matter*
Editors: N. Boccara and M. Daoud

6 *Aerogels* Editor: J. Fricke

7 *Nonlinear Optics: Materials and Devices*
Editors: C. Flytzanis and J. L. Oudar

8 *Optical Bistability III*
Editors: H. M. Gibbs, P. Mandel, N. Peyghambarian, and S. D. Smith

9 *Ion Formation from Organic Solids (IFOS III)*
Editor: A. Benninghoven

10 *Atomic Transport and Defects in Metals by Neutron Scattering*
Editors: C. Janot, W. Petry, D. Richter, and T. Springer

11 *Biophysical Effects of Steady Magnetic Fields*
Editors: G. Maret, J. Kiepenheuer, and N. Boccara

12 *Quantum Optics IV*
Editors: J. D. Harvey and D. F. Walls

13 *The Physics and Fabrication of Microstructures and Microdevices*
Editors: M. J. Kelly and C. Weisbuch

14 *Magnetic Properties of Low-Dimensional Systems*
Editors: L. M. Falicov and J. L. Morán-López

15 *Gas Flow and Chemical Lasers*
Editor: S. Rosenwaks

16 *Photons and Continuum States of Atoms and Molecules*
Editors: N. K. Rahman, C. Guidotti, and M. Allegrini

17 *Quantum Aspects of Molecular Motions in Solids*
Editors: A. Heidemann, A. Magerl, M. Prager, D. Richter, and T. Springer

18 *Electro-optic and Photorefractive Materials*
Editor: P. Günter

19 *Lasers and Synergetics*
Editors: R. Graham and A. Wunderlin

20 *Primary Processes in Photobiology*
Editor: T. Kobayashi

21 *Physics of Amphiphilic Layers*
Editors: J. Meunier, D. Langevin, and N. Boccara

22 *Semiconductor Interfaces: Formation and Properties*
Editors: G. Le Lay, J. Derrien, and N. Boccara

23 *Magnetic Excitations and Fluctuations II*
Editors: U. Balucani, S. W. Lovesey, M. G. Rasetti, and V. Tognetti

24 *Recent Topics in Theoretical Physics*
Editor: H. Takayama

25 *Excitons in Confined Systems*
Editors: R. Del Sole, A. D'Andrea, and A. Lapiccirella

26 *The Elementary Structure of Matter*
Editors: J.-M. Richard, E. Aslanides, and N. Boccara

27 *Competing Interactions and Microstructures: Statics and Dynamics*
Editors: R. LeSar, A. Bishop, and R. Heffner

28 *Anderson Localization*
Editors: T. Ando and H. Fukuyama

29 *Polymer Motion in Dense Systems*
Editors: D. Richter and T. Springer

30 *Short-Wavelength Lasers and Their Applications*
Editor: C. Yamanaka

31 *Quantum String Theory*
Editors: N. Kawamoto and T. Kugo

32 *Universalities in Condensed Matter*
Editors: R. Jullien, L. Peliti, R. Rammal, and N. Boccara

33 *Computer Simulation Studies in Condensed Matter Physics: Recent Developments*
Editors: D. P. Landau, K. K. Mon, and H.-B. Schüttler

34 *Amorphous and Crystalline Silicon Carbide and Related Materials*
Editors: G. L. Harris and C. Y.-W. Yang

35 *Polycrystalline Semiconductors: Grain Boundaries and Interfaces*
Editors: H. J. Möller, H. P. Strunk, and J. H. Werner

36 *Nonlinear Optics of Organics and Semiconductors*
Editor: T. Kobayashi

37 *Dynamics of Disordered Materials*
Editors: D. Richter, A. J. Dianoux, W. Petry, and J. Teixeira

38 *Electroluminescence*
Editors: S. Shionoya and H. Kobayashi

39 *Disorder and Nonlinearity*
Editors: A. R. Bishop, D. K. Campbell, and S. Pnevmatikos

40 *Static and Dynamic Properties of Liquids*
Editors: M. Davidović and A. K. Soper

41 *Quantum Optics V*
Editors: J. D. Harvey and D. F. Walls

42 *Molecular Basis of Polymer Networks*
Editors: A. Baumgärtner and C. E. Picot

43 *Amorphous and Crystalline Silicon Carbide II: Recent Developments*
Editors: M. M. Rahman, C. Y.-W. Yang, and G. L. Harris

MIX
Papier aus verantwortungsvollen Quellen
Paper from responsible sources
FSC® C105338
FSC
www.fsc.org

If you have any concerns about our products,
you can contact us on
ProductSafety@springernature.com

In case Publisher is established outside the EU,
the EU authorized representative is:
Springer Nature Customer Service Center GmbH
Europaplatz 3, 69115 Heidelberg, Germany

Printed by Libri Plureos GmbH
in Hamburg, Germany